Laboratories of Faith

Laboratories of Faith

Mesmerism, Spiritism, and Occultism in Modern France

JOHN WARNE MONROE

CORNELL UNIVERSITY PRESS

Ithaca & London

Publication of this book was made possible, in part, by a Publication Subvention Grant from Iowa State University.

First published 2008 by Cornell University Press

Printed in the United States of America

Library of Congress Cataloging-in-Publication Data

Monroe, John Warne, 1973–
Laboratories of faith: Mesmerism, spiritism, and occultism in modern France / John Warne Monroe.
p. cm.
Includes bibliographical references and index.
ISBN-978-0-8014-4562-0 (cloth : alk. paper)
1. Mesmerism—France—History. 2. Spiritualism—France—History. 3. Occultism—France—History. 4. France—Religion—19th century. 5. France—Religion—20th century. I. Title.

BF1125.M66 2008
130.944—dc22 2007031128

Cornell University Press strives to use environmentally responsible suppliers and materials to the fullest extent possible in the publishing of its books. Such materials include vegetable-based, low-VOC inks and acid-free papers that are recycled, totally chlorine-free, or partly composed of nonwood fibers. For further information, visit our website at www.cornellpress.cornell.edu.

Cloth printing 10 9 8 7 6 5 4 3 2 1

For my parents,
John and Margaret Monroe

And for my brother,
Charles

Oui, *je le sais,* nous ne sommes que de vaines formes de la matière—mais bien sublimes pour avoir inventé Dieu et notre âme. Si sublimes, mon ami! que je veux me donner ce spectacle de la matière, ayant conscience d'elle, et, cependant, s'élançant forcènement dans le Rêve qu'elle sait n'être pas, chantant l'Ame et toutes les divines impressions pareilles qui se sont ammassées en nous depuis les premiers âges, et proclamant, devant le Rien qui est la vérité, ces glorieux mensonges!

Stéphane Mallarmé, letter to Henri Cazalis, April 28, 1866

Contents

Acknowledgments

The idea for this book came to me in early June 1996. After a year of graduate school, I had managed to find a summer job in Paris as a translator of instruction manuals for medical equipment. On one of my first weekends in the city, prowling the Latin Quarter for jazz records, I happened upon Editions Leymarie, an old bookstore-publisher on the rue Saint-Jacques. Its interior remains much as it must have been in the belle époque: dark, permeated with the smell of tobacco, packed floor to ceiling with works on Spiritism, Occultism, esotericism, and related topics. The upper shelves held hundreds of titles from the nineteenth and early twentieth centuries, some well-worn, others with spines crisp and pages still uncut, all unfamiliar to me. There were two other customers that day, men in tweed jackets; both were talking with the proprietor, Michel Leymarie, a redoubtable gentleman who looked to be in his mid-fifties and smoked short, pungent cigarillos. He spoke emphatically about the importance of science and the shortcomings of the Catholic Church. After an hour or two of browsing, I left the shop intrigued, with a bag full of books and a plan for the next semester's research.

As that initially modest project has developed, I have received help from many quarters. John Merriman, Frank Turner, and Thomas Kselman have all been invaluable mentors. This book would have taken a very different form if I had not met Sofie Lachapelle, who has been a remarkably generous friend and colleague. At several crucial stages I have also benefited from the insights of other scholars of French heterodoxy: Naomi Andrews, Matthew Brady Brower, David Allen Harvey, Lynn L. Sharp, Wiktor Stoczkowski, and Ann Taves. André Aciman, Ellen Amster, Susan Ashley,

Jonathan Beecher, Michael Behrent, William Christian, Esther da Costa-Meyer, John Demos, Stéphane Gerson, Anthony Grafton, Suzanne Kaufman, Kelly Maynard, Mark Micale, Philip Nord, Kevin Repp, and Robin Walz have all offered useful advice, comments, and critiques. In the course of my work, I have also drawn heavily on the linguistic knowledge that Simone Lewis inculcated with such remarkable enthusiasm. While the late Robert Amadou, an eminent esotericist and scholar, would probably still have had reservations about the approach I take here, his pointed admonitions nevertheless helped shape the book that follows.

Financial support for the bulk of my research came from a *Bourse Chateaubriand*, a Gilbert Chinard Research Grant from the *Société Française de Washington*, a Millstone Travel Grant from the Western Society of French History, and a Yale University Enders Research Grant. Grants from the Iowa State University College of Liberal Arts and Sciences and Center for Excellence in the Arts and Humanities both underwrote supplemental research trips, as did a summer stipend from the National Endowment for the Humanities. A Yale University Leylan Fellowship and a summer grant from the Institute for the Advanced Study of Religion at Yale financed a year of writing; the Theron Rockwell Field Prize, also from Yale, paid for one more summer. In spring 2005, a semester's fellowship at the University of Iowa's Obermann Center for Advanced Study allowed me to conclude my revisions in peaceful surroundings. Finally, I owe a particular debt of gratitude to Ingrid Walsøe-Engel, who played a crucial role in securing financial support for me early in my graduate studies.

This project would have been impossible to complete without the help of librarians, archivists, and booksellers. The staff in *salle* M of the Bibliothèque nationale de France, where I conducted most of the research for this book, deserve special mention for unflagging patience and enthusiasm. Jacques Pernet and Patrick Fuentès, volunteer conservators of the Fonds Camille Flammarion, distinguished themselves both as guides to the riches of their collection and as lunch companions. Father Philippe Ploix, conservator of the Archives de l'Archevêché de Paris, and Claude Charlot of the Archives de la Préfecture de Police de Paris provided invaluable assistance. The bookseller and publisher Jean Meyniel gave me important research leads. André Braire and his wife, Nicole, proprietors of Editions Traditionnelles, were extremely generous, both with their time and with their matchless collection of nineteenth-century Spiritist and Occultist publications.

Over the last several years, a host of unsuspecting colleagues and friends have consented to let me bend their ears about French heterodoxy, and have offered excellent suggestions in return. At Iowa State, this group includes many current and former members of the junior faculty: Kevin Amidon, Michael Bailey, Patrick Barr-Melej, Brett Bowles, Christopher

Curtis, Sara Gregg, Paul Griffiths, David Hollander, Bernhard Rieger, Leonard Sadosky, and Matthew Stanley. By asking a few well-timed and searching questions, David B. Wilson and his wife, Julia, have done much to hone the approach to heterodoxy I develop here. Hamilton Cravens and Alan Marcus have also read and commented insightfully on portions of this manuscript. Andrejs Plakans and Charles Dobbs, as my department chairs, came through many times in a pinch. I was lucky to have attended Yale with a group of sociable and intelligent Europeanists, including Jennifer Boittin, Kate Cambor, Mark Choate, Rachel Chrastil, Jay Geller, Maya Jasanoff, Charles Lansing, Richard Lofthouse, Kieko Matteson, Elisa Milkes, Leon Sachs, Michael Shurkin, George Trumbull, Stephen Vella, and Cherie Woodworth. My friends Jeffrey Atteberry, Dallas Dickinson, Julian Fisher, Michael Harrison, Jean Laurens, Andrew Lewis, Michael McCabe, Timothy O'Reilly, Lisa-Marie Priddy, Maryse Reynolds, Timothy Robertson, Julie Simpson, Terence Strick, Joe Teltser, Florence Thomas, Carey Wallace, and Katy Walden have all contributed as well. On the nonhuman side, Mocha and Suleyman have been both furry and tolerant, albeit notably unreliable judges of prose style.

At Cornell Press, I would like to thank John Ackerman for his wise editorial advice and Ange Romeo-Hall for seeing this book through to publication. The copyeditor, Cathi Reinfelder, with her eagle eye for adverbial excess, saved me from myself on all too many occasions. Michael Morris has been a useful source of administrative assistance. Iowa State University has also supported this project with a Publication Subvention Grant. Portions of chapters 3 and 4 originally appeared in *History of Religions* and *French Historical Studies*; I thank both journals for permission to include this material here.

My parents, John and Margaret Monroe, and my brother, Charles, have been unflagging in their support. When seen in terms of the *longue durée*, this book is in many ways the product of their efforts, and is dedicated to them, with love and the greatest respect. My final thanks are reserved for my wife, Wendie Schneider. While busy with projects of her own, she nevertheless has found time to be a steadfast intellectual companion, a source of emotional sustenance, and an extraordinary reader. The chapters that follow could not have been written without her.

Laboratories of Faith

Introduction

Now faith is the substance of things hoped for,
the evidence of things not seen.
Hebrews 11:1

In the summer of 1859, the pharmacist and retired army officer P. F. Mathieu submitted a long paper to the Académie des sciences. In it, he presented the results of a series of experiments he had undertaken with the help of a medium named Honorine Huet. One of these, which occurred in a chapel off the nave of the Parisian basilica of Notre Dame des Victoires, involved a sheet of ordinary office paper Mathieu had removed from his desk the day before. After verifying that the sheet was unmarked, he folded it into quarters and placed it on a step in the chapel. Huet meditated in silence for a few moments, and then, while reciting a prayer, touched her gloved fingertips to the folded paper. When Huet pulled her fingers away, Mathieu opened the sheet, and, to his "great stupefaction, discovered the word *faith* traced, as if by a pencil, on one of the interior leaves!" The leadlike substance that formed the script, he hypothesized, could perhaps have reached the paper "in an invisible molecular state, as in electro-galvanic transport."[1]

Mathieu called this uncanny manifestation "direct writing," and argued that it constituted nothing less than a "Providential" effort to "combat the increasingly materialistic tendencies of our age." Direct writing, he maintained, provided irrefutably tangible proof that spirits existed and could make their presence felt in the world. In the face of such sound empirical evidence, the scientific conviction that the human soul and its divine creator were merely superstitious illusions became "blind and unjust." By

[1] Reprinted in *La Revue spiritualiste* 2 (1859): 147, 152. Italics in the original. These translations, and all those that follow, are mine unless otherwise noted.

transforming metaphysics from a matter of philosophical speculation to one of rigorous experimental study, Mathieu declared, this data gave humanity the first inklings of a "science of God." If pursued with suitable commitment, this science would eventually arrive at an understanding of the beyond as clear and precise as a chemist's understanding of the novel process of electroplating.[2]

Two years later, in 1861, the young astronomy student Camille Flammarion wrote a letter to the Abbé Berillon, a priest who had been his childhood confessor. In his letter, he described some of his own efforts to contact the other world. Through automatic writing, he had received messages from the spirits of Fénélon and Galileo, who had provided numerous cosmological insights, including an account of the origin of the universe.[3] Flammarion had also witnessed other, more spectacular manifestations, which he only alluded to in his letter, perhaps in deference to the old priest's sensibilities. At one séance, with Huet acting as medium, he had seen a table hanging in the air, suspended as if by an invisible force; at another, he had been present when the spirit of a murdered man revealed his body's resting place.[4] Foreseeing his confessor's anxieties, Flammarion ended his letter with a reassurance: "I should warn you in advance that I am not in the presence or under the influence of any evil spirit: I study Spiritism as I study mathematics."[5] He did not see himself as an ecstatic visionary but rather as a sober observer, an astronomer who contemplated the beyond with the same impartial rigor he devoted to the stars.

Déodat Roché, a young lawyer in the southern city of Toulouse, adopted a similarly assured tone in a series of letters, written in 1898 and addressed to Gérard Encausse—known as Papus—the head of an Occultist secret society called the Ordre Martiniste. To fulfill one of the order's requirements for advancement, Roché provided his initiator with a spiritual autobiography. In it, he described his own efforts to develop a more thorough knowledge of the other world. Like Flammarion, Roché saw his metaphysical pursuits in mathematical terms. Geometry had kindled his interest in higher things: The contemplation of theorems had gradually led him to the contemplation of "pure essences."[6] His magical experiments began after he read Papus's best-selling *Traité de magie pratique*. After much practice, he had succeeded in having "a seriously verifiable

[2] Ibid., 147, 152, 153.

[3] Fonds Camille Flammarion de l'observatoire de Juvisy-sur-Orge [FCF], ms. copybook marked *Miscellanées 1861*, letter to the Abbé Berillon, Dec. 31, 1861.

[4] Ibid., letters to Charles Burdy, Nov. 1 and Oct. 15, 1861.

[5] Ibid., letter to Berillon, Dec. 31, 1861.

[6] *Fonds Papus de la Bibliothèque municipale de Lyon* [FP], letter from Deodat Roché, March 4, 1898.

vision and even perhaps in producing an apparition of myself at a distance."[7] At the same time, however, he found that the incantations and visualization techniques he learned from books had an ever-diminishing effect, "much as the same dose of coffee would."[8] As a member of the Ordre Martiniste, he hoped to learn more effective methods for harnessing these strange powers, exploring the depths of his psyche to cultivate a richer understanding of "the Holy Light."[9]

Though the details of their stories differ in ways that reflect broader developments in intellectual and cultural history—from a positivistic focus on objective phenomena to a psychologically conceived notion of inner power—each of these people was engaged in a similar project. Mathieu, Huet, Flammarion, and Roché were intrigued by phenomena that seemed to reveal the action of intelligence beyond the limits of the human body. In addition, they all viewed studies of such phenomena as contributions to a new form of metaphysics based on tangible experience rather than philosophical speculation. These seekers were by no means isolated eccentrics; their efforts to reach a new understanding of the beyond were part of a wave of innovative religious thought and practice that first emerged with Mesmerism in the late eighteenth century and developed into a widespread cultural phenomenon after 1850. During this period, discussions of novel forms of heterodoxy became a surprisingly common feature of middle-class urban life.

Whether through study of the strange phenomena mediums produced, communication with disembodied souls, or a rehabilitation of ancient magical practices, figures such as Mathieu, Huet, Flammarion, and Roché sought to resolve what they perceived to be a central problem of their time: a *crisis of factuality* in religious life. As scientific knowledge grew more refined, the precepts religion affirmed seemed increasingly difficult to defend. Metaphysical concepts like the immortality of the soul could not be tested in the laboratory; they could be accepted only intuitively, as articles of faith. The more prestige scientific standards of empirical proof acquired, these thinkers and believers argued, the less convincing faith seemed as a way of knowing. The religious seekers this book investigates sought to resolve this crisis by inventing new "sciences of God," approaches to the beyond capable of turning faith into fact by providing empirical evidence for metaphysical propositions. In the process, they helped cause a spiritual sea change, the ramifications of which are still visible today in France and throughout the West.

[7] Ibid., March 22, 1898.
[8] Ibid., March 4, 1898.
[9] Ibid., March 22, 1898.

Fact and Faith in Modern France

Over forty years ago, the anthropologist Clifford Geertz described the role of factuality in religious discourse in a way that sheds considerable light on the problem these heterodox innovators faced. A *religion*, Geertz wrote, is

> a system of symbols which acts to establish powerful, pervasive, long-lasting moods and motivations in men by formulating conceptions of a general order of existence and clothing those conceptions with such an aura of factuality that the moods and motivations seem uniquely realistic.[10]

Religion, in other words, is an interpretive framework that allows people both to imbue their experience with meaning and to elaborate a guide for right conduct. By projecting themselves into a cosmic order—a particular constellation of symbols—believers discover a way to bear the most painful, chaotic-seeming aspects of life. Religious systems do not deny suffering, absurdity and evil; instead, they explain them, and by placing them in a larger interpretive framework, prove that despite any temporary appearances to the contrary, the universe is a just and orderly place.

For an explanatory system of this kind to function properly, it needs to have an "aura of factuality," to strike believers as "uniquely realistic." In addition to articulating a particular vision of symbolic order, a given religious system must make that order seem authoritative. Numerous factors can work together to create this sense of authority—language, strategies of argument, ritual, institutional structures, a privileged relationship to political power, and so on. All of these factors, in turn, can be considered aspects of a coherent discourse. They are ways of talking, acting, and organizing that work to convey a clear message to believers: "the vision of cosmic order espoused here is undeniably real." Among other things, then, a religion is a discursive machine for the production of truth. It puts forward a vision of cosmic order in a manner shaped to seem irrefutable when judged according to the standards of a particular audience.

The discursive techniques for creating this "aura of factuality" are historically contingent. Where a medieval French Catholic may have seen his belief system made real in the king's ritual postcoronation laying on of hands, for example, a late-nineteenth-century believer might have found similar affirmation by contemplating the profusion of cast-off crutches at Lourdes. What might seem to be an irrefutable sign of authority in one

[10] Clifford Geertz, "Religion as a Cultural System," in *The Interpretation of Cultures* (New York: Basic Books, 1973), 90. The essay was originally published in 1966.

context could appear decidedly inadequate or old-fashioned in another. The period this book explores was one in which the historical contingency of religious symbol systems became particularly apparent. France in the late nineteenth and early twentieth centuries was a society engaged in a deep and thoroughgoing reconception of what, exactly, constituted an "aura of factuality" in the first place.

In many respects, the defining characteristics of this period's heterodox discourse derived from a broader view of modernity and its epistemological challenges that had acquired the force of common sense by the mid-nineteenth century.[11] During the period from 1848 to 1914, journalists and other commentators tirelessly proclaimed the radical difference and novelty of the era they inhabited. The modern age, according to late-nineteenth-century conventional wisdom, was characterized above all by an ever-growing knowledge and mastery of the forces of nature. Max Weber provided perhaps the most influential formulation of this idea in his work, which presented "rationalization" as one of the primary characteristics of modernity.[12] New technologies, which appeared to be both causes and consequences of this rationalization, dramatically altered the material fabric of society: On the Continent, this was the age of railways, telegraphs, and rapidly expanding industrial production. These technological changes also had far-reaching social and cultural effects—as seen in the emergence of a mass press, the expansion of cities, the depopulation of the countryside, and the gradual growth of a culture of consumption.

Like Weber, many others assumed that this new reality entailed a particular way of seeing the world, one based on "scientific" objectivity and empirical rigor. For the modern mind, as many nineteenth-century commentators saw it, valid knowledge was knowledge based above all on facts, impartially accumulated and presented in ways that made them appear to "speak for themselves."[13] This self-conscious concern with the presentation and accumulation of fact pervaded nineteenth-century discourse—Utilitarianism, Positivism, literary realism, and even the discipline of history itself testify to its influence. Despite its seemingly "scientific" neutrality, however, this epistemological ideal had clear moral overtones: not only was it the path to truth, it was also the path to virtue. For many

[11] For a useful account of the early development of this epistemological ideal, see Mary Poovey, *A History of the Modern Fact: Problems of Knowledge in the Sciences of Wealth and Society* (Chicago: University of Chicago Press, 1998).

[12] For a discussion of this concept in relation to religion, see Max Weber, "The Social Psychology of the World Religions," in *From Max Weber: Essays in Sociology*, trans. and ed. Hans Heinrich Gerth and Charles Wright Mills (Oxford: Oxford University Press, 1958).

[13] Lorraine Daston and Peter Galison, "The Image of Objectivity," *Representations* 40 (Fall 1992): 81–128.

nineteenth-century commentators, the solid empirical information that a modern reverence for fact produced seemed to provide the basis for an understanding of moral truth with the potential to be equally substantial.[14]

This conception of objectivity and its moral value, as Lorraine Daston and Peter Galison point out, also entails a particular conception of subjectivity and its shortcomings.[15] In this framework, conventional forms of religion, with their emotional and intuitive approach to knowledge, seemed to be "subjective"—and therefore inferior—ways of understanding the world. Auguste Comte's famous "three stage" model of history exemplifies this pattern of thought. In Comte's conception, mankind gradually moved toward epistemological and social perfection by successively abandoning two forms of "subjectivity," the invented worlds of myth and of metaphysics, in favor of an exclusive, objective focus on empirical phenomena. The world of the future, in Comte's view, would be characterized by a "Positive spirit," one that had no place for old-fashioned "fictions" like the notion of the immortal soul.[16]

In the French context, this critique of subjectivity—and by extension of religion—also assumed a political dimension, which stemmed from the Catholic Church's support of antidemocratic conservatism. Beginning in the 1850s, and increasingly after the 1870s, French advocates of democracy and reform used Comte's vision of history to affirm their own sense of political destiny. As these commentators and statesmen saw it, secular republicanism was a rational political system characterized by an objective reliance on democratic decision making—and hence was perfectly suited to the demands of the modern age. The triumph of the Republic, in this view, would necessarily entail the withering away of religious life.

These philosophical and political developments placed French seekers of religious consolation in a difficult position. Given what seemed to be a revolution not only in the criteria of epistemological authority but also in the very assumptions on which society was based, how could religious systems continue to maintain their "auras of factuality"? They could do so by reshaping religious experience, ritual, and even doctrine—either subtly, or more dramatically. Since the 1980s, scholars have acknowledged the vitality of nineteenth-century religious life by abandoning an older equation of modernity and secularization for a new focus on the resilience and

[14] See Daston and Galison for a discussion of the moral aspects of objectivity in the specific context of the sciences; for a more general discussion of the nineteenth-century moralization of "realism," see Hayden White, *Metahistory: The Historical Imagination in Nineteenth-Century Europe* (Baltimore: Johns Hopkins University Press, 1973), 45–46.

[15] Daston and Galison, 82.

[16] Auguste Comte, *Discours sur l'esprit positif* (Paris: J. Vrin, 1995 [1844]).

mutability of religious forms.[17] Historians of France, for example, have already shown that the second half of the nineteenth century was a time of marked Catholic ferment, which saw the emergence of new forms of piety focused on tangible religious experience.[18] This was an era of mass pilgrimages; of popular journals describing communications with souls in purgatory; of devotion to the Sacred Heart of Jesus; of Marian visions, stigmatics, and miraculous cures. While these phenomena are deeply rooted in Catholic tradition, their expression in the nineteenth century also had elements of novelty. First, they relied on material innovations: Pilgrimage sites like Lourdes thrived because of the railway; statues of the Virgin proliferated thanks to advances in mass production; and popular books and images became cheaper with improvements in printing technology. Second, the nationwide expansion of education, and the consequent growth in literacy, allowed believers to become engaged in religious life in a very different way than they had in the past; the devout were now also avid readers of a new Catholic mass press. Third, the Church hierarchy proved more willing to endorse these intense, exuberant expressions of piety than it had been previously. Fourth, and most important, there was the sheer *scale* of this interest in tangible religious experience: This surge of enthusiasm was truly a phenomenon of mass culture.

Mathieu, Huet, Flammarion, and Roché, then, were hardly alone in their quest for ways to make the truths of religion more palpable. In the French context, heterodox efforts to create "sciences of God" were part of a broader current in nineteenth-century religious life. Of course, the paths Mathieu and other Spiritists, Occultists, and Mesmerists followed also differed markedly from the one orthodox Catholics trod in such great numbers. Where Catholic innovations in this period emerged in a context

[17] See Lisa Abend, "Specters of the Secular: Spiritism in Nineteenth-Century Spain," *European History Quarterly* 34, no. 4 (2004): 507–534; David Blackbourn, *Marpingen: Apparitions of the Virgin Mary in a Nineteenth-Cenutry German Village* (New York: Vintage, 1993); William B. Christian Jr., *Visionaries: The Spanish Republic and the Reign of Christ* (Berkeley: University of California Press, 1996); Olav Hammer, *Claiming Knowledge: Strategies of Epistemology from Theosophy to the New Age* (Leiden: Brill, 2001); Thomas Laqueur, "Why the Margins Matter: Occultism and the Making of Modernity," *Modern Intellectual History* 3, no. 1 (2006): 111–135; Alex Owen, *The Place of Enchantment: British Occultism and the Culture of the Modern* (Chicago: University of Chicago Press, 2004); Helen Sword, *Ghostwriting Modernism* (Ithaca: Cornell University Press, 2002); Corinna Treitel, *A Science for the Soul: Occultism and the Genesis of the German Modern* (Baltimore: The Johns Hopkins University Press, 2004).

[18] Gérard Cholvy and Yves-Marie Hilaire, *Histoire religieuse de la France, 1800–1880* (Paris: Privat, 2000), esp. the discussion of Ultramontane piety, 177–225; Ruth Harris, *Lourdes: Body and Spirit in the Secular Age* (London: Penguin, 1999); Raymond Jonas, *France and the Cult of the Sacred Heart, an Epic Tale for Modern Times* (Berkeley: University of California Press, 2000); Suzanne Kaufman, *Consuming Visions: Mass Culture and the Lourdes Shrine* (Ithaca: Cornell University Press, 2005); Thomas Kselman, *Death and the Afterlife in Modern France* (Princeton: Princeton University Press, 1993), and *Miracles and Prophecies in Nineteenth Century France* (New Brunswick: Rutgers University Press, 1983).

of always-affirmed continuity, heterodox innovations involved a self-conscious break with the past. For these believers, the only way to guarantee the continued validity of religion in the modern world was to radically change the texture of the human experience of the sacred. "Unique realism," in this view, had to come from a religious system that made the contemplation of the beyond into a scientific, empirical project.[19]

Despite this desire to create a new basis for belief, however, the religious systems these nineteenth-century innovators proposed bore marked structural similarities to the Catholicism they were intended to supersede. Religious symbol systems, no matter how novel they may seem, cannot emerge from a vacuum—to convey the necessary "aura of factuality," they must hew to structures that strike potential believers as credible. Stephen Prothero has suggested a useful way of thinking about this problem by likening religious innovation to the "creolization" of a language.[20] In creole languages, the grammatical structure remains relatively stable, while the vocabulary, imported from a wide array of other languages, changes dramatically. Similarly, in religious systems, a "grammar" of deep structures can be separated from a "vocabulary" of specific practices, doctrines, and institutional arrangements.

Nineteenth-century French religious innovators used drastically new vocabularies, many of American origin—holding "séances," using "mediums" to produce "direct writings"—but also adhered in many crucial ways to the grammar of the Catholicism with which the vast majority of them had grown up. Ideas of Christian morality, the importance of charity and the soteriological value of repentance, the necessity of doctrinal uniformity, and the fundamental role of centralized authority in the legitimation of religious teachings all remained crucial, in varying degrees, to their spiritual projects. In this respect, Spiritism and Occultism are similar to the Positivist "Religion of Humanity" Comte created in the 1850s. Even as he sought to replace Catholicism, Comte readily adapted distinctively Catholic forms, substituting the figure of his deceased *grand amour* Clotilde de Vaux for the Virgin Mary, structuring devotion around "sacraments," and envisioning an ecclesiastical hierarchy led by a decidedly papal *Grand*

[19] For other works addressing aspects of French heterodoxy in the nineteenth and early twentieth centuries, see Matthew Brady Brower, "Fantasms of Science: Psychical Research in the French Third Republic, 1880–1935" (Ph.D. diss., Rutgers University, 2005); Nicole Edelman, *Voyantes, guérisseuses et visionnaires en France, 1785–1914* (Paris: Albin Michel, 1995); David Allen Harvey, *Beyond Enlightenment: Occultism and Politics in Modern France* (DeKalb: Northern Illinois University Press, 2005); Sofie Lachapelle, "A World outside Science: French Attitudes toward Mediumistic Phenomena, 1853–1931" (Ph.D. diss., University of Notre Dame, 2002); Lynn L. Sharp, *Secular Spirituality: Reincarnation and Spiritism in Nineteenth-Century France* (Lanham MD: Lexington Books, 2006).

[20] Stephen Prothero, *The White Buddhist: The Asian Odyssey of Henry Steel Olcott* (Bloomington: Indiana University Press, 1996), 8.

Prêtre.[21] Acknowledging the pervasiveness of Catholicism, then, both as something to react against and as a paradigm for the creation of religious authority, makes it possible to perceive a distinctively French brand of nineteenth-century religious innovation.[22]

Laboratories of Faith

French heterodoxy continues to flourish today, and the phenomena that attracted spiritual seekers more than a century ago retain their appeal. In the period between 1982 and 2000, for example, a relatively constant 47 to 54 percent of the French population professed belief in the curative powers of Mesmerism, a practice based on the manipulation of an invisible "universal fluid;" a similarly constant 10 to 16 percent agreed that it was possible to contact the beyond in séances.[23] According to the sociologist Daniel Boy, these beliefs "prosper in a kind of religious 'betwixt-and-between' " inhabited by occasionally practicing Catholics and people who identify themselves as adherents of unspecified "other religions."[24] For many in France, in other words, religion has not disappeared with the twentieth-century decline of organized churches; it has instead become a matter of independent *bricolage*—a personal quest for meaning to be pursued by autonomous individuals in their private lives.[25] Boy's findings indicate that, for a surprisingly large percentage of these individuals, certain ideas and practices first introduced by nineteenth-century innovators remain attractive and vital commodities in this new spiritual marketplace.

Many French commentators have explained the late twentieth century prominence of heterodoxy by seeing it as the product of a fairly recent socio-cultural rupture, usually either the tumult of 1968 or the emergence

[21] Andrew Wernick, *Auguste Comte and the Religion of Humanity: The Post-Theistic Program of French Social Theory* (Cambridge: Cambridge University Press, 2001).

[22] For material on related developments in the Anglo-American world, see Ann Braude, *Radical Spirits: Spiritualism and Women's Rights in Nineteenth-Century America* (Boston: Beacon Press, 1989); R. Laurence Moore, *In Search of White Crows: Spiritualism, Parapsychology, and American Culture* (Oxford: Oxford University Press, 1977); Janet Oppenheim, *The Other World: Spiritualism and Psychical Research in England, 1850–1914* (Cambridge: Cambridge University Press, 1985); and Alex Owen, *The Darkened Room: Women, Power, and Spiritualism in Late Victorian England* (London: Virago, 1989).

[23] Daniel Boy, "Les Français et les para-sciences, vingt ans de mesures," *Revue Française de sociologie* 43, no. 1 (January–March 2002): 35–45.

[24] Ibid., 41–42.

[25] A number of sociologists and philosophers have theorized this development in different ways. For two approaches that have informed this study, see Marcel Gauchet, *The Disenchantment of the World*, translated by Oscar Burge (Princeton: Princeton University Press, 1997); Thomas Luckmann, *The Invisible Religion: The Problem of Religion in Modern Society* (New York: Macmillan, 1967).

of a new mass culture of consumption after the Second World War.[26] As Wiktor Stoczkowski has observed, these attempts to ascribe the emergence of innovative religious systems to some kind of distinctively contemporary crisis suffer from a serious logical inconsistency.[27] If these new forms of heterodoxy are indeed the consequence of a Weberian "disenchantment of the world," as scholars have tended to agree, then the crisis to which they respond far predates either 1968 or 1945—it is as old as modernity itself. Those seeking to explain the new forms of religious expression that seem so characteristic of contemporary culture, therefore, should look not only to the present but also to the past.[28]

This book lays the groundwork for a new, broader conception of modern French religious life that allows us to see the gradual emergence of the heterodox ideas and practices that remain such durable features of contemporary French society. It accomplishes this task by charting the development of this current of innovation from the mid-nineteenth century, when it first emerged as a widespread cultural phenomenon, to the early decades of the twentieth, when it assumed new forms in the wake of the First World War. During this period, three movements dominated the French heterodox scene: spiritualist Mesmerism, Spiritism, and Occultism. The chapters that follow examine the institutions, ideas, and practices that characterized each of these movements and reconstruct the specific ways in which these new religious systems functioned as structures of meaning for those who adopted them.

The story this book tells, in broad terms, is one of growing individualism and diversity. The earliest heterodox movements it investigates tended to be highly structured social groups, with formal organizations and an expectation that members would share a fixed system of beliefs. As the century came to an end, however, these organizations fragmented, and believers became increasingly free to construct eclectic, personal creeds. By 1930, this

[26] See, e.g., the anticult judge Georges Fenech, who attributes the continuing prominence of heterodoxy in France to the post-1968 "decline of traditional religions, the collapse of family structures, and the crisis of Western values." Fenech, *Face aux sectes: politique, justice, état* (Paris: Presses Universitaires de France, 1999), 4.

[27] Wiktor Stoczkowski, *Des Hommes, des dieux et des extraterrestres, Ethnologie d'une croyance moderne* (Paris: Flammarion, 1999), 19–37.

[28] The tendency to perceive this current of heterodox innovation as distinctively contemporary is an artifact of the models historians and sociologists have used to determine the contours of "religious life" in France. When these observers consider the period before 1945, they see religion in the terms French historians have conventionally used to describe it: as a social and cultural field entirely defined by the partly overlapping spheres of "institutional" and rural "popular" religion. When these observers turn their attention to the contemporary scene, however, they define the field more broadly, including not only the shrinking spheres of "institutional" and "popular" religion but also a third sphere, which the Catholic scholar Jean Vernette has called "the new religiosity." See, e.g., Cholvy and Hilaire, vol. 3; Vernette, *Jésus dans la nouvelle religiosité* (Paris: Desclée, 1987).

change had nearly run its course: where the world of heterodoxy had once been characterized by commitment, it came to be characterized instead by independence. Meanwhile, beginning in about 1885, experimental psychology took shape as an autonomous discipline. During this early period, many psychologists considered psychical research—the scientific study of paranormal phenomena—to be a legitimate, if controversial, branch of their field. By 1930, however, psychical research had lost much of the intellectual prestige it had once enjoyed. As this connection to mainstream science weakened, heterodoxy moved further and further toward the margins of intellectual life. These twin currents of increasing individualism and increasing marginality place us at the origins of the social phenomenon Colin Campbell has identified as the "cultic milieu," with its distinctive "ideology of seekership" and "underground" ethos.[29]

The religious innovation described in the chapters that follow does not exist as a thing apart, but is instead integral to the intellectual, cultural, and social history of the nineteenth and early twentieth centuries. During the period covered here, French heterodoxy changed in ways that reveal broader preoccupations. Figures like Mathieu, Huet, and Flammarion, for example, saw the soul as an unproblematic entity, capable of manipulating physical objects by means of an imponderable substance similar to electricity or the "universal fluid" that Mesmerists posited in the 1770s; for this reason, they often envisioned the séance table as a kind of battery or wireless telegraph, a device activated by the "coil" a group of experimenters created when they joined hands. The cosmologies they espoused drew heavily on the various forms of Romantic Socialism that emerged in France during the first half of the century and on Comte's vision of historical progress. Figures like Roché, on the other hand, considered the soul and the cosmos to be altogether more complex. Instead of deriving their approach from the physical sciences, late-century seekers tended to look to the nascent science of psychology, especially the new conceptions of multivalent human consciousness elaborated by thinkers such as Pierre Janet, Frederic W. H. Myers, and Theodore Flournoy. These innovators also expressed a distinctively fin-de-siècle ambivalence about the narrative of progress that had seemed so authoritative in previous decades: They self-consciously rejected Positivism, making new claims for the value of ancient texts that Comteans would have dismissed as useless relics of a primitive age.

Religious innovators, then, were not mere eccentrics fulminating to an invisible audience; they were active participants in some of the period's

[29] Colin Campbell, "The Cult, the Cultic Milieu, and Secularization," in *The Cultic Milieu: Oppositional Subcultures in an Age of Globalization*, ed. Jeffrey Kaplan and Heléne Lööw (Oxford: AltaMira Press, 2002), 12–25.

most important debates. The best-known of them were among the most widely read popular philosophers of their time, and they were acknowledged—if sometimes ridiculed—participants in a rich, contentious dialogue on the nature of faith that involved journalists, clergymen, literary writers, philosophers, psychologists, and scientists. These heterodox thinkers and believers left a deep imprint on the French cultural landscape. For both their advocates and their adversaries, they introduced an array of intriguing new identities—the spirit medium, the researcher of paranormal phenomena, the modern mage. The novel conduct and thought of religious innovators also provided commentators with distinctive opportunities to assess the subjective impact of modernity's numerous material and social changes. To do justice to the richness of the subject, therefore, this book seeks to understand heterodoxy as a broad socio-cultural phenomenon, constituted not only by the ideas and practices of believers but also by the responses those ideas and practices provoked. The analysis I present encompasses a wide array of voices, from artisans and shopkeepers to well-known figures such as Victor Hugo, Louis Veuillot, Emile Littré, and the Nobel Prize–winning physiologist Charles Richet.

This book comprises five chronological chapters and an epilogue. Chapter 1 describes the 1853 arrival of American Modern Spiritualism in France. News of this movement and the practices of its adherents triggered a short-lived but intense French fascination with such curiosities as *tables tournantes*—tables that seemed to turn of their own accord under the hands of séance participants. Each section of the chapter explores the specific ways in which a particular group—journalists, Catholic thinkers, Academic scientists, and leftists disillusioned by the fall of the Second Republic—grappled with the questions these strange phenomena raised. Chapter 2 is a history of Mesmerism's development in the first half of the nineteenth century. In 1853, Mesmerism was a form of occult science with a long French pedigree; the incursion of American spiritualism changed it in a variety of revealing ways. Above all, the new religion from across the Atlantic introduced the practice of the séance, and with it, a novel and disconcerting social type: the spirit medium.

By the mid-1860s, a movement called Spiritism had emerged as the dominant force in French heterodox religious life. Chapters 3 and 4 trace its rise and gradual decline and explain how this trajectory was tied to broader social and political developments. Founded in 1857 by a former mathematics teacher who called himself Allan Kardec, Spiritism sprang directly from a strain of Mesmerism but far surpassed its predecessor in popularity. Chapter 3 provides an account of Kardec's ideas and an analysis of the practices of those who adopted them. Spiritism succeeded, this chapter argues, because it placed consoling dialogue with the souls of the deceased in a philosophical and ritual context that struck believers as familiar, modern, and serious. Chapter 4 describes a turning point in the history of Spiritism—a highly publicized trial that took place in 1875, in which several Spiritists were convicted of producing and marketing false spirit photographs. The events leading up to this trial and the widespread polemic between republican and Catholic journalists that followed it show the revealing role Spiritism played in the "war of religion" that shaped the political, cultural, and intellectual history of France after the fall of the Second Empire in 1870.

During the period from 1880 to 1914, heterodox beliefs and practices attracted greater interest than ever before. At the same time, however, the landscape of French religious innovation changed dramatically: The domination of a single doctrine gave way to pluralism. Chapter 5 and the Epilogue provide an account of this shift, in which old organizations fragmented and believers became increasingly willing to develop their own distinctive approaches. Chapter 5 relates the fin-de-siècle proliferation of heterodox ideas to two linked developments: the growing prestige of psychical research and the emergence of a novel conception of the psyche that stressed the importance of intellectual activity beyond the compass of ordinary waking awareness. The Epilogue is a brief overview of the years following the First World War and the emergence of the fluid, diverse, and individualistic "cultic milieu" that now exists in France and so many other Western countries.

A Note on Methodology

In France, the period from 1853 to 1930 witnessed a wide ranging search for a modern faith, for new forms of "evidence of things not seen." While this book makes no claims for the empirical reality of the phenomena so enthusiastically endorsed by the heterodox seekers it discusses, it also does not present their beliefs as signs of a persistent, atavistic irrationality. These believers, after all, did not see themselves as either backward-looking or superstitious. On the contrary, they considered their approach utterly rational, and, in elaborating their views, they drew self-consciously on their knowledge of scientific discourse and method. Indeed, the multifarious visions of a "factual" metaphysics that heterodox thinkers advanced during this period were as much a part of the emerging landscape of modernity as the railway or the telegraph.

The goal of this book, then, is not to unmask delusion and expose the "truth" behind these phenomena. Instead, it seeks to explain and analyze the conflicting—and often contradictory—ways in which individuals sought out, described, and made sense of a particular type of experience, one that appeared to provide empirical confirmation for metaphysical concepts. As a result, unless evidence of fraud would have been clear to the historical actors I discuss, I have sought to maintain a stance of epistemological agnosticism. What is important here is not whether these phenomena were authentic but instead how they appeared to particular people at particular times. The uncanny manifestations described in the pages that follow have a way of changing shape to reflect the preconceptions of those observing them. Charting these shifting forms reveals aspects of religious experience—and subjectivity more generally—that might otherwise elude the historian's grasp.

Similarly, to consider these nineteenth-century ideas and practices to be mere curiosities, marginal expressions of anxiety in the face of change, is to ignore a fact already familiar to sociologists: that heterodoxy of a kind very similar to this continues to play an important role in contemporary Western religious life. In France and elsewhere, the vitality of the New Age movement, UFO religions, channeling, and other forms of heterodox belief and practice demonstrates that the desire to make metaphysics concrete in novel ways—like the notion of self-expression through consumerism or the idea of artistic bohemia—is an enduring conceptual strategy the long nineteenth century has bequeathed to the present.[30] The social, economic, and political changes we associate with modernity turn out to have provided astonishingly fertile terrain for the development of religious thought and practice. The nineteenth-century crisis of factuality did not render religion obsolete; it caused it to transform. The heterodox movements discussed in the following chapters were at the forefront of this transformation and therefore provide a key to understanding its dynamics.

[30] Michael F. Brown, *The Channeling Zone: American Spirituality in an Anxious Age* (Cambridge, MA: Harvard University Press, 1997); Wouter J. Hanegraaff, *New Age Religion and Western Culture: Esotericism in the Mirror of Secular Thought* (Albany: State University of New York Press, 1998); T. H. Luhrmann, *Persuasions of the Witch's Craft: Ritual Magic and Witchcraft in Present Day England* (Oxford: Blackwell, 1989); Sarah Pike, *New Age and Neopagan Religions in America* (New York: Columbia University Press, 2004).

CHAPTER ONE

Interpreting the *Tables Tournantes*, 1853–1856

Early in 1848, Kate and Maggie Fox, two young sisters in rural New York State, began to receive mysterious communications from the beyond. These took the form of "spirit raps," sharp sounds that emanated from walls, furniture, or any other hard surface. Shortly after the raps first occurred, members of the Fox family started to ask questions of the unseen force that produced the noises. Initially, the answers were simple: a single tap for "yes," silence for "no." Using this unwieldy method, the Foxes and their growing circle of intrigued guests determined that the soul of a murdered peddler produced the "raps." The sisters eventually found they could summon these noises at will, in any location, and that a tremendous variety of spirits heeded their calls. A Quaker from the nearby town of Rochester discovered that these spectral visitors could also communicate by means of a "spirit telegraph:" the human questioner recited the alphabet slowly, over and over, while the raps spelled their dispatches by marking each appropriate letter. Word of these novel manifestations spread quickly, and the two Fox girls, under the shrewd management of their sister Leah Fox Fish, embarked on spectacular careers as interlocutors with the other world.[1]

[1] Retellings of this origin-story appear frequently in histories of Anglo-American spiritualism. For the earliest version, see E. E. Lewis, "A Report of the Mysterious Noises, Heard in the House of Mr. John D. Fox, in Hydesville, Arcadia, Wayne County, Authenticated by the Statements of the Citizens of that Place and Vicinity" (Canandaigua, NY: E. E. Lewis, 1848). For other versions, see Ann Braude, *Radical Spirits: Spiritualism and Women's Rights in Nineteenth-Century America* (Boston: Beacon Press, 1989), 10–19; Eliab Wilkinson Capron, *Modern Spiritualism: Its Facts and Fanaticisms, Its Consistencies and Contradictions* (Boston: Bela Marsh, 1855); Barbara Weisberg, *Talking to the Dead: Kate and Maggie Fox and the Rise of Spiritualism* (New York: HarperCollins, 2004).

American believers in these strange phenomena founded a new religion, which came to be called Modern Spiritualism. Spiritualism's cornerstone was the belief that the living could contact the souls of the dead. Only certain gifted individuals, like the Fox sisters, could establish these connections, spiritualists believed. The faithful called these privileged intermediaries "mediums." A medium's contacts with the spirit world tended to take place only under special circumstances. Usually, spirits appeared during "séances," rituals in which a number of believers, seated in darkness around a table, often holding hands, reinforced the medium's powers with their concentration.

During the late 1840s and early 1850s, Modern Spiritualism rapidly grew into an influential social and intellectual movement. Mediums and spiritualist societies sprang up across the United States. While the opinions of individual groups varied, the spirits that communicated tended to be advocates of social reform, including the abolition of slavery; they elaborated cosmologies inspired by the eighteenth-century mystic Emmanuel Swedenborg and the contemporary visionary Andrew Jackson Davis, known as the "Poughkeepsie seer."[2] Ritual changed with this growth and development: rather than conjuring raps from distant places, some mediums spoke for spirits or served as vehicles for "spirit writing." Others concentrated their attention on the table around which séance participants sat, having spirits turn it, levitate it, or cause it to produce the now-familiar alphabetic raps.

News of the marvels said to occur in American séances spread to Britain, and by the middle of 1852 had also reached the Continent. German newspapers began to publish reports of séances held on European soil in early 1853. In late April, French journalists turned their attention to the subject. Within weeks, burgeoning coverage had helped create a full-fledged vogue for these phenomena, which commentators dubbed *tables tournantes* (turning tables). Publishers quickly took advantage of this widespread interest: a flurry of articles, pamphlets, and books appeared, providing instructions for holding séances, attempts to explain the new phenomena, visionary speculation about their possible significance, and a wide variety of polemical diatribes. Popular enthusiasm dwindled during the summer of 1853, but these uncanny manifestations proved too compelling to disappear entirely and remained a subject of passionate debate.

French discussions of the *tables tournantes* owed much of their intensity to their social and political context, which was deeply marked by the

[2] For discussions of the ideas and political concerns of American Spiritualists, see Braude; Barbara Goldsmith, *Other Powers: The Age of Suffrage, Spiritualism, and the Scandalous Victoria Woodhull* (New York: Knopf, 1998); R. Laurence Moore, *In Search of White Crows: Spiritualism, Parapsychology and American Culture* (New York: Oxford University Press, 1977), 3–129.

Revolution of 1848. In February of that year, the July Monarchy collapsed in a wave of popular demonstrations, giving way to a democratic government that came to be called the Second Republic. During the heady months after the revolution, it seemed to many as if France were on the verge of full-scale social transformation. Under the provisional government that replaced the deposed King Louis-Philippe, advocates of a wide array of socialist ideologies took an unprecedented role in political life, thrilled by the prospect of turning theory into practice. This period of idealism did not last long, however. Radical programs of social restructuring appealed strongly to urban populations, but worried the rural majority. As a result, the first elected government of the new republic was decidedly conservative; in June 1848, its repeal of earlier reforms triggered several days of bloody rioting in Paris and other cities, ending the idealistic phase of the revolution.

Several months later, in December, Louis-Napoleon Bonaparte, nephew of the former emperor, was elected president of the republic. On December 2, 1851, after assiduously cultivating the allegiance of conservative factions—the Catholic Church in particular—he ended the republic with a coup. This triggered a second wave of uprisings, more widespread than the first, but quickly suppressed. A year later, Louis-Napoleon Bonaparte became Emperor Napoleon III, and the Second Republic gave way to the Second Empire. During his first decade in power, the new emperor took drastic steps to repress the currents of dissent that had burst forth in 1848. His regime rigorously censored the press and regulated political discussion of all kinds; an augmented police force kept social clubs and periodicals under strict surveillance.

In this tense atmosphere, the *tables tournantes* became a flash point for debate. As the four case studies presented in this chapter demonstrate, these new phenomena had a disconcerting way of changing shape to reflect the preoccupations of those observing them. For popular journalists, Catholic clergy and laypeople, members of the scientific community, and disenchanted leftists, *tables tournantes* became ciphers that addressed an array of often urgent intellectual, social, and political concerns. Journalists established the conceptual vocabulary that would dominate all subsequent discussions of the new phenomena, but their primary interest in the subject seems to have been pragmatic: *Histoires de table* provided a supply of diverting, inoffensive material in a period when censorship had dramatically limited the range of acceptable topics. Catholic priests and writers simultaneously justified and assuaged post-1848 conservative anxieties by presenting the Devil as a frighteningly tangible presence in the séance room. These accounts made the *tables* agents of moral turpitude and spreaders of revolution, but they also transformed them into tangible proof of the rightness and power of Catholic dogma. For members of the

Académie des sciences and their allies, explaining the *tables tournantes* became a way of linking science with the authoritarian, technocratic ethos of the Second Empire. Debunking the new phenomena allowed members of the Academy (*académiciens*) to elaborate an image of the scientist as objective guardian of rationality and, by extension, as protector of the social stability that 1848 had threatened. Finally, for socialists and republicans demoralized by the collapse of the Second Republic, these new phenomena were a source of hope and metaphysical consolation. In their séances, the entities that spoke through the *tables* explained the left's political failures while reaffirming its transcendent aspirations. The séance vogue was not simply a revealing cultural phenomenon, however. By providing what many considered to be a distinctively modern, factual corroboration for metaphysical principles, it also touched off a far-reaching process of religious change.

The *Tables Tournantes* Vogue in the Press

While a few reports of American spiritualism had appeared sporadically in 1852, the French fascination with *tables tournantes* began in earnest on April 20, 1853, when the *Constitutionnel* published a story describing a strange German fad. People in Bremen had discovered that they could cause tables to move "without visible impulsion."[3] A doctor named Karl André published the first account of this mysterious phenomenon in the *Augsburger Allgemeine Zeitung*; others followed rapidly. By the first week of May, French newspapers major and minor began publishing similar letters from local readers, who described various instances of this strange new "rotatory phenomenon."[4] Pamphlets giving instructions for the holding of séances appeared shortly thereafter. Playwrights, humorists, *feuilletonnistes* (writers of popular newspaper serials), and science journalists were also quick to comment on these phenomena; the daily *La Patrie* even instituted a regular "Occult Science Bulletin."[5]

Though sensationalistic and often light-hearted, these letters, newspaper stories, pamphlets, songs, plays, and caricatures established many of the terms and concepts that would recur in subsequent discussions of the *tables tournantes*. Most strikingly, the early accounts presented these phenomena as novel and quintessentially modern. Often, writers couched their accounts in a neutral, self-consciously precise "scientific" language.

[3] Quoted in *Le Journal du magnétisme* 12 (1853): 201. For more on the German beginnings of the Continental séance vogue, see Corinna Treitel, *A Science for the Soul: Occultism and the Genesis of the German Modern* (Baltimore: Johns Hopkins University Press, 2004), 30–32.

[4] *Le Siècle*, May 13, 1853, 3.

[5] The first of these columns appeared in *La Patrie*, May 11, 1853, 2.

When analyzing the broader significance of the phenomena, they tended to use a rhetoric of discovery. Many of the strongest advocates were already students of Mesmerism and, crucially for the subsequent career of the *tables*, readily drew on its vocabulary in their accounts. This improbable fad was also a boon for satirists, who quickly noted the myriad ways in which séances seemed to foster credulity and upend social convention.

A typical letter appeared on May 13, 1853, in *Le Siècle*, the most widely circulated daily newspaper of the period. In it, Achille Cherreau, a man the editor assured his readers was "a well known and respected doctor," described his own curious experiment with *tables tournantes*. The previous Sunday, Dr. Cherreau had gathered a group of friends to attempt to move "a miniature pedestal table of the kind that hawkers sell in the street for 29 *sous*." His guests took their seats around the table, then placed their hands on it, palms down, fingers spread, "the little finger of the right hand touching its palmary surface to the dorsal surface of the neighboring little finger, and so on, sequentially." Once the group had assumed this position, after a mere eight minutes of waiting, the table "underwent what appeared to be a kind of molecular tremor." After this "vibratory movement" had continued for several minutes, accompanied by muted cracking noises, the table began to turn, slowly at first, and then with steadily increasing speed. Cherreau's guests were obliged to rise from their seats to follow its movements. Eventually, the table appeared to take on a life of its own: it began to move about rapidly, forcing its way over uneven spots in the *parquet* and leading the experimenters in a circuitous path around the room.[6]

Cherreau insisted on the authenticity of the strange events he described, even as he acknowledged that they would have seemed incredible to readers who had not yet witnessed the *tables tournantes*. Whatever skeptics might think, Cherreau wrote, "this phenomenon exists; it is there, it can be seen, it can be touched, its character leaves no room for serious illusions." Readers were wrong to consider the *tables tournantes* to be farfetched, Cherreau argued. The nineteenth century had already proved to be remarkably "fertile in brilliant discoveries"—this new human ability to move objects spontaneously, with negligible physical effort, was merely the latest in the long string of scientific advances that had come to characterize the age. If mankind had been able to create the steam engine and the hot-air balloon, why should it not have the capacity to animate "inert bodies" with the force of will alone?[7]

The range of these dramatic new phenomena was not limited to simple movement. Most accounts, particularly those appearing in May, described

[6] *Le Siècle*, May 13, 1853, 3.
[7] Ibid.

hats, tables, salad bowls and other objects that seemed to turn under their own power when touched by a group of people.[8] Some experimenters, however, discovered that animated tables also had the ability to communicate, either by tapping the floor with their legs or by making mysterious cracking noises during recitations of the alphabet. These new manifestations, in which the phenomena appeared to convey messages, were named *tables parlantes* (speaking tables).

In the weeks after the *tables parlantes* made their debut, the phenomena journalists and readers described became increasingly uncanny. In the first stories that appeared, animated tables had simply repeated information already known to the séance participants; in subsequent accounts, they became considerably more prescient. Etienne Mouttet, author of the "Occult Science Bulletin" in *La Patrie*, for example, described a remarkable experiment he had witnessed. In a dining room with no clock, he and two other people caused a table to turn, then asked the table to tap out the time, first giving the number of hours, then of minutes. The table tapped nine times for the hour, and another nine for the minutes. Then, the "master of the house" went into the adjoining *salon* and discovered that the table's response was "perfectly exact."[9]

As uncanny stories like this one became more common, writers leapt to provide explanations. The most devoted advocates of the new phenomena, drawing on the vocabulary of Mesmerism, tended to ascribe them to an invisible "fluid" emitted by the human will, which seemed to function in ways similar to electricity. In what was probably the most widely read of all pamphlets on the *tables tournantes*, Dr. Félix Roubaud argued that this "magic fluid" was an entirely "positive" and "indisputable" force, the properties of which anyone could observe in experiments of their own.[10] The object to be turned had to be light: on the average, a single individual's fluid could move no more than 120 grams (4.2 ounces). If the object was a table or other piece of furniture, smoothly functioning rollers and an even parquet guaranteed more rapid results. In his own experiments, Roubaud claimed, objects tended to turn faster in rooms with northern

[8] See, e.g., *L'Estafette*, May 11, 1853, 3; *La Gazette de France*, May 9, 1853, 3; May 13, 1853, 3; *Le Pays*, May 5, 1853, 3; *La Presse*, May 11, 1853, 3, May 13, 1853, 3, May 15, 1853, 2, May 21, 1853, 2; *Le Siècle*, May 20, 1853, 3.

[9] *La Patrie*, May 27, 1853, 2.

[10] Félix Roubaud, *La Danse des tables, phénomènes physiologiques démontrés*, 2e éd. (Paris: Librairie Nouvelle, 1853), 8. This book went through two editions in 1853. Roubaud also made his instructions available in a smaller, more cheaply printed pamphlet, which also appeared in two editions. See Roubaud, "La Danse des Tables dévoilée, expériences de magnétisme animal pour s'amuser en société, manière de faire tourner une Bague, un Chapeau, une Montre, une Table, et même jusqu'aux têtes des expérimentateurs et celles des Spectateurs." (Paris: l'auteur, 1853).

Fig. 1. The explanatory engraving Roubaud included in his 1853 instruction book *La Danse des tables.* Note the contact between pinkies and the alternation of men and women. (Image courtesy of the Bibliothèque nationale de France.)

exposure. All the people forming the circle had to be in contact with one another and with the object being turned; each needed to have one little finger beneath their neighbor's, and the other on top, as demonstrated in an explanatory engraving (fig. 1). Healthy people between the ages of twenty-five and forty with "nervous temperaments"—by which Roubaud meant excitable natures—tended to produce these phenomena with particular ease. Men, because they had stronger wills, elicited spontaneous rotations more rapidly than women could. The best results of all, however, came from circles in which men and women alternated; this mixed composition also served to "dispel boredom," since normally twenty or thirty minutes passed before the table began moving. If the men and women in the circle felt attraction for one another, the phenomena would start considerably more quickly. In this way, Roubaud handily transformed table-turning from a pastime for amateur scientists into a titillating party game.[11]

The precision of Roubaud's instructions belied the remarkable mutability of these new phenomena. In the early stages of the *tables tournantes* vogue, every commentator described the behavior of the tables, and the conditions necessary to make them move, in a manner that justified his own pet theories. How "activated" tables, hats, and salad bowls might actually

[11] Roubaud, *Danse des tables,* 31–33, 68, 34–35, 42, 51.

have behaved remains unclear; it appears, in fact, that objects turned differently for different observers. A writer from Bordeaux who signed himself A. T., for example, produced a pamphlet that contradicted Roubaud's on a number of points. Where Roubaud maintained that the table would turn only if the people seated around it placed their hands in a particular position, one little finger covering that of their neighbor, the other covered, A. T. emphatically stated that such contact "IS NOT NECESSARY," and did nothing to facilitate the experiment.[12] In addition, unlike Roubaud, A. T. asserted that if the experimenters lifted their fingers from the table's surface one by one, the rotation would slow; it would accelerate again when the experimenters put their fingers back in place.

The peculiar blend of inconsistency, earnestness, and outright credulity that seemed to characterize many of these newspaper and pamphlet discussions provided a rich vein of material for satirists. In May and June of 1853, a torrent of humorous commentary appeared in the *grande presse,* in popular song, in the theaters, and in satirical newspapers like the *Charivari.* Perhaps the most striking aspect of this voluminous material is its repetitiveness; humorists used a surprisingly limited array of techniques and tropes in their sallies against the new fad. The satirical commentary took two primary forms: (1) direct mockery of the advocates of the phenomena and (2) more nuanced observation of the ways in which *tables tournantes* functioned in society. The *Charivari,* for example, published a letter signed by "An Amateur Physician," who prepared for his experiment by drinking several bottles of wine "for purely scientific purposes." Thus fortified, he returned to his house:

> Then—O prodigy most likely caused by my own electricity—my house suddenly began to turn with great speed, and not only the house, but the street itself, such that it was soon impossible for me to recognize myself or to locate my door.[13]

Satirists who wished to critique the new craze more subtly advanced the notion that these simultaneously banal and marvelous phenomena had figuratively turned society's tables, suspending the rules that normally governed relationships between people in public and in private. In the *feuilletons*—entertaining essays and serials that newspapers included to attract readers—and on the stage, ardent young men used experiments with turning tables or hats as opportunities to caress the little fingers of the women they courted; unscrupulous rakes held séances to

[12] A. T., "Les Tables tournantes, origine et découverte de ce phénomène" (Bordeaux: l'auteur, 1853), 11. Emphasis in original.

[13] *Le Charivari,* May 5, 1853, 2.

distract old pantaloons and seduce their pretty wives; indiscreet *tables parlantes* revealed the true ages of old maids; and workers ignored their bosses to attend to the higher imperatives of scientific discovery, ideally in collaboration with a pretty maid or shopgirl.[14] The series of lithographs Honoré Daumier devoted to the new phenomena—aptly titled "fluidomanie"—ran the gamut of these social-satirical tropes, presenting them with particular effectiveness (figs. 2–5, 6).

For all their formulaic qualities, these satirical commentaries were as revealing as the more serious accounts they sought to debunk. Both showed the *tables*' remarkable capacity to disrupt and challenge taken-for-granted ideas in ways that could seem either liberating or disturbing. Socially, the séance was a novel situation in which anything could happen: in the dark, as participants sat around the table, both the laws of physics and the rules governing interaction between the sexes were temporarily suspended. More profoundly, the *tables parlantes* seemed to contradict an array of fundamental assumptions about the nature of subjectivity, the power of the mind, and the limits of human knowledge. Whether real or imagined, these strange taps and movements demanded to be explained.

Tables Tournantes and French Catholics

As reports of séances became more common in the *grande presse*, explicitly Catholic commentary on the subject began to appear. This current of opinion developed gradually, moving through three phases. During the first, clerics and laypeople voiced a surprising variety of ideas: some dismissed experimentation with the *tables* as a mere party game, while others saw it as a potential source of moral edification. A few went so far as to argue that the new phenomena could serve as a valuable link between living believers and souls in purgatory. In the second phase, clerical and lay commentators proved increasingly willing to ascribe the new phenomena to demonic intervention, and did so in alarmist, highly colored language. The *tables*, according to these writers, were a rebuke from God—a spiritual plague intended to confound the arrogant certainty of scientific materialists. The third phase emerged gradually, as French bishops and archbishops began publishing pastoral letters against the practice of séances in late

[14] For courtship, see a short verse pamphlet, Jacquet, "Les Tables tournantes" (Nantes: Busseuil, 1853), and two *feuilletons* by Eugène Guinot in *Le Pays*, May 15, 1853, 1–2, and May 22, 1853, 1–2; for unscrupulous rakes and pantaloons, see a "Revue des Théâtres" recounting the plots of three unsuccessful new comedies, all called *La Table tournante*, in *Le Siècle*, May 30, 1853, 1, and H. Lefebvre, "La Danse des Tables, pochade en un acte" (Lyon: Vingtrinier, 1853); for negligent workers, see the first installment of Louis Desnoyers' *feuilleton* on the *tables*, *Le Siecle*, May 22, 1853, 1.

Fig. 2. A Daumier lithograph from 1853. The caption reads "An Indiscreet table.—What! This table dares to say that I am forty-eight years old! . . . . oh! horrors . . oh! my nerves! (The lady feels poorly, which does nothing to prove that she is under forty-eight.)" (Image courtesy of the Beinecke Library, Yale University.)

Fig. 3. Daumier presenting the *tables tournantes* as a bourgeois folly: "—Is it turnin' yet, sir? —Not yet . . . but we have only reached the sixty-third minute . . . leave us, Madeleine, do not disturb our fluid." (Image courtesy of the Bibliothèque nationale de France.)

Fig. 4. Daumier satirizing Catholic critics of the *tables*: "AN EXPERIMENT THAT HAS SUCCEEDED TOO WELL. Stop, stop, I say . . . you see, there is now no means of stopping this table . . . it has gone to the devil!" (Image courtesy of the Bibliothèque nationale de France.)

Fig. 5. "What all the different people of the Earth are doing at present." (Image courtesy of the Bibliothèque nationale de France.)

Fig. 6. Charles-Edouard de Beaumont presents the *tables tournantes* as romantic ruse: "—Look at that! My wife has also taken an interest in Mesmerism . . . she is trying to turn a hat all by herself . . . I am relived that she has adopted this scientific hobby. Now I may go out without any concerns about the use she makes of her free time!" (Image courtesy of the Bibliothèque nationale de France.)

1853. Though these prelates generally accepted the diabolical origin of the phenomena, they sought to moderate the rhetorical excesses that characterized the more popular critiques, proposing an alternative explanation that mixed psychology with demonology.

This evolving response to the *tables* was the product of a particular moment in French religious history. In the decades since the 1820s, the nature of French Catholicism had gradually been changing. As Gérard Cholvy has noted, Catholic preaching and practice increasingly emphasized the emotional aspects of religious experience; at the same time, French priests became more permissive in their approach to rural popular religion.[15] Visions, cases of demonic possession, and miraculous cures figured more and more prominently in French Catholic life. The 1850s saw the growth of the Cult of Mary, the visions at Lourdes, and a surge in pilgrimages. The willingness of so many commentators to accept the authenticity of the *tables tournantes*, then, reflected both a growing interest in tangible religious experience—whether positive or negative—and a greater openness to conceptions of the spirit world that resonated with popular beliefs.

Catholic books, pamphlets, and periodicals devoted to the new phenomena found an enthusiastic audience. The most successful publications were those that presented the *tables tournantes* as products of demonic intervention. The stories they told appear to have generated a *frisson* readers found appealing. The diabolists' arguments also tended to offer an indirect source of reassurance by emphasizing the effectiveness of exorcism, holy symbols, and faith as proof against evil. The Devil, as he appeared in these accounts, not only reaffirmed the moral framework of Catholic orthodoxy but also posed a threat that the Church could easily address—in striking contrast to the less tractable political and social tensions the events of 1848 had revealed.

Skepticism and Enthusiasm

During the spring of 1853, many Catholic journalists ignored the *tables* altogether. The few stories that discussed the phenomena tended to treat them as a mere popular diversion with no relevance to religious concerns. In the *Ami de la religion*, one of the most important Catholic journals of the period, for example, Léon Desdouits voiced his skepticism about the uncanny "*histoires de table*" that had begun to appear in the daily press. While he was willing to admit that a table could rotate spontaneously, Desdouits stopped well short of acknowledging the reality of *tables parlantes*. The

[15] Gérard Cholvy and Yves-Marie Hilaire, *Histoire religieuse de la France, 1800–1880* (Paris: Privat, 2000), 177–178.

phenomena in séances that seemed to indicate otherworldly intervention, he asserted, "must be attributable to some sort of hidden trickery."[16] The Abbé F. Moigno, who served both as science correspondent for *Le Pays* and as editor of the popular science journal *Cosmos,* took a similarly skeptical position. If tables, hats, and salad bowls seemed to move of their own volition under the hands of séance participants, he wrote, it was entirely the result of a "robust faith born of imagination, illusion, and preoccupation of mind."[17]

Other Catholic priests and journalists, however, were less willing to dismiss these new phenomena so abruptly. A few looked on them with enthusiasm, organizing and attending séances of their own.[18] For these early experimenters, the *tables parlantes* seemed to provide an intriguing new way of learning about the beyond. The Abbé Almignana, for example, a parish priest in the Parisian neighborhood of Batignolles, published a pamphlet in which he presented the *tables parlantes* as a powerful sign that the age of miracles had not yet passed. He carried out his experiments with the authorization of the archbishop of Paris—who had probably not anticipated the conclusions his parish priest would reach—and saw nothing untoward in the new phenomena.[19] "Apparitions of the dead to the living," after all, figured prominently in the Bible, and Ecclesiastes reminded believers that "the thing that hath been is that which shall be." Since for Almignana the phenomena occurred even when séance participants spoke the name of Jesus, touched the table with a crucifix, or sprinkled it with holy water, he also argued that these new manifestations were essentially benign.[20]

Henri Carion, a journalist for the ultraconservative *Gazette de France,* took a similar stance, though he pushed the point further. In a short book he argued that spirit communication could be useful to Catholics both as a political tool and as an aid to religious devotion. To bolster his case, Carion cited the results of his own experiments. He had begun by receiving messages from the beyond in the typical way, using a *table parlante* that struck the floor with one of its legs. He soon abandoned this cumbersome method, however, and switched to a *planchette,* a small piece of wood with a pencil attached, which could trace spirit messages directly onto sheets of paper. Once he adopted this new device, a host of renowned figures

[16] *L'Ami de la religion* 160, no. 5535 (Jun. 2, 1853): 544.

[17] *Le Pays,* May 13, 1853, 3.

[18] In addition to the texts by Almignana and Carion, cited in notes 20 and 21, see *L'Ami de la religion* 163, no. 5632 (Jan. 14, 1854): 113.

[19] See the letter from Almignana, dated June 17, 1852, in the Archives historiques de l'Archevêché de Paris, Carton 4E21, dr. M12 bis.

[20] Abbé A. Almignana, *Du Somnambulisme, des tables tournantes, et des médiums, considérés dans leurs rapports avec la théologie et la physique* (Paris: Dentu, 1854), 29, 11.

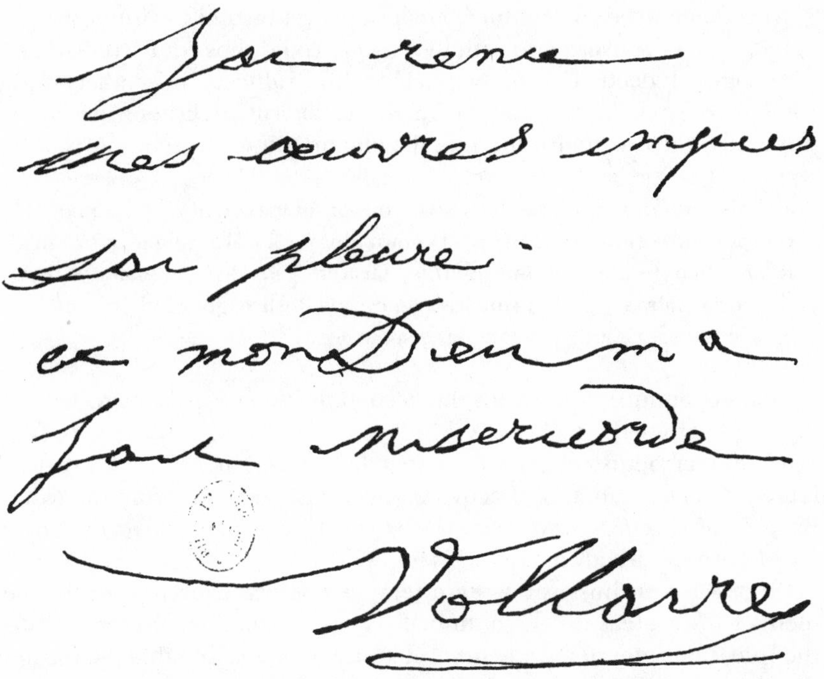

Fig. 7. Posthumous autograph attributed to Voltaire, received by the Catholic journalist Henri Carion via planchette and published in 1854. The text reads "I have renounced my irreligious works, I have cried, and my God has had mercy upon me." (Image courtesy of the Bibliothèque nationale de France.)

favored Carion with visits. Joan of Arc informed him that Louis XVI and his family, "martyred on the scaffold of '93," were in paradise. Charles X, the arch-conservative monarch unseated by the Revolution of 1830, sent a spiritual emissary to thank Carion for his support. Most remarkably, the spirits of Voltaire and Rousseau declared themselves to be in purgatory and to have recanted their previous ideas. At Carion's request, Voltaire produced an autograph (fig. 7), which he preceded with a brief statement: "I have renounced my irreligious works, I have cried, and my God has had mercy upon me." Rousseau showed a similar willingness to change his mind; he told Carion that the text he suffered for most was *Emile* because it was "the one in which I parody the ministry of my God." For Carion, the posthumous conversions of these heroes of the secular left were powerful political ammunition for the conservative devout.[21]

[21] Henri Carion, *Lettres sur l'évocation des esprits* (Paris: Dentu, 1853), 7, 66, 99, 100.

When conducted "within the bounds of discretion and Christian charity," the practice of conversation with the beyond could thus confer remarkable advantages. Indeed, Carion maintained, the value of this otherworldly commerce extended well beyond mere ideological vindication because it provided a "new demonstration" of the immortality of the soul,

> and whatever one might say, I have seen this proof make a more profound impression on worldly people than all known philosophical arguments. The sufferings and rewards of the afterlife, Heaven and Hell, Purgatory—that temporary abyss where divine Clemency and Justice go hand in hand—appear to us, in a way, with tangibly evident signs.[22]

Carion's communications with the beyond, in other words, provided empirical proof for the reality of Catholic teachings and, as such, were powerful inducements to piety. These new phenomena also provided forgotten souls a notably effective way to request prayers from the living. By animating a *planchette*, even the most obscure denizen of purgatory could now act directly to better his lot.

Despite his enthusiasm, Carion tempered his affirmative pronouncements with a measure of caution. Even as he extolled the benefits of the "pleasant interviews" the new phenomena made possible, he warned that summoning spirits entailed significant risks. "From both the physical and the moral point of view," he wrote, "it is, for the majority of mortals, gravely imprudent to indulge in these interviews with the Spirits." Carion claimed to have seen many women with excitable temperaments succumb to "veritable attacks of nerves" after entering into contact with spirits; excessive study of these phenomena, he maintained, had cost a number of "very learned men" their sanity. In addition to these physical dangers, Carion argued, these experiments inevitably entailed occasional brushes with the Devil. While saying the name of Christ always caused demons to flee, even brief contacts with the forces of evil could have serious consequences. He therefore urged his readers to abstain from replicating his experiments, out of "Christian prudence."[23]

Confronting the Demons of Revolution

Almignana and Carion aside, Catholic advocates of the new phenomena seem to have refrained from making their opinions public. The same was not true for a growing number of other commentators who expressed

[22] Ibid., ix, viii–ix.
[23] Ibid., 3, vii, viii, 3, 76.

an emphatically negative view of the *tables tournantes.* As early as the summer of 1853, pamphlets, books, and newspaper articles attributing the new phenomena to demonic intervention began to appear. By 1854, this point of view dominated Catholic discourse on the subject, even in the initially skeptical *Ami de la religion.* These accounts proved so popular among French readers that they quickly became a cultural phenomenon in their own right. The most famous book espousing this point of view, Jules Eudes de Mirville's *Pneumatologie, des esprits et de leurs manifestations diverses,* enjoyed a "prodigious success" after its appearance in 1853; it was widely reviewed and went through several editions.[24] In mid-1854, the prominent publisher Henri Plon, building on Mirville's success and an ever-increasing pamphlet literature, established a periodical called *La Table parlante,* devoted entirely to articles describing the new phenomena as products of demonic intervention.[25]

For these Catholic commentators, attacks on the séance vogue provided a point of departure for more sweeping critiques of Second Empire French society. Their countrymen had left traditional faith behind, these writers argued, and had succumbed to the easy charms of crass materialism. The *tables tournantes* were a celestial rebuke to those who had previously denied the reality of the supernatural. According to these critics, unseen spirit forces clearly caused the phenomena that took place in séances. It was foolish, however, to think such forces could be benign, since—in the eyes of these commentators, at any rate—the *tables tournantes* encouraged immoral conduct. Not only did the new phenomena pose an array of serious threats to the physical and mental health of individual experimenters, they also had grave political implications: the Devil, these writers argued, was using the *tables parlantes* to destabilize French society by reigniting the revolutionary fires of 1848.

The only way to overcome the multiple threats the new phenomena posed was a return to the Catholic faith. Fortunately, these critics believed, the widespread vogue for séances had paved the way for just such a revitalization of French religious life. In this respect, the diabolists were not so far from Almignana and Carion. The *tables tournantes,* which so clearly demonstrated the reality of the supernatural, they argued, had permanently disproved the philosophical materialism so many scientists and nonbelievers espoused. Even as the *tables* posed a threat, therefore, they offered hope for the future. The demons that appeared in séances were powerful adversaries, but they were also forces that the Catholic Church was uniquely suited to explain and conquer.

[24] *Le Journal du magnétisme* 14 (1855): 464.
[25] *La Table parlante, journal des faits merveilleux* (1854–1855).

These Catholic critics often made their case for the diabolical nature of séances by likening the *tables tournantes* to older, more apparently sinister forms of necromancy. In the process, they transformed these phenomena into something that bore little resemblance to the often-benign manifestations Almignana and Carion had described. According to the anonymous author of an 1853 pamphlet, for example, the true nature of the *tables tournantes* was most clearly visible in America, the sole country that had made them the basis of a new religion. To prove his point, the pamphleteer quoted a description of a séance in the "*pays des* Yankees" that had appeared in *Le Siècle*:

> A few of the new sectarians await the coming of the spirit immobile or absorbed in profound thoughts. But these silent meetings are not the most frequent. More often, the ceremony begins with a kind of witches' sabbath. At first, it is a nameless dance, which sweeps you away in its whirl, surrounded by confused sounds and inarticulate cries. When the faithful have reached a sufficient level of exaltation—high enough to make the spirits hear and obey, and to let the believers endure direct conversation with them—the dance stops. Then it seems as if the walls resound strangely with the sounds of repeated blows. These blows are the language of the souls who have just been called, and come running.[26]

This diabolical ritual was a long way from the innocent parlor games being played in French *salons*, but the two involved identical phenomena, the pamphleteer warned, and it was only a matter of time before diversion became Devil worship. Accounts like these transformed the new phenomena into the stuff of Gothic fantasy. Events that took place in American séances marked themselves as "diabolical" by assuming forms already familiar from history, literature, and folklore.

When these demonic manifestations appeared in French drawing rooms, and not far-off America, they tended to be more prosaic but no less patently evil. The *Ami de la religion*, for example, cited a letter that had appeared in a provincial newspaper called the *Abeille de la Ternoise*, in which a reader described an experiment carried out by a priest from a village near Versailles. He and his fellow experimenters began their séance by praying for the true nature of the *tables parlantes* to be revealed to them. After several minutes, the table began to move, and one of the experimenters asked it: "Who are you?" The table responded by tapping out the word "DEMON."[27] The Abbé Louis Bautain had an array of similar facts to report in his 1853 pamphlet on the *tables*. In his experiments,

[26] "Examen raisonné des prodiges récents d'Europe et d'Amerique, notamment des tables tournantes et répondantes, par un philosophe" (Paris: Vermot, 1853), 29.

[27] *L'Ami de la religion* 162, no. 5624 (Dec. 27, 1853): 760. Emphasis in the original.

Bautain wrote, animated tables abruptly stopped rotating if touched by holy water or a crucifix. Making the sign of the cross "with faith" had a similar effect. Dialogues with the spirits led to equally clear conclusions, Bautain argued. When the *tables* consented to answer metaphysical questions, they generally did so in a decidedly unorthodox manner, denying the existence of purgatory and hell or even claiming "that there is neither God nor Providence, and that the universe is ruled by fate."[28]

To bolster their case for the Satanic nature of the phenomena, Catholic commentators also often stressed the physical, moral, and social dangers of séance practice, even when no demons explicitly revealed their presence. According to this view, the séance was particularly dangerous because it created a situation that fostered uncontrolled behavior. This lack of restraint posed threats on two levels. First, individuals who participated in séances could lose control of themselves, succumbing to convulsions, trances, or madness. Second, the diabolical spirits that communicated through the *tables* allegedly fostered a larger, political elimination of restraint, spreading revolutionary ideas among their auditors.

The ways in which mediums behaved when in contact with spirits provided Catholic critics with powerful examples of the dangerous wildness these otherworldly entities fostered. Conservative newspapers and tracts were full of accounts describing strange physical symptoms and nervous ailments.[29] The Chevalier Henri Roger Gougenot des Mousseaux, for example, described a séance he attended at the home of a respectable family in Paris, where a sixteen-year-old girl served as medium. Mousseaux observed that "the *natural* spelling of our young medium is pure, and her penmanship has all the facile and uniform elegance of English writing." When she became possessed by spirits and performed automatic writing, however, a sinister change occurred:

> The medium's hand immediately seizes the pencil, while her eyes, raised or wild, seem to seek a higher sphere; she lets the pencil run irregularly over long sheets of paper, and seems to be nothing but the Spirit's mechanical instrument. The path the pencil takes is convoluted, but supple and full of fancy, certain admirable bits of calligraphy excepted; for regularity and rectitude seem to fit neither the habits nor the tastes of these Spirits.

[28] Abbé Louis Eugène Marie Bautain (pseud. Un Ecclésiastique), "Avis aux chrétiens sur les tables tournantes et parlantes" (Paris: Devarenne, 1853), 21, 11.

[29] For nervous ailments, see *La Gazette de France*, May 14, 1853, 2–3, May 19, 1853, 2; *L'Union*, May 15, 1853, 3, May 23, 1853, 3; for trembling, see *L'Union*, May 4, 1853, 3 and Jules Eudes, marquis de Mirville, *Pneumatologie, des esprits et de leurs manifestations diverses* (Paris: H. Vrayet de Surcy 1863 [1853]), 1: 426.

Here was a well brought up and pretty girl, "who acts very much her age," voluntarily becoming a passive vessel to be occupied by unknown, unruly spirits. Once she was in contact with the beyond, as Mousseaux pointed out, the "regularity" and "rectitude" that had previously made her so charming fell by the wayside—her grace gave way to a mechanical stiffness, and her carefully trained writing became a wild, fanciful scrawl. In accounts like these, the séance was a dangerous moment, in which normal boundaries between good and evil, virtue and lasciviousness, reality and unreality, disappeared. Many people, women above all, Catholic critics argued, lacked the moral strength to recover from this awful liberation. Since these writers generally agreed that women figured disproportionately among séance participants, this argument carried particular force.[30]

For religious polemicists, séances posed the same threats to social well-being as they posed to that of the individual. In both cases, the chief damage came from the way in which the intervening spirits eliminated restraints usually kept firmly in place. Catholic writers generally described the dangers of this pathological license by directly or indirectly evoking memories of 1848. Mirville, for example, gave a tellingly shrill description of the political opinions that spirits expressed in séances:

> In the case of social institutions, it is necessary, they say, *to remake everything*, do you understand? And remake *everything from the foundations*, to divide land equally, abolish the death penalty, do away with all laws regulating debts, and above all, never extend tolerance to the Roman Catholic Church, *mother of all superstitions.*[31]

The democratization of religion propounded by visionary advocates of the new phenomena had clear political implications, which proved extremely disturbing for these conservative polemicists. In a religion based on the *tables tournantes*, anyone who cared to contact spirits was free to receive all sorts of subversive ideas, which would have the resonance of revealed truth. Like the novel religious systems propounded by the Romantic Socialists of 1848, the ideas the *tables* communicated did more than threaten the hegemony of Catholicism; they also jeopardized the position of anyone who enjoyed a privileged position in Second Empire society.

A propaganda conduit of this kind, these Catholic critics believed, was exactly what the demonic forces of disorder needed to bring about the

[30] Chevalier Henri Roger Gougenot des Mousseaux, *La Magie au dix-neuvième siècle, ses agents, ses vérités, ses mensonges* (Paris: Henri Plon, 1864 [1853]), 15, 8. Italics in the original.
[31] Mirville, 1:422. Italics in the original.

downfall of French society. Indeed, according to the Abbé Cognat, a frequent contributor to the *Ami de la religion*, the spirits already seemed to be laying the groundwork for a new network of radical anti-Christian secret societies, disconcertingly similar to the *montagnards* that had coordinated the 1851 republican revolt against Louis-Napoleon's coup d'état:

> We know, and can guarantee, that in several of our cities, there are self-proclaimed prophets who dream of a radical change in the order of things, the substitution of a new religion, new dogma, and new rituals for the symbols and rituals of Christianity. Now, the tables have revealed these prophets to others, who did not know them previously in any way, and who sought them out only after being advised to do so by *the spirit*!

The grand plans of Second Republic socialists had failed because they were ill-conceived human inventions. A new ideology, imbued with all the charms the Devil's ingenuity could muster, spread from city to city on demon wings, would prove considerably more difficult for the forces of order to combat. The number of believers in *tables parlantes* might as yet be small, the Abbé concluded, but the damage they had already done gave concerned observers "every reason to fear . . . the gravest disorders, the most monstrous aberrations."[32]

Dangerous as the *tables* might have seemed to them, these critics tended to argue that the popular fascination with séances also had a beneficial aspect. In a long review of Mirville's book, for example, the critic and novelist Jules Barbey d'Aurévilly supported the author's contention that the *tables tournantes* were a divinely sanctioned rebuke of scientific materialism. These strange phenomena, Barbey argued, tangibly demonstrated the limits of scientific knowledge and as a result paved the way for a new social order in which the Catholic Church would regain its authority:

> While Philosophy grows inconclusively muddled or . . . remains silent; while Pantheism, Eclecticism, and Rationalism respond to this question of spirits—as much a part of tradition everywhere on Earth as God and the fall of man—by denying history and closing their eyes to the contemporary phenomena, Catholicism rises up before a perplexed science and addresses the eternal problem, a problem it has solved in the same way at every point in its history. Once suspected of being the enemy of progress, today it is Catholicism, unperturbed by the advances of the natural sciences, that demands them to account for their worldly observations.[33]

[32] *L'Ami de la religion* 162, no. 5624 (Dec. 27, 1853): 759. Italics in the original.
[33] *Le Pays*, July 6, 1853, 4.

Science and secular philosophy, Barbey believed, were powerless to explain the *tables tournantes*, while the Church, with its venerable, unchanging knowledge of the ways of God, could account for the new phenomena handily. Scientists, when attacking the legitimacy of the Church, routinely discounted its doctrine by claiming that it lacked an empirical basis; the *tables* had been sent by God to provide that empirical foundation, humiliating arrogant rationalists in the process. In the face of such powerful evidence, Barbey argued, scientists would finally be forced to abandon their reductive materialism and acknowledge the irrefutable truth of Catholic teachings.

Ecclesiastical Condemnations

Despite the steadily increasing number of critics who proclaimed the diabolical nature of the new phenomena, official pronouncements from the ecclesiastical hierarchy did not appear until the séance vogue was six or seven months old. The influential Catholic pedagogue Félix Dupanloup, bishop of Orléans, denounced the *tables* from the pulpit in the fall of 1853.[34] The bishop of Viviers issued the first published pastoral letter on the subject in late November. He opened his letter by explaining why he had waited so long to express his opinion. At first, he wrote, experiments with the *tables tournantes* had struck him as a faddish "recreational exercise," too frivolous and ephemeral to merit ecclesiastical commentary. The situation changed, however, with the emergence of *tables parlantes*, and the growth of visionary speculation that accompanied it. Once experimenters began to approach their task "in a spirit of seriousness" and to view the new phenomena as "a means of tearing down the barrier that separates us from the invisible world," the Church could no longer remain mute. What had begun as "an amusing physics experiment," the bishop wrote, had transformed itself into something that closely resembled the "mysterious operations of magic, divination, or necromancy," all of which the Church forbade. This new turn of events demanded that he explicitly prohibit the faithful from engaging in such activities.[35]

Though he readily connected the *tables tournantes* to "spirits of the abyss," the bishop stopped short of fully endorsing the lurid claims other Catholic commentators had made in less formal contexts. In his view, it was more likely that "these marvelous phenomena exist only in the imaginations

[34] *Le Journal du magnétisme* 13 (1854): 150.

[35] Joseph-Hippolyte Guibert, évêque de Viviers, "Lettre pastorale de monseigneur l'évêque de Viviers, au clergé de son diocèse, sur le danger des expériences des Tables Parlantes." (Privas: Guiremand, 1853), 1.

of people who take part in the operations as agents or witnesses." Numerous reputable observers had testified to the reality of the phenomena, the bishop conceded, but even dignitaries could have their perceptions warped if they were "under the influence of enthusiasm." The absence of direct spirit intervention, however, did not make the *tables parlantes* any less diabolical. On the contrary, the chimerical nature of the phenomena testified to the "skillful and sly methods" the "infernal serpent" used in his exploitation of the "natural disposition that attracts mankind to all that is marvelous." By introducing this strange phenomenon in the form of an innocuous parlor game and then slowly influencing the imaginations of those experimenting with it, the Bishop argued, the Devil had succeeded in coaxing unsuspecting souls into the "abyss" while sparing himself the trouble of direct manifestation. Table-moving séances probably did not summon demons, but they rendered experimenters unusually susceptible to demonic inspiration. The belief that spirits could communicate through the *tables,* for the bishop, was thus as dangerous as the act of communicating itself.[36]

After the bishop of Viviers' pastoral letter appeared, many other prelates issued injunctions of their own. Between December 1853 and March 1854, the archbishops of Paris, Besançon, Rouen, and Albi, along with the bishops of Autun, Le Mans, Marseille, Verdun, and Dijon all condemned experiments with the new phenomena.[37] These denunciations took very similar approaches, adopting the balance the bishop of Viviers had struck between skepticism and rhetorically compelling, but potentially alarmist, denunciations of demonic intervention. In these clerical pronouncements, the possibility that the demons summoned in séances were only psychological made them no less threatening, and no less powerful as a justification for the continuing importance of the Catholic clergy's role in the regulation of human relations with the beyond. In July 1856, Pope Pius IX supported these proclamations with the encyclical *Adversus Magnetismi Abusus,* which prohibited Catholics from undertaking any experiments that involved, or seemed to involve, conversation with the spirit world.[38]

[36] Ibid., 10, 11.

[37] Most of these condemnations are reprinted as appendices to Ambroise Matignon, S. J., *Les Morts et les vivants, entretiens sur les communications d'outre tombe* (Paris: Adrien Le Clere, 1862), 106–139. Excerpts also appear in *Le Journal du magnétisme* 13 (1854): 113–122, 147–153. *L'Ami de la religion* printed all of them in their entirety. See *L'Ami de la religion* 163 (Dec. 1853—March 1854): 59–60, 173–174, 500–504, 519–524, 645–652, 701–702, 763.

[38] A broadside of this document, along with related correspondence, has been preserved in the Archives historiques de l'Archevêché de Paris, Carton 4E18, dr. E25.

Tables Tournantes and the *Académie des Sciences*

Like the Catholic response to the new phenomena, the scientific one moved through several stages, culminating in a synthesis that dealt with the phenomena in at least partially psychological terms. During the spring of 1853, the Académie des sciences quickly agreed on a hypothesis to account for the *tables tournantes*—that their rotation was the product of imperceptibly tiny muscular tremors produced by séance participants. A vocal group of advocates for the new phenomena, led by the well-known engineer Marc Séguin, protested the insufficiency of this hypothesis, which did little to explain the more spectacular manifestations that many observers reported. These advocates tended to present the question as a matter of scientific integrity with metaphysical implications. If approached with a suitably open mind, they argued, the new phenomena that appeared in séances promised to transform human knowledge, resolving the crisis of factuality that beset modern religious life: the *table tournante* would make it possible to conduct empirical studies of what had previously been the exclusively speculative realms of metaphysics and morality. Academic scientists foreclosed this advance in human knowledge by refusing to take the new phenomena seriously.

As reports of spectacular events in séances multiplied, the insufficiency of the initial Academic hypothesis became impossible to ignore. In response, *académiciens* and others who shared their views elaborated a more nuanced approach to these phenomena. The commentators who developed this new line of argument made few attempts to prove or disprove the empirical reality of particular manifestations. Instead, they turned their attention to the mental states of those who believed they had perceived them.

Shifting perspective in this way helped advance a conception of the scientist's role in society that was eminently suited to the technocratic ethos of the Second Empire. By adding a psychological dimension to their analysis, the second wave of *académiciens* and their allies transformed the terms of the debate about the *tables tournantes.* The conflict was no longer one between intolerance and openness to innovation; instead, it became a struggle between virtuous objectivity and destructive subjectivity, in the form of an all-too-human *amour du merveilleux* (love of the marvelous). When confronted with a dispute of this kind, these writers argued, the scientist's role was to guarantee future progress—to continue shepherding France toward the Positive Age—by protecting the nation's subjects from their own worst inclinations, using cool rationality to calm the excesses of fanaticism and superstition that psychology so effectively laid bare. Here, the stern, objective scientist, rather than the empathetic priest, became the expert best equipped to navigate the spiritual shoals of the modern age.

The First Academic Hypothesis

In the first months of the séance vogue, the Académie des sciences received several reports describing remarkable phenomena obtained during experiments with the *tables tournantes.* Three were from amateurs who had little or no standing in the Académie: Dr. Kaeppellin, a physics teacher at the Collège de Colmar; Vauquelin, a bailiff from Mortagne; and Bonjean, a pharmacist from Chambéry.[39] The obscurity of these writers made their accounts easy to dismiss. The same rule did not apply to the fourth report, however; it came from one of the most eminent corresponding members of the Académie, Marc Séguin, inventor of the wooden railway tie and the suspension bridge. François Arago, the *secrétaire perpetuel,* read Séguin's paper to the Académie during the meeting of May 19, 1853, and designated a commission of three members, Ernest Chevreul, Jean-Baptiste Boussingault, and Jacques Babinet, to study the new phenomena.[40]

The manifestations Séguin described were surprising indeed. He performed the bulk of his experiments with his friend Eugène de Montgolfier, a relative of the famous balloonist, who joined him in forming a human "chain" around a small, light walnut table. After a few hours, he wrote, "beating time along with the piano; indicating ages, numbers of people, and facts known to the person or persons in communication with [the table]; were experiments repeated a thousand ways, all with equal success." Séguin and Montgolfier even succeeded in causing a beaver hat to lift from the table top and hang briefly suspended in the air, so that only "a few hairs on the convex part of the crown" remained in contact with the surface. Taken together, Séguin wrote, these phenomena appeared to indicate that "the laws of gravity, in this circumstance, are completely inverted and subjected to a cause that has temporarily acquired superiority over them." He refrained from speculating about what caused this strange inversion, but the title he gave his *mémoire* provisionally ascribed it to an "electricity of a particular nature" emitted by the nervous system and controlled by the

[39] For Kaeppellin, see *Le Journal des débats,* May 19, 1853, 1; for Vauquelin, see *Le Siècle,* May 31, 1853, 3 and *Cosmos, Revue encyclopédique hebdomadaire des progrès des sciences* 2 (May 29, 1853): 662; for Bonjean, see *Cosmos* 2 (May 29, 1853): 664.

[40] *La Patrie,* May 24, 1853, 2.

human will.[41] According to the *Gazette de France*, hearing these conclusions from such a reputable author left many *académiciens* "powerfully moved."[42]

The week after he read Séguin's report to the Académie, Arago presented a detailed response that probably did much to put the disconcerted *académiciens* at ease. He began with an outright dismissal of Séguin's most disturbing claims. The report's accounts of tables guessing people's ages, tapping rhythmically to music, and otherwise behaving in a manner that seemed to manifest "the presumed action of the will upon inert matter," Arago declared, were "inadmissible."[43] No serious scientist, in his view, could accept the reality of phenomena that so manifestly contradicted the known laws of physics and physiology. Any attempt to explain such bizarre manifestations, or to study them experimentally, Arago concluded, would be futile. The simple rotation of hats and tables, however, was an unquestionably authentic phenomenon and deserved careful scientific scrutiny. To initiate this process, he proposed an explanatory hypothesis: The seemingly spontaneous rotation of hats and tables was "only the accumulation, summation, integration of small impulses that have built to produce a quantity of movement intense enough to provoke the displacement of even large masses."[44] Tables turned, in other words, because séance participants pushed them. The fact that each individual push was imperceptibly tiny made their cumulative effect seem to reveal the operation of a mysterious external force.

The *académiciens* seized this hypothesis as their official explanation for the *tables tournantes*. In addition to accounting for the phenomena in a manner that seemed rational, Arago's theory had the added advantage of building on the work of Ernest Chevreul, who had made similar arguments about the uncanny movement of the dowsing rod and divining pendulum in the 1830s.[45] In late June, this hypothesis received further confirmation when the English physicist Michael Faraday published an account of his own experiments with the *tables tournantes* and concluded that the physical pressure séance participants unwittingly exerted on the table-top caused the mysterious rotation.[46]

[41] Quoted in *La Gazette médicale*, June 4, 1853, 350.
[42] *La Gazette de France*, June 14, 1853, 3.
[43] *La Presse*, July 5, 1853, 2.
[44] *Cosmos* 2 (May 29, 1853): 665.
[45] Chevreul published a letter on this subject in 1833 in the *Revue des deux mondes*; it was reprinted in the *Journal des débats*, May 13, 1853, 2.
[46] See *L'Illustration*, July 9, 1853, 27; Alison Winter, *Mesmerized: Powers of Mind in Victorian Britain* (Chicago: University of Chicago Press, 1998), 276–305.

Objectivity and the Limits of the Possible

Arago's "imperceptible movement" hypothesis, even after it had received the confirmation of a scientist as illustrious as Faraday, struck many observers as insufficient because it failed to account for all the phenomena séance participants described. In the *Gazette médicale,* for example, journalist Jules Guérin mentioned the case of Dr. Prévost of Alençon, who reported that he and a group of twenty-one fellow experimenters had successfully caused a heavy billiard table to rotate a half-turn.[47] It seemed implausible that such spectacular results could be the work of tiny, imperceptible hand movements. Many other commentators took similar issue with the Académie's abrupt refusal to consider any of the more spectacular phenomena Séguin and other observers had described.

For these critics of the Académie, the crucial question was how responsible scientists ought to approach phenomena that appeared to be impossible—that is, patently at odds with principles widely accepted as true. The debate hinged on the question of objectivity, which all agreed was the sine qua non of valid scientific knowledge. One group asserted that a truly objective attitude entailed absolute open-mindedness. To deem even the most implausible-seeming phenomenon impossible without first subjecting it to thorough experimental study, in this view, was a violation of proper experimental method. The other group, which followed Arago's lead, held that current scientific knowledge was sound enough to furnish certain irrefutable principles. These accepted principles, in turn, could be used to make an objective distinction between the possible and the impossible in evaluations of particular phenomena.

The most widely publicized exchange on this subject began when Séguin sent a letter announcing his discoveries to the Abbé Moigno at the popular science journal *Cosmos.* Initially, Séguin wrote, he had shared Moigno's skepticism of the *tables tournantes.* When his experiments first began to yield their strange results, Séguin believed he was "under the influence of a hallucination." Since subsequent experiments duplicated the initial ones, however, he wrote, "it is impossible for me to deny the evidence, even when it overturns and confounds all my ideas." As a responsible scientist, Séguin felt obliged to publicize his findings and to assert their authenticity. It made no difference that these findings contradicted widely accepted physical laws, he argued, since "laws are the slaves or the humble followers of facts . . . , nothing but their empirical and scientific expression." To refuse a fact a priori, purely because it contradicted cherished nostrums, therefore, was an inherently unscientific act. The moment scientists began to use invented laws to discount inconvenient pieces of empirical evidence,

[47] *La Gazette médicale,* May 28, 1853, 335.

Séguin argued, the process of scientific discovery that had produced such astonishing results throughout the nineteenth century would cease.[48]

Séguin's plea for scientific open-mindedness left Moigno unconvinced. The famous engineer's emphatic language did not compensate for the ambiguity of the facts at issue. Séguin claimed that the mysterious phenomena he had witnessed could not be explained with any existing scientific concepts and required a new hypothesis: the notion that the human will could emit a kind of "electricity" capable of moving inert objects. Moigno contested this assertion by examining the conditions under which this novel "electricity" appeared. The new phenomena, he noted, were notoriously resistant to systematic experimental scrutiny. Many people, Moigno included, could not produce them at all; even successful experimenters could not replicate them reliably. Scientists, Moigno argued, rightly used this instability as justification for discounting the "electricity" hypothesis and looking skeptically upon data that seemed to support it.

In the *Gazette médicale*, Guérin strongly criticized this rejection of the "particular electricity" hypothesis. Moigno's argument had long been used against Mesmerism, which involved similarly unstable phenomena, and Guérin's defense was precisely the one Mesmerists had invoked for decades.[49] The elusiveness of these new phenomena in the laboratory, he argued, did not tell against them. Indeed, by placing such a heavy emphasis on experimental replicability, he believed, Moigno and the *académiciens* had fundamentally misconstrued the nature of the manifestations. *Tables tournantes* were not physical phenomena, like chemical reactions; they were *physiological* ones, like aches that appeared only when the weather was sufficiently cold. Hence, Guérin maintained, these novel manifestations depended on a host of extremely complex biological conditions that no scientist could ever hope to control completely in a laboratory.[50]

While Séguin and Guérin presented this debate over scientific open-mindedness as above all an issue of experimental methodology, the comments of others who shared their views indicate that larger issues were also at stake. The Protestant count and former Second Republic deputy Agénor de Gasparin expressed this position most clearly in his two-volume study of the *tables*, which supported Séguin's contention that an invisible force emitted by the will caused the phenomena.[51] A thorough scientific

[48] *Cosmos* 2 (May 19, 1853): 612, 614, 612–613. This correspondence also appeared in *Le Pays*.

[49] For a detailed analysis of this question's role in the Academic career of Mesmerism, see Bertrand Méheust, *Somnambulisme et médiumnité* (Le Plessis-Robinson: Synthélabo, 1999), 1: 412–469.

[50] *La Gazette médicale*, May 28, 1853, 336.

[51] Comte Agénor de Gasparin, *Des Tables tournantes, du surnaturel en général et des esprits*, 2 vols. (Paris: Dentu, 1854).

investigation of this force, Gasparin argued, would cure what he saw as the primary intellectual malaise of his time, a reductive materialism caused by the "despotism of the positive sciences." For him, the Académie's skeptical reaction to the *tables tournantes* was a classic example of the orthodox scientific community's pathological tendency to reject any phenomenon that seemed to entail the contemplation of the transcendent. The fact that a table could rotate under the sole impulsion of the human will, in Gasparin's view, demonstrated that the soul was as tangible as any other phenomenon in nature, and that "there are other phenomena than those the telescope perceives or the scalpel exposes." By proving the empirical reality of the soul in such a striking way, the *tables tournantes* would trump the materialism of the scientific community and restore philosophy to its rightful place as the most prestigious of all intellectual disciplines.[52]

Gasparin's optimistic view was implicit in many other arguments for an extension of the realm of the possible to include these new phenomena. By bringing the soul into focus as an object of objective, empirical study, the *tables tournantes* seemed to provide the starting place for a novel approach to metaphysics. They would be the tool that would allow human beings to achieve the ultimate, utterly solid knowledge of the cosmos and their place in it that so many French commentators of the period saw as the gleaming end-point of all scientific progress. In this new world, the existence of the soul would cease to be a philosophical postulate and become a simple fact. When humanity had reached this final goal, science would become a moral force far more powerful than Catholicism had ever been, creating the future that Victor Hugo—another experimenter with *tables tournantes*—poetically envisioned as an airship's ascent "toward the religious and holy truth."[53]

A Cold Eye for the Inclinations of the Heart

In 1854, Babinet and Chevreul, two members of the three-man commission Arago had appointed to study these phenomena, addressed critics like Séguin and Gasparin by advancing more nuanced explanations for the *tables tournantes.* Both agreed with Arago's hypothesis that tiny unconscious movements contributed to these strange manifestations, but they also assigned a crucial role to the imagination, aided by an all-too-human desire to be amazed. The presence of the imagination, in their view, was clearest in séances where the animated table, or a medium, seemed to receive communications from the beyond. For these

[52] *Le Journal des débats,* Aug. 30, 1853, 1.

[53] Victor Hugo, "Plein Ciel," section LVIII of *La Légende des Siècles* (Paris: Gallimard, 1955), 729.

manifestations, it made more sense to examine the minds of the séance participants themselves.

When Babinet and Chevreul shifted their attention from phenomenon to observer in this manner, they changed the terms of the debate. What had previously been a conflict between two different conceptions of objectivity became a conflict between objectivity and its opposite—a subjective, atavistic *amour du merveilleux.* This conceptual shift marked a revealing stage in a broader process: the emergence of a new way of envisioning the scientist's role as producer of knowledge. Beginning in the 1830s, as Lorraine Daston and Peter Galison have argued, objectivity acquired a powerful moral significance to go along with its epistemological meaning.[54] Students of the sciences began to speak of their task in terms of virtue and self-denial, making the practice of detached, careful observation into an ascetic discipline. An objective scientist had to restrain his all-too-human desire to see particular theories confirmed in order to let nature "speak for itself." Objectivity, here, became a matter of negating unruly human desires for understanding and certainty in the name of truth and social order.

Babinet and Chevreul argued against advocates of the new phenomena by asserting that any signs of intelligence a *table tournante* might manifest—like an ability to communicate using a spirit alphabet—came entirely from the direct physical and intellectual action of the séance participants, not from otherworldly entities or mysterious psychic forces. Chevreul intended his study to dispel these illusions:

> I hope to show definitively and precisely how people of light-hearted temperament, under the influence of that love of the marvelous so natural to man, cross the boundaries of the known, the finite; and how, unwilling to bring a considered assessment to bear on new opinions that have the allure of the marvelous and supernatural, they precipitously accept what, examined with a cold eye, would be phenomena of a kind amenable to human explanation.[55]

The desire to be in contact with a world beyond the material was an ancient human weakness, Chevreul argued, and it clouded the judgment of otherwise rational people when they attempted to make sense of the *tables.* Here, objectivity did not entail a willingness to consider the authenticity

[54] Lorraine Daston and Peter Galison, "The Image of Objectivity," *Representations* 40 (Fall 1992): 81–128; Galison, "Judgment against Objectivity," in *Picturing Science, Producing Art,* ed. Caroline A. Jones and Galison (London: Routledge: 1998).

[55] Ernest Chevreul, *De la baguette divinatoire, du pendule dit explorateur et des tables tournantes, au point de vue de l'histoire, de la critique et de la méthode expérimentale* (Paris: Mallet-Bachelier, 1854), 26.

of these manifestations; instead, it became a way to resist their fascinating influence. Chevreul's carefully cultivated scientific detachment and self-control—his "cold eye"—allowed him to conquer any irrational inclinations and see the new phenomena for what they were.

Babinet cast the struggle between scientific objectivity and the innate human weakness for the marvelous in similar terms. Using the *tables tournantes* as his case in point, he argued that one of the central missions of scientists and educated bureaucrats should be to protect society from the perennial resurgence of human irrationality. The séance vogue proved that even the French could occasionally behave like unenlightened beings, and that the state must therefore never relax its vigilant efforts to protect and extend the domain of reason. There were only two ways to control humankind's tenacious love of the marvelous: scientific education and a strong police force, which Babinet insisted should punish unscrupulous mediums for the abuses they would inevitably perpetrate. It was a clear case of the intellect against "the inclinations of the heart," which were atavistic but impossible to eliminate. Babinet maintained that a similar frenzy had overtaken France during the revolutions of 1830 and 1848, but with the *tables*, the unreason had become spiritual rather than political, and would therefore be easier to control.[56]

Though they tended to describe the innate human *amour du merveilleux* with an indulgent condescension, these scientists were not entirely sanguine about the possible social consequences of the séance vogue. Babinet worried primarily about the corrupting influence frequent participation in séances might have on otherwise well-bred young women.[57] Chevreul, thinking more broadly, linked the rise of the *tables tournantes* to a decline in orthodox religious faith and a thirst for novelty among the young, both of which he believed were products of a distinctly modern malaise. The allure of the *tables tournantes*, if not dispelled by reason, Chevreul argued, would provide the basis for a new religion. The specious foundations of this innovative belief system, totally without roots in tradition, made it extremely dangerous. The exaltation of something both irrational and unprecedented, Chevreul fretted, could impede the "march of civilization," perhaps even endangering the remarkable scientific and intellectual progress that had begun during the Enlightenment.[58]

Writing in the *Revue des deux mondes* two years later, the lexicographer and philosopher Emile Littré took the arguments of Babinet and Chevreul a step further, making them the basis for a sweeping critique of religious

[56] Jacques Babinet, "Les Sciences occultes au XIXe siècle, les tables tournantes et les manifestations prétendues surnaturelles considérées au point de vue des principes qui servent de guide dans les sciences d'observation," *La Revue des deux mondes* 24, no. 4 (April 1854): 527.

[57] Ibid., 529.

[58] Chevreul, 255–256.

faith and a consequent justification of a new social role for scientists. Both *académiciens* had joined their moral cases for objectivity to a conception of human development rooted in the Positivist philosophy of Auguste Comte. This vision of progress made faith, and the mystical impulse more generally, into an atavistic "love of the marvelous" that would have to be eliminated—or at least strictly controlled—if the "march of civilization" was to continue. Babinet and Chevreul had absorbed this view secondhand, and indeed it had become conventional wisdom among many mid-century French men of science.[59] Littré, however, had been one of Comte's most prominent disciples in the 1840s, and his analysis of the tables represented a more direct response to the older philosopher's ideas, particularly his controversial attempt to create a new "Religion of Humanity" in the years after 1848.

Like Babinet and Chevreul, Littré saw the *tables tournantes* as a telling example of the danger of religious belief and a powerful demonstration of the prophylactic social role scientific objectivity could play in a Positive Age.[60] For Littré, the persistent French interest in the phenomena of Modern Spiritualism was a manifestation of an ancient, perennial, and insidiously contagious human disease. To prove this point, he cited several instances in French history when large numbers of people had become attracted to strange phenomena that appeared to have otherworldly causes. The behavior of people tormented by witches in the Middle Ages, the ecstatic visions of seventeenth-century Protestants in Cevennes, and the trances of Jansenist *convulsionnaires*, he argued, all resembled the strange conduct of present-day mediums:

> The range of these unhealthy manifestations is quite narrowly limited. They always involve disturbances of the senses that cause individuals to see, hear, or touch; ecstasies that provoke very singular conditions in the nervous system; grave modifications of sensitivity; energetic convulsions that give the muscular system an incalculable power. In addition, elements provided by the ideas and beliefs of the time join with these general circumstances.

The specific significance attached to these "convulsions" and nervous perturbations might have changed in relation to the intellectual climate of each period, but the basic physical attributes of such "unhealthy manifestations" remained constant, Littré argued. By presenting an account of mediums' behavior that strongly emphasized its resemblance to earlier

[59] For a discussion of various aspects of this "Cult of Science" and the notion of progress that accompanied it, see D. G. Charlton, *Secular Religions in France, 1815–1870* (Oxford: Oxford University Press, 1963), 38–64, 155–199.

[60] See Emile Littré, *Auguste Comte et la philosophie positive* (Paris: Hachette, 1864), iv.

forms of trance and ecstasy, he sought to turn their physical actions into fixed symptoms, and the social movement they represented into a form of pathology. In the end, he wrote, the *tables parlantes* and related phenomena, like the demonic apparitions that had appeared in witchcraft cases, were merely collective hallucinations caused by "the combination of mental lesions with the predominance of an order of ideas familiar to all minds at the time."[61]

While the physical "symptoms" of this malady were timeless, Littré argued, the circumstances that caused their recent appearance were products of a distinctive historical context. The interpretations believers in the new phenomena advanced, he wrote, could have emerged only in the mid-nineteenth century:

> Our age is an age of revolutions. Considerable upheavals have troubled our society at short intervals, inspiring incredible terror in some, unlimited hope in others. Under these conditions, the nervous system has become more vulnerable than ever before. Also, in the absence of an apparent social foundation, many souls have anxiously sought refuge in religious ideas; this return, however, has been mixed with other elements. It has occurred in the presence of opposing ideas, which have retained their ascendancy, and in the presence of scientific ideas, which have inspired a great respect, even among those who deplore their influence. Here, then, we have a combination of circumstances clearly favoring the contemporary explosion.

The widespread interest in séances and communications received through the *tables*, then, was the pathological byproduct of a society in transition. As such, Littré believed it offered a powerful object lesson in the complexity of human social development and the need to keep from taking progress for granted. Over time, civilization inevitably tended toward "a progressive improvement," he observed, but nevertheless remained subject to "perturbations and disorders that slow, block and divert the overall movement." The vogue for *tables tournantes*, like witchcraft scares before it, exemplified this type of dangerous "perturbation."[62]

In the face of such uncertainty, Littré argued, scientists played a crucial role: by supplying objective, empirical information about the world, they rendered a "noble and brilliant service," dispelling illusions and curing outbreaks of pathological religious enthusiasm that might divert humanity from its course of perpetual improvement.[63] Doctors had eliminated

[61] Emile Littré, "Des Tables parlantes et des esprits frappeurs," *La Revue des deux mondes* 26, no. 3 (Feb. 15, 1856): 869, 867.
[62] Ibid., 869, 851.
[63] Ibid., 864.

witchcraft by proving that physical signs thought to be caused by demonic possession were in fact symptoms of an organic disease; Littré suggested that the strange beliefs the séance vogue had inspired could be destroyed in a similar manner. A scientist could occupy this crucial regulatory role only if he denied and controlled his own innate tendency to respond irrationally when confronted with an uncertain and rapidly changing social landscape, however—and this was precisely where Comte had failed, when he made his fateful turn to religion. Following the path of true Positivist objectivity required self-discipline; this prodigious restraint, in turn, became one of the key signs that the scientist's authority was legitimate.

By advancing this view of the scientist as prophylactic explainer, Littré linked his argument to those of Chevreul and Babinet. For these writers, the proponents of scientific open-mindedness were fundamentally misguided. When confronted with such a large number of dramatically implausible accounts, a truly rational analyst turned first to the study of those who claimed to have observed the phenomena, not to the phenomena themselves. According to this new explanatory approach, commentators like Séguin, Guérin, and Gasparin, who claimed to have witnessed inexplicable manifestations firsthand, ceased to be valid observers. All three had abdicated their positions as legitimate scientists by succumbing to an all-too-human willingness to abandon the steady ground of reason for the shifting terrain of the marvelous.

The image of the scientist presented by Babinet, Chevreul, and Littré embodied a new moral vision, rooted not in faith but in an ascetic commitment to objectivity. The role of the trained scientist, these writers argued, was to act as a bulwark against sentiment—not to provide new forms of consolation, as Comte had attempted to do, but rather to serve as a living example of the virtue that came from denying the desire for consolation itself. Rationality, for these writers, was the engine of progress, but it was also fragile, a recent and artificial product of human ingenuity. The scientist's role was to protect this delicate attainment from the ancient, disruptive inclinations of the heart.

Tables Tournantes and the Post-Revolutionary Left

Despite the criticisms of priests and scientists, a vocal minority of French men and women saw the phenomena of American spiritualism as sources of an utterly new kind of religious experience. These believers responded most strongly to the *tables parlantes*, which they presented as a distinctively modern means of receiving consolation and metaphysical insight from the beyond. The otherworldly communications this group of religious seekers received did not simply echo Catholic orthodoxy, but instead elaborated an array of alternative cosmologies. For these thinkers and believers, the *tables parlantes* became a novel technology of revelation, a telegraph connected to the beyond not by wires but by an invisible stream of "fluid." Resolving religion's crisis of factuality in this case involved abandoning orthodoxy in favor of innovative systems that seemed better suited to the demands of a changing society.

Many of those who regarded the *tables* as a source of revelation came from the political left. For many advocates of democracy and social reform, the rapid, violent collapse of the Second Republic provoked a crisis of faith. This was especially true among exponents of the theories of Romantic Socialists such as Henri de Saint-Simon, Etienne Cabet, and Charles Fourier; thinkers who, in the earlier decades of the century, had elaborated vast, totalizing schemes for the reorganization and perfection of society. In the heady days of February 1848, many reformers believed that a wide-ranging social transformation, inspired by one or another of these great systems, was imminent. After the catastrophic Parisian revolt in June—known as the June Days—and the conservative backlash that followed, the implausibility of these dreams of paradise on Earth became clear. *Tables tournantes* and *parlantes* became a powerful source of reassurance in this atmosphere of disappointment and thwarted aspiration. The visionary ideas of the Romantic Socialists might have fallen short in the realm of practical politics, but the revelations of the *tables* seemed to indicate that they remained empirically true nevertheless and retained their power to effect social change.

Left-wing approaches to the new phenomena varied widely. For Victor Hugo, his close friend Auguste Vacquerie, and the visitors they received in exile on the Isle of Jersey, the *tables tournantes* were sources of both religious and political consolation. For a group of newly unemployed contributors to *La Démocratie pacifique*, a Fourierist newspaper shut down by the censors after Louis-Napoleon Bonaparte's coup d'état, the tables seemed to redeem the grand theories of Romantic Socialism. The former socialist lecturer, deputy, and journalist Victor Hennequin derived a very different message from these phenomena. His decision to heed the critique of Fourier that his *table parlante* advanced made him a *cause célèbre* of the early Second Empire. Damaging as Hennequin's example may have been, the *tables tournantes* proved too tempting a source of revelation for leftist visionaries to ignore. Throughout the later 1850s, an array of idiosyncratic thinkers continued to use communications from séances as sources of authority for broader speculations about society and the cosmos.

The *tables tournantes*, then, were both a boon and a burden for disappointed republicans and socialists. On the one hand, they consoled, seeming to prove the rightness of the transcendent principles that so many leftists of the period embraced. On the other, even as these strange manifestations affirmed cherished beliefs, they served to push socialist advocates for democracy even further toward the margins of political life. During the 1850s and 1860s, the emergence of a new kind of visionary thinker—the medium who elaborated theories of social organization based on communications with the beyond—played an important role in the broader series of developments that caused utopian theories to lose the credibility they had once enjoyed.

Hugo's Consolation

In 1853, Victor Hugo and Auguste Vacquerie were in exile on the Isle of Jersey, having been forced to leave France in the wake of Bonaparte's coup d'état. Their experiments with *tables tournantes* began at the urging of the *salonnière* and writer Delphine Gay de Girardin, who, during a visit to the island in September, asked Hugo's family to hold a séance.[64] Their first experiment, performed one evening after dinner, failed; several other attempts met with equally poor results. Hugo himself had refused to participate in these gatherings.

Finally, the day before she was to leave, Mme de Girardin persuaded Hugo to place his hands on the table with her. After a few minutes, the piece of furniture made a cracking sound and began to tremble. Various members of the party—Mme Hugo, her adult children Charles and Adèle, and Vacquerie—began to ask questions, which the table answered handily by tilting and tapping a leg on the floor. Then the table stopped moving and, when asked, declared that another spirit was present. Slowly, it rapped the name of Hugo's favorite daughter, Léopoldine, who had drowned in a boating accident several years before. In a memoir of the séance, Vacquerie wrote:

> Here, disbelief was no longer possible: no-one would have had the heart or the effrontery to turn this tomb into a stage before our eyes. A mystification was already difficult to admit, but an infamy! Suspicion would have mistrusted itself. The brother questioned his sister, who had come from death to console the exile; the mother cried; an inexpressible emotion welled up in our breasts; I distinctly felt the presence of the girl, who had been torn from us by a dire blast of wind. Where was she? Did she still love us? Was she happy? She responded to all the questions, or responded that she was forbidden to respond. The night melted away, and we remained there, souls fixed on the invisible apparition. Finally, she told us: Farewell! And the table stopped moving.[65]

[64] For more on Girardin's life and remarkable career, see Whitney Walton, *Eve's Proud Descendants: Four Women Writers and Republican Politics in Nineteenth-Century France* (Stanford: Stanford University Press, 2000).

[65] Auguste Vacquerie, *Les Miettes de l'histoire* (Paris: Pagnerre, 1863), 408.

Though Vacquerie, out of a desire for discretion, left specific names out of his 1863 account, the identities of the participants in this séance appear in private correspondence and journals.[66]

The journal Adèle Hugo kept during this period gives a slightly different, and in some ways more revealing, account of Léopoldine's first manifestation. In Adèle's telling, the spirit revealed her identity by spelling out the Latin word *soror* (sister), and then answering "yes" to Victor Hugo when he tearfully asked if she was "happy when I include your name in my prayers every evening." In addition to describing her felicitous state in the afterlife, Léopoldine offered a consoling political prophecy, which Adèle transcribed:

> *Victor Hugo*: Do you have a commentary for us?
> *The Table*: Republic.
> *VH*: When? Strike the floor as many times as there are years from now until the Republic.
> The table struck two blows.

The animated table, here, became a powerful source of reassurance, not only in matters of metaphysics but in worldly affairs as well. Manifestations like this one seemed to prove both that a divine order ruled the universe and that a French republic, even if momentarily thwarted, was an integral part of that order. Despite recent setbacks, the table indicated, Victor Hugo was indeed on the side of the angels.[67]

This conversation deeply moved Hugo and Vacquerie, inspiring them to devote long hours to séances for more than a year. In addition to contacting deceased relatives, the men summoned the spirits of great writers—among them Racine, Dante, and Molière—philosophers, and even religious figures, not stopping short of Christ himself. Through the Jersey séance table, Shakespeare even dictated the first act of a new play—in French.[68] These communications took hours because the spirits spoke in much the same way as they had to the Fox sisters, causing the table legs to strike the ground during repetitions of the alphabet, spelling long disquisitions letter by letter. The séances stopped in 1855, for reasons that remain unclear. The experiments might have ended after a fellow French

[66] See Jean Gaudon, ed., *Ce que disent les tables parlantes: Victor Hugo à Jersey* (Paris: Pauvert, 1963), which includes excerpts from correspondence, along with transcripts from a series of séances held by Hugo and Vacquerie; Gustave Simon, ed., *Chez Victor Hugo: les tables tournantes de Jersey* (Paris: Stock, 1980), which includes numerous transcripts of other sessions, often expurgated.

[67] Adèle Hugo, *Le Journal d'Adèle Hugo, deuxième volume, 1853,* ed. Frances Vernor Guille (Paris: Minard, 1971), 275–276, 277.

[68] This fragment appears in Gaudon, 33–58.

exile suffered a psychological breakdown at Hugo's table or because the highly publicized madness of Fourierist Victor Hennequin seemed to reveal that *tables tournantes* could be dangerous when used to excess. Some also see the end of the séances as a consequence of Hugo's expulsion from the island.[69]

To the end of his life, Hugo retained a guarded belief in the otherworldly origin of these phenomena, though as early as September 1853 he questioned whether the spirits that appeared were actually the illustrious personages they claimed to be.[70] In his will, he stipulated that transcripts of the séances be published after his death. Vacquerie also confessed his belief in the spiritual aspect of the *tables*:

> As for the existence of what we call spirits, I do not doubt it; I have never shared the human complacency that insists the scale of being ends with man; I am persuaded that we have as many levels above our heads as beneath our feet, and I believe as firmly in spirits as I do in evening primroses.[71]

For these men, the séance was not simply a novel game *pour tuer le temps* after dinner. On the contrary, Hugo and Vacquerie saw the *tables* as a legitimate and powerful source of religious consolation, ideally suited to the modern world. Unlike Catholicism, which they believed had become a prop for the social order, the *tables tournantes* were a vital spiritual force. Hugo, in fact, had lost his Catholic faith even before his daughter's death and had dabbled in things occult since the 1830s. His Kabbalistic and alchemical experiments had none of the visceral impact of the séances at Jersey, however.[72]

Around the séance table, Hugo and his circle seemed to enter into palpable contact with the beyond: Vacquerie had *felt* Léopoldine's presence in the room. Where traditional religion was based on philosophical speculation, the *tables tournantes* were empirical. Of course, Hugo's cultural stature makes his case extraordinary by definition, but the author of *Les Misérables* was hardly alone in his conviction—many others in this period reacted to the *tables* with the same excitement and credulity. For seekers like Hugo and Vacquerie, this new American parlor game was a way to regenerate dying sources of faith, transforming them to meet the intellectual and political demands of the modern believer.

[69] See Jann Matlock, "Ghostly Politics," *Diacritics* 30, no. 3 (Fall 2000): 68.

[70] Gaudon, 253.

[71] Vacquerie, 412.

[72] For more on Hugo's religious beliefs and their relation to the Jersey séances, see Auguste Viatte, *Victor Hugo et les illuminés de son temps* (Montreal: Editions de l'Arbre, 1942), 101–131; André Maurois, *Olympio: The Life of Victor Hugo,* trans. Gerard Hopkins (New York: Harper and Bros., 1956), 299–321.

Fourier Redeemed

In the Paris offices of the Fourierist newspaper *La Démocratie pacifique*, other defeated proponents of the Second Republic looked to the *tables tournantes* for less personal, more explicitly political forms of consolation. As early as March 1853, the journalist Allyre Bureau came upon a description of the new phenomena in an American newspaper and translated it for several of his colleagues. The writers Eugène Nus, Méray, Brunier, and Franchot responded with particular enthusiasm, holding their own séances in the *Démocratie pacifique's* offices on the rue de Beaune. Soon, they were spending hours seated at their table, producing mysterious communications. Like Hugo's, the table at the rue de Beaune spelled out its messages by tapping the floor during recitations of the alphabet.

In a memoir written many years later, Eugène Nus attributed the strength of these writers' interest to the political situation. After the coup d'état of 1851, the new government had suppressed the *Démocratie pacifique*. Its editor, Victor Considerant, a prominent member of the National Assembly during the Second Republic, had fled to Texas, where he planned to found a Fourierist colony.[73] These events, Nus dryly observed, "left us with a good deal of leisure time." Unemployment came with a healthy dose of frustration. The writers continued to gather in their old newspaper's offices "out of idleness, out of habit, out of friendship above all, out of the natural need to soothe our disappointments and outrages." Since their hopes for creating a perfect society in France had been thwarted, they regretfully turned their attention to Texas, but could not help "looking sadly across the Seine from our windows on the quai Voltaire, as the first plumes of the Empire wafted from the Tuileries to parade in the *Bois* [*de Boulogne*]." In this atmosphere, the *tables tournantes* provided both a welcome diversion and, for nearly a year, a powerful source of hope.[74]

Initially, the journalists had been content to cause tables to rotate. The motion appeared to prove the reality of Mesmerism—an idea the *Démocratie pacifique* had long supported—in a striking manner.[75] Their experiments changed dramatically, however, after they received a visit from a doctor named Arthur de Bonnard, who told them about a new type of manifestation. Bonnard's family had begun to use their table to produce messages, which often bore the signature of a spirit named Jopidiès. Inspired by this information, the journalists began to produce communications of their

[73] For more on Considerant and the early years of the *Démocratie pacifique*, see Jonathan Beecher, *Victor Considerant and the Rise and Fall of French Romantic Socialism* (Berkeley: University of California Press, 2001).

[74] Eugène Nus, *Choses de l'autre monde* (Paris: Dentu, n.d. [1880]), 3–4.

[75] For an article discussing the *Démocratie pacifique's* support for Mesmerism, see *Le Journal du magnétsme* 12 (1853): 247.

own. At first, these messages appeared to come from the souls of the deceased. The experimenters received visits from a variety of obscure and eminent individuals, including Pythagoras, Confucius, Saint Paul, and Socrates, but remained skeptical of these uncanny manifestations.[76]

For the journalists, all of whom were convinced believers in the immortality of the soul, Nus wrote, these manifestations seemed too good to be true. The fate of the soul after death, Nus maintained, was "the greatest mystery of life, the most profound stimulator of thought." It was a conundrum human beings ardently wished to resolve; the impossibility of resolving it, in turn, spurred their intellects to otherwise unreachable heights. To suggest that a single, banal-seeming phenomenon could settle the question "brutally, by means of an almost grotesque procedure," Nus argued, was to make a mockery of all the previous achievements of philosophy and, perhaps more disturbingly still, to render any future speculation on the subject unnecessary.[77] In response to this objection, one of the experimenters suggested that perhaps the table claimed to be speaking for spirits because that was what the people around it expected.

Armed with this new hypothesis, the journalists questioned the table about the origin of the communications it produced. This time, it provided a different explanation, which the experimenters found considerably more congenial. Nus reproduced it verbatim:

> The phenomenon results from the association of your souls among themselves, and with the spirit of life. The manifestation emanates from human forces and the universal force. The Being that your souls form, associated with the spirit of life, OVER TIME, immaterial, connected to your senses and sentiments, is merely the expression of your animic solidarity: a message half-human, half-divine, when your souls are in harmonic vibration with the universal order, which is to say, the beautiful, the true, the just.

The experimenters' feelings of solidarity, coupled with the "fluids" they emitted, produced a "harmonic vibration" that fused their souls into a collective entity, which in turn had the power to contact a transcendent and impersonal "universal intelligence." This vision of community as a metaphysical "battery" appealed strongly to the journalists.[78]

Perhaps most important, the new theory appeared to provide empirical evidence for their Fourierist convictions, which recent political events had shaken. These new phenomena seemed to prove Fourier's claim that true solidarity would allow human beings to develop remarkable,

[76] Nus, 9, 112.
[77] Ibid., 13.
[78] Ibid., 13–14. Emphasis in the original.

world-transforming powers. Any political defeat, therefore, would be nothing more than a temporary setback; the divine force of human solidarity, which the moving table made so compellingly palpable, would eventually come into its own and overcome all selfish efforts to prevent the realization of a truly harmonious society. The table confirmed this view, noting that, when they did nothing but "overturn the established government," revolutions "had no utility." Political upheavals of this kind, the table wrote, "encourage bad ambitions," and "overexcite the minds of intelligent and generous men, dulled by excessive repose."[79] Social transformation would come, but it could not be brought about through the alteration of political institutions because such reforms did nothing to change citizens' attitudes. Only unanimous and sincere feelings of solidarity, fostered by the widespread practice of association, could unleash the forces of true social transformation. Around their table at the rue de Beaune, therefore, these Fourierist journalists were setting the stage for mankind's brilliant future.

After the group had adopted this new theory of causation, the journalists' experiments with the *tables* gradually became literary games. Their collective soul revealed itself to be both witty and artistically gifted: It could communicate in nearly correct English, define any term or concept in twelve words, and even compose melodies. News of these feats spread quickly through Parisian literary circles; as the popular interest in *tables tournantes* grew, the séances at the old *Démocratie pacifique* offices became fashionable events. Gérard de Nerval witnessed some experiments there, as did Victor Meunier, the science columnist for *La Presse,* who proved an outspoken advocate for the new phenomena. In her *salon,* Delphine de Girardin held a special concert comprised entirely of melodies dictated by the table, the "cloudy" quality of which impressed the composer Félicien David. The journalists' behavior around the table created an atmosphere of lighthearted urbanity. During their séances, they would smoke "irreverently, one his pipe, another his cigarette, a third his cigar," chuckling at their collective soul's jokes and chiding it when communications failed to adhere to the word limit they imposed.[80]

After about a year, Nus wrote, the table began to move less frequently, and the experimenters became less enthusiastic, largely because of the repetitiveness of the communications. The table's movements themselves also reflected the waning interest of those whose collective will had once been a source of vivacious animation. Where the table had previously spelled out words "with a well accented tap at the end of each letter," it now produced an enervated "*cottony* sound." Eventually, Nus wrote, the "holy table, as it sometimes modestly dubbed itself," ceased moving entirely and

[79] Ibid., 18.
[80] Ibid., 112, 90, 106.

resumed its previous function, supporting a backgammon board in the newspaper office. Some of the experimenters left to join Considerant's colony in Texas, and others found work elsewhere.[81]

Hennequin and the Decline of the Visionary Left

While Nus's memoir is unclear on this point, the journalists' diminishing enthusiasm for the new phenomena probably had to do not only with boredom but also with an unforeseen consequence of their experiments—the widely publicized defection of the politician and journalist Victor Hennequin from the Fourierist cause. This episode proved to be a serious humiliation for those who believed in the utopian strain of leftism that played such a crucial role in the events of 1848. The widespread journalistic reaction to Hennequin's strange pronouncements drove home the extent to which visionary Romantic Socialism had been discredited after the collapse of the Second Republic.

In the spring of 1853, the rue de Beaune séances attracted Hennequin's attention. The son of a well-known Legitimist lawyer, he had been a contributor to the *Démocratie pacifique,* a Fourierist lecturer, and deputy for the *département* of Saône-et-Loire in the dissolved National Assembly. The coup d'état was a catastrophe for him. He had been imprisoned along with many other deputies of the left on December 2, 1851. Once released, in order to support his wife and daughter, he abandoned journalism and resumed his practice as a lawyer, which he detested.[82] Hennequin freely expressed his disappointment with this situation; he was "ambitious" by nature and loved material "signs of dignity," like his deputy's rosette.[83]

In the face of these disappointments and humiliations, Hennequin undertook his own experiments with the phenomena he had witnessed at the offices of his old newspaper. He and his wife Octavie began by causing tables to move; soon, they received communications from Hennequin's deceased mother and brother.[84] As Hennequin continued his experiments, his connections to the beyond became closer. Soon, the former deputy wrote, "the table became a voice." [85] Instead of being tapped letter by letter, Hennequin's communications now came to him directly, transmitted through an invisible "aromal trumpet" affixed to his head.[86] This

[81] Ibid., 109–110.

[82] Ibid., 137.

[83] Quoted in Alexandre-André Jacob [Alexandre Erdan, pseud.], *La France mistique, tablau des excentricités religieuses de ce tems* (Paris: Coulon-Pineau, 1855), 2:597. Jacob's title reflects his interest in orthographic reform.

[84] *La Presse,* Oct. 4, 1853, 2.

[85] Victor Hennequin, *Sauvons le genre humain* (Paris: Dentu, 1853), xxviii.

[86] Jacob, 2:600.

new method of communication, Hennequin wrote, allowed him to enter into direct contact with the "*âme de la terre*" (soul of the Earth), the transcendent intelligence that, according to Fourier, God had placed in charge of human society.

The *âme de la terre* provided Hennequin with a surprising explanation for the left's failure in 1848. Above all, the entity claimed, the debacle stemmed from the personal shortcomings of Charles Fourier. In a letter to a friend that was later published in *La Presse,* Hennequin wrote that the "stars of the vortex" had long been disgusted with humanity's inability to improve itself, and prevailed upon the *âme de la terre* to take matters into her own hands. Unfortunately, she bungled the task. "Still young," and "distracted from work by a romance," the *âme de la terre* let centuries pass without using the "seeds of inspiration" the Creator had given her. Eventually, when pressed, she hurriedly affixed "organs of intuition" to Charles Fourier, whose soul, while "honest," proved too "limited" and "trivial" to fully comprehend the divine messages it received. In his writings, therefore, Fourier misconstrued a variety of crucial points, reaching several conclusions that were "immoral or ridiculous."[87] Once the revolution of 1848 began, these errors became grave indeed, Hennequin wrote, because Fourier's system was the only available theory of social organization that was truly "serious." Inevitably, the upheaval would have led to its triumph. "The creator," realizing that the implementation of Fourier's ideas would have been "fatal" for humanity, therefore, had no choice but to intervene, and "suddenly stopped the democratic movement."[88]

While the failure of the revolution had saved mankind from disaster, Hennequin wrote, it had also stalled the *âme de la terre*'s inept but sincere efforts to perfect human society. According to Hennequin, these events had placed the Earth and its inhabitants in grave danger. The creator was prepared to destroy the planet, consigning "the fragments of all souls" to the "abyss," and would already have carried out this painful but necessary task if the *âme de la terre* had not requested a reprieve. To save humanity, the *âme de la terre* had decided to inspire a new work of social theory, which would correct Fourier's errors while conserving the aspects of his system that stemmed from authentic divine inspiration. Once this work appeared, it would provide the impetus for a second, far more successful social transformation. The vogue for *tables tournantes,* according to Hennequin, was a precursor of these impending developments.[89]

The *âme de la terre* chose Hennequin to receive the new revelation and began dictating a book to be entitled *Sauvons le genre humain*—"Let Us

[87] *La Presse,* Oct. 4, 1853, 2.
[88] Hennequin, 168.
[89] *La Presse,* Oct. 4, 1853, 3, 2.

Save Humankind". In the course of these initial communications, which took place in early August, 1853, the *âme de la terre* prophesied that in seven days Hennequin would receive a visit from the publisher Adolphe Delahays, who would offer him 100,000 francs for his manuscript. Hennequin, at the urging of his otherworldly interlocutor, responded to this news with a series of rash actions. First, he decided to resign from the court case he was currently preparing. He turned the project over to a colleague, whom he also informed of his divine mission, thereby bringing his legal career to an abrupt end. Hennequin then sent a letter to his sister, renouncing his claim to his recently deceased mother's estate. Finally, he wrote to the treasurer of the *Démocratie pacifique.* The *âme de la terre* had instructed Hennequin to use the bulk of his advance to settle the paper's substantial debts; his letter offered this money on the condition that the Fourierist movement dissolve itself completely.[90]

Hennequin's letter, Nus wrote, "hit us like a bolt from the blue." Everyone on the paper's staff was astounded first by the implausibility of Hennequin's prediction, and second, by the sudden defection of one of their most active collaborators. Ferdinand Guillon, the journalist who knew Hennequin best, went to his friend's apartment and asked him to explain the letter. Hennequin replied by reading him a part of his manuscript, and asking, "Do you believe I did this?"[91] Throughout this period, Hennequin emphasized his role as the passive vessel of forces beyond his control. Often, he noted, the words the *âme de la terre* made him write ran counter to his own sympathies. Recent events had caused his "intelligence" to be "recast." In his new role as messenger for the divine, Hennequin wrote, "we no longer write what once pleased us; we write what we see."[92]

The former deputy retained this attitude of passivity even after Delahays and his 100,000 francs failed to materialize. Hennequin wrote a letter to his friend Mme de Curton—which he eventually published as part of the preface to his book—explaining that he had been "completely misled."[93] This deception did not lead him to renounce his belief in his own divine mission, however. Instead, he chose to view it as a test of his faith, and he readily acquiesced when the *âme de la terre* instructed him to write more letters, this time to Napoleon III and the press.

In his letter to the Emperor, Hennequin rejected his earlier views and declared his support for the new order. After requesting the Emperor's authorization to publish his manuscript, he wrote:

[90] See Nus, 137–138; Hennequin, xii–xiv.
[91] Nus, 138.
[92] Hennequin, 212.
[93] Ibid., xix.

> The second motive for this missive is to announce that God has overturned all my political convictions; that my book attacks the principles most important to democracy; that it supports the cause of established power in general, despite my own strong distaste for such opinions; and that I have been ordered to tell you, personally, that you have a providential mission.[94]

Hennequin sent copies of this letter to a variety of major newspapers, and it generated a flurry of rumors among journalists. Copies of it appeared first in the *Indépendance Belge* and then in *La Presse*, one of the most widely read newspapers of the day.[95] In the coming months, Hennequin would publish several other letters in the *Presse*, further explaining his role as vehicle for the *âme de la terre*, and elaborating on the crucial significance of the manuscript he had written.[96]

Despite Hennequin's very public proclamation of support for the established regime, the *Directeur de la librairie*—the government official in charge of book censorship—initially denied him authorization to publish. Once permission had been granted, however, Hennequin had no trouble placing his manuscript, which his letters to *La Presse* had turned into a *cause célèbre*. In November 1853, Edouard Dentu—who, one journalist acidly observed, could always be counted on to print "the fulminations of the unhinged"—brought out *Sauvons le genre humain*.[97] Less than a year later, as Dentu was printing *La Religion*, Hennequin's second, considerably more abstruse volume, the author's family had him committed to an insane asylum. He died shortly thereafter.[98]

Sauvons le genre humain, as many commentators noted, was a straightforward and coherent summary of Fourier's doctrine, with some crucial modifications.[99] Hennequin's central criticism had to do with Fourier's concept of the "passions" and their role in human behavior. According to Fourier, Hennequin asserted, human beings were incapable of resisting their passions, and therefore, the perfect model of social organization was one that would allow all human beings to gratify themselves freely.[100] The evils that existed in society as it was currently constituted, Fourier argued, all stemmed from its tendency to block or deflect what would otherwise be constructive desires and impulses. This vision of human behavior, while valid in some respects, Hennequin argued, was nevertheless fundamentally misguided. Most important, Fourier's radical trust in the rightness of

[94] Ibid., xxix.
[95] *La Presse*, Sept. 16, 1853, 2.
[96] See *La Presse*, Oct. 4, 1853, 2–3, Oct. 15, 1853, 3.
[97] Jacob, 2:601.
[98] Nus, 140.
[99] See, e.g., *La Presse*, Nov. 13, 1853; Jacob, 2:602.
[100] Hennequin, 16.

all passions led him into dubious moral territory, particularly in matters of sex and love: In his 1808 *Théorie des quatre mouvements*, he had condemned monogamy, and in a manuscript known to an inner circle of followers, he had expressed tolerance for such practices as male homosexuality, lesbianism, and incest.[101] Hennequin sought to remedy what he perceived to be the moral weakness of Fourier's system by making the monogamous couple the fundamental unit of an ideal society.

Alongside his critique of Fourier, Hennequin included a scathing analysis of the events of 1848. The revolution, he believed, had been a God-given opportunity for social transformation, which the provisional government that took power in February had squandered by foolishly entrusting French citizens with the responsibility of choosing their own leaders. For a democracy to function properly, Hennequin asserted, the voters needed to have "enough understanding and virtue to recognize and to elevate those whom nature made to be their guides." The conservative outcomes of so many of the Second Republic's elections, coupled with the "infernal horror" of the June Days, proved that the bulk of French citizens lacked this necessary intellectual and moral refinement and inspired the Creator to thwart the revolution he had instigated earlier. Given the shortcomings of its citizens, Hennequin argued, France could achieve real social progress only under an authoritarian government. Napoleon III's system of dictatorship periodically bolstered by plebiscites was the sole arrangement that would allow France any hope of continuing down the path of social progress. This, then, was the Emperor's "providential mission": He had been sent to take control of France and to force it to accept the social reforms its citizens were too foolish and egotistical to implement of their own accord.[102]

Hennequin's well-publicized defection proved humiliating for those who remained loyal to the Fourierist cause. Moderate commentators tended to praise Hennequin's critique of Fourier even as they mocked his claims of otherworldly inspiration; this approach had the implicit result of making loyal Fourierists appear even further out of touch with reality than their eccentric critic. Adolphe Garnier, professor of philosophy at the Sorbonne, for example, published a pamphlet in which he argued that Hennequin's peculiar mental state stemmed from a dissonance between his utopian "enthusiasm" and the practical realities of his situation. This conflict, coupled with a high self-regard, Garnier argued, had led Hennequin to misconstrue the source of his new, anti-Fourierist ideas:

[101] Ibid., 106–115. Fourier's manuscript on the erotic passions was not published until 1967. See Charles Fourier, *Le Nouveau monde amoureux*, ed. Simone Debout-Oleskiewicz (Paris: Anthropos, 1967).

[102] Ibid., 128, 166, 132.

> The proof, M. Hennequin says, that a mind different from my own speaks to me, is that it gives me ideas that are entirely contrary to those I previously formulated. But, we would respond, you no longer live in the same milieu as before; you are no longer surrounded by fellow believers; good sense has prevailed over the *esprit de système.* You are married, you have a child: you can no longer practice the promiscuity the Master preached; it is your own mind that complains from within, not another.

A residual enthusiasm from Hennequin's visionary past had led him to mistake the natural process of outgrowing youthful illusions for a direct revelation from the beyond. This disillusionment, Garnier believed, was inevitable; the danger and peculiarity of Fourier's doctrine stemmed from the credulity and susceptibility to the "marvelous" that it seemed to foster in temperaments as ardent as Hennequin's.[103]

Nus, who had a considerably more charitable view of Fourier and the aspirations of his followers, explained Hennequin's change of heart quite differently. Experiments with the *tables* drove Hennequin mad, Nus argued, because he had undertaken them alone. Even in his days as a Fourierist, Hennequin "was a solitary person," Nus wrote; "he rarely or, to say it better, never took part in our good effusions, our serious or mad conversations." This tendency to avoid social interaction would prove to be Hennequin's downfall, according to Nus. The table itself had warned the journalists of the dangers of solitary dialogue with the universal intelligence. According to the communications the experimenters at the *Democratie pacifique* had received, the table was a force for good only when under the power of a group of souls fused by feelings of mutual affection and solidarity. When an individual activated a table without the support of a group, he gave free reign to his own selfish illusions, and risked madness if he succumbed to them. Hennequin, Nus argued, was a victim of this phenomenon; the mysterious manifestations he produced had led him to mistake his own thoughts for the emanations of a transcendent intelligence. In Nus's assessment, then, Hennequin's madness seemed to affirm Fourier's idea of the power of solidarity; it was thus not a fatal blow to the cause but an exception that proved the rule.[104]

From Revolution to Revelation

Hugo, the *Démocratie pacifique* circle, and Hennequin were not the only disappointed advocates of social reform to look to the new phenomena

[103] Adolphe Garnier, "Sauvons le genre humain, par Victor Hennequin" (Paris: Paul Dupont, 1854), 4, 3, 4.

[104] Nus, 136, 19.

for reassurance and inspiration. Indeed, the *tables parlantes* inspired an exuberant variety of left-leaning metaphysical and political speculation. For those who published texts on the subject, communications from the beyond seemed to promise an alternative means of social transformation that would succeed where political revolution had failed. The pamphleteer Paul Louisy, for example, proclaimed that "the phenomenon of the *tables tournantes* is a prelude to scientific and social regeneration."[105] By acting as the conduit for an irrefutably true, purely rational new revelation, Louisy argued, the *tables* would achieve what the revolutionaries of 1848 had failed to accomplish. The *quarante-huitards* (forty-eighters, or leftists of 1848) had attempted to change society by imposing a new order from above; the *tables*, by provoking individual moral transformations, would cause reform to occur organically, as each newly enlightened French subject independently came to desire a society based on justice and fellow-feeling.

The writer Louis Goupy shared this vision of organic transformation, but made his predictions more specific. The triumph of the *tables parlantes* as a source of empirical metaphysical authority, he argued, would create a religion that could be explained "by the phenomena of nature." This innovation would lead French subjects to question a host of irrational ideas they had previously accepted as indisputably necessary and true, ranging from "the monarchical principle" to the use of gold and silver as the bases of currency.[106] Though censorship rules prohibited Goupy from discussing broad questions of social justice, he nevertheless strongly implied that a dramatic, and beneficial, restructuring of society would necessarily result from this process of self-criticism.

For these Second Empire visionaries, the *tables* provided a new source of hope: a sign that the forces of reform enjoyed a cosmic sanction powerful enough to overcome even the most total political defeat. At the same time, however, their arguments tended to make visionary leftism seem increasingly marginal. Political theories developed by human beings had a long and august history; political theories tapped out by animated tables, on the other hand, struck many observers as being of dubious provenance. For Louisy and Goupy, the audience was no longer all rational French readers; it was all rational French readers willing to believe in revelations from the *tables parlantes*.

Optimistic pronouncements like those of Louisy and Goupy found their readiest audience among Mesmerists, a group fully prepared to

[105] Paul Louisy, *Lumière! Esprits et tables tournantes, révélations médianimiques* (Paris: Garnier Frères, 1854), 11.

[106] Louis Goupy, *L'Ether, l'électricité et la matière, deuxième édition de quæare et invenies* (Paris: Ledoyen, 1854 [1853]), 197–199, 146–147.

accept the reality of the new phenomena. The Mesmerists, however, linked these phenomena to an already highly developed set of theories and practices of their own, which the *tables* seemed to reaffirm and undercut simultaneously. Chapter 2 investigates the conflicting ways French Mesmerists approached these strange phenomena and the important consequences of their disagreement, which set the stage for a dramatic new attempt to resolve the crisis of factuality: a distinctively French version of Modern Spiritualism.

CHAPTER TWO

Mesmerism and the Challenge of Spiritualism, 1853–1859

At a séance held in the summer of 1859, a distinguished committee of French Mesmerists attempted to reach a definitive conclusion about the reality of spirit phenomena. These eight doctors, journalists, and idealistic *bourgeois* considered themselves uniquely qualified to answer the vexed metaphysical and psychological questions such manifestations continued to pose. The committee members, chosen months before at a special meeting of the Société philanthropico-magnétique de Paris, had devoted their lives—or at least a substantial portion of their spare time—to the field of *magnétisme animal*. All considered themselves experts in the experimental study of the human mind, the impalpable forces it exuded, and the unique states of consciousness those forces provoked. Where the Académie des sciences had refused to accept the *tables tournantes* and *parlantes*, declaring them to be the hallucinatory product of *amour du merveilleux* and involuntary muscular tremors, these Mesmerists took the new phenomena seriously.

The *commission d'enquête* (commission of inquiry) gathered at the home of Honorine Huet, who had acquired a reputation as France's most gifted medium. This meeting had not been easy to bring about. France's leading Mesmerists had spent months persuading Huet to subject herself to the experiments they wished to conduct. She was well aware of the suspicion with which many members of the committee viewed the more spectacular manifestations said to occur in séances: Several had written articles expressing disbelief at the highly colored newspaper reports that made their way from America to France. While these *magnétiseurs*—as French Mesmerists called themselves—may have been receptive to the notion that the mind had an array of mysterious powers, they were not necessarily willing to believe the deceased capable of producing material wonders in the world of the living. Huet's repertoire of prodigies was humbler than those of the most famous mediums on the other side of the Atlantic, but it was impressive enough, and she shared the American conviction that the presence of skeptics was inimical to an effective séance.

Even so, hands flat on the table, feet resting daintily on a small stool, she began in the usual way, with a look of intense concentration. Soon, the men seated around her heard tapping noises, which appeared to come from under the table. In sequence, each person in attendance beat a different rhythm. The spectral raps responded in kind, seeming to move about, first occurring near the medium, then at the table's center, then close to its edges. Huet's friend and publicist, the pharmacist P. F. Mathieu, placed a large piece of cardboard, marked with the letters of the alphabet, on the table. He produced a wooden pointer, which he moved slowly across the alphabet; the raps spelled out messages by indicating the appropriate letters. The first message was not encouraging: "science pas ce soir" (no knowledge—or science—this evening).[1]

Despite this warning, the committee members insisted on trying a new experiment, which they believed would conclusively determine whether these noises were the work of an autonomous otherworldly intelligence. One of the committee members took the cardboard and held it vertically on his lap, so Huet could not see the alphabet. He then moved the pointer at random across the letters, and asked the invisible force producing the raps to spell out a message. A series of noises followed, though more slowly, and without their former confident snap. When the taps stopped, the committee member read out the letters they had indicated, which "did not form any word it would even be possible to pronounce." Huet produced further raps, but this did little to redeem her previous failure. At a meeting after the séance, the Mesmerists drafted their formal report, which ended with a declaration that "the phenomena the committee witnessed were not conclusive." Six members voted in favor of the committee's decision to publish the report, and two voted against. A bitter polemic raged for months afterward.[2]

This unsuccessful séance and the polemic that followed it marked a decisive moment in the history of French heterodoxy, one in which advocates of an older way of thinking about the mysterious powers of the mind made their last serious attempt to counteract the emergence of a newer approach. The phenomena of American spiritualism precipitated this transition by posing a knotty challenge to students of *magnétisme animal.* On the

[1] *L'Union magnétique, journal de la Société philanthropico-magnétique de Paris* (Aug. 25, 1859): 3.
[2] Ibid.

one hand, spiritualist manifestations seemed to corroborate theories about the mind's power to act outside the body that Mesmerists held dear, and had fought zealously to prove since their discipline's heyday late in the eighteenth century. On the other, the spiritualist tendency to ascribe these phenomena to the intervention of otherworldly beings exacerbated a fundamental difference of opinion that had become increasingly pronounced in Mesmerist circles during the 1840s. Some *magnétiseurs*, accustomed to the notion that certain people, in a trance state, could explore the beyond and communicate with its inhabitants, readily embraced the idea that disembodied souls caused these new manifestations. Others saw this willingness to posit the existence of spirits as an unscientific departure from Mesmerism's primary mission, which was to serve as a uniquely effective, inexpensive form of medical therapy.

The tension between these schools of thought had roots deep in Mesmerism's past, but emerged with particular intensity in the years after the advent of the *tables tournantes*. Initially, the phenomena that captured public attention in 1853 struck many *magnétiseurs* as a form of vindication: spontaneously rotating tables and hats seemed to confirm their theory that the will emitted a quasi-electric "universal fluid." As simple rotation gave way to telegraphic dispatches from the beyond, however, many practitioners who saw themselves primarily as healers—therapeutic Mesmerists—began to grow skeptical.[3] Spirit communications seemed to imperil the stance of objectivity that, in their eyes, was a crucial aspect of Mesmerism's claim to medical and scientific legitimacy. The skepticism of these therapeutic Mesmerists increased in 1857, when the visiting American medium Daniel Dunglas Home caused a sensation in the French press by introducing a new array of phenomena. Table-tipping and automatic writing had given way to luminous spirit hands, levitations, instruments that played of their own accord, uncanny revelations about deceased loved ones, and written messages that appeared as if from nowhere. Spiritualist Mesmerists embraced these astonishing manifestations as signs of a new era in human history—one in which mankind would finally resolve the crisis of factuality that scientific progress had created in religious life. Therapeutic Mesmerists, for their part, argued that their discipline would never be accepted by the scientific community if its practitioners surrendered so readily to metaphysical aspirations that, in their view, were nothing but superstitious fanaticism.

[3] In the *Journal du magnétisme* and the *Union magnétique*, the *magnétiseurs* focused on medical practice described their approach with a variety of terms, such as "pure," "true," or "clinical" Mesmerism. Those who preferred to address metaphysical questions usually called themselves "spiritualists"—even though their ideas and practices often differed from the ones English-speaking readers associate with the term. To make the polemic between these factions easier for the reader to follow, I refer to the first as "therapeutic Mesmerists" and the second as "spiritualist Mesmerists."

The skeptics lost this debate. Therapeutic Mesmerism lapsed into obscurity after 1859, not to revive until the early twentieth century. Spiritualist Mesmerism, on the other hand, served as the foundation for Spiritism, a new movement that would become an enduring part of French religious life. As we will see, American spiritualism, in the form of *tables tournantes*, articles from American newspapers, and the publicity surrounding Home's visit, reshaped Mesmerist theory and practice. This transformation laid the groundwork for the emergence both of a new social type—the spirit medium—and of a new kind of religious experience.

From Science to Superstition

Mesmerism began in 1778, when Franz Anton Mesmer, a German doctor trained in Vienna, arrived in Paris with a new technique for treating disease, which he called *magnétisme animal*. This therapy, Mesmer claimed, owed its efficacy to an invisible force, roughly analogous to electricity, which he called the "universal fluid." In healthy people, this fluid circulated freely; in the diseased, it encountered obstacles. Trained *magnétiseurs* could eliminate these obstructions by manipulating and concentrating the fluid. Their techniques for doing so involved a variety of methods, including massage, special gestures called "magnetic passes," and the use of an apparatus known as the *baquet*, a water-filled tub containing iron rods to which patients were connected by ropes. Once the *magnétiseur* had successfully restored free circulation of the fluid, the patient would experience what Mesmer called a "crisis," which usually involved a trance state and convulsions. Mesmer's theories and treatments generated enthusiasm among amateur scientists and learned men during the decades before the French Revolution. Even after 1784, when a government-appointed commission issued a report denying the reality of Mesmer's fluid, French interest in these new ideas remained strong, though the Revolution changed this situation by scattering Mesmer's mostly aristocratic followers. After 1789, Mesmer severed his own ties with France, and died in Germany in 1815.[4]

During the First Empire and Restoration, French Mesmerism returned, but in a different form. This shift was largely due to the influence of a provincial aristocrat, Armand Marie Jacques de Chastenet, marquis de Puységur. He began his experiments with *magnétisme animal* in the 1780s. Unlike many of his fellow nobles, he not only remained in the country during the Revolution but also succeeded in surviving its most tumultuous years

[4] Robert Darnton, *Mesmerism and the End of the Enlightenment in France* (Cambridge, MA: Harvard University Press, 1968), 3–4, 62, 67, 135.

while preventing the expropriation of his property. A term of service as an artillery general in 1791 put the marquis in good enough graces with the Empire to earn him an appointment as mayor of Soissons, a post he abandoned in 1807 to devote himself to the propagation of Mesmerist ideas. With the financial support of his ample rents, he quickly emerged as the most influential figure in the revitalized movement, a position he enjoyed until his death in 1815. Puységur approached the study of Mesmer's universal fluid in a manner that differed markedly from that of its originator. Rather than focusing on dramatic "crises," Puységur directed his attention to the study of "somnambulism," a more placid—though no less remarkable—variety of trance that subjects under the influence of the fluid often entered.[5]

Many scholars present Puységur's studies of "magnetic sleep," the first of which appeared in 1786, as key precursors of what later came to be called hypnosis.[6] Certainly, the trance state he studied had similarities to those that would later be induced in some forms of psychiatric therapy, but it also differed in crucial respects. Somnambulism, as it appeared in the works of Puységur and the numerous *magnétiseurs* who built on his example, was a complex and polymorphous phenomenon that expressed itself in a variety of uncanny ways—not only psychologically, but also physically. As Bertrand Méheust observes, the state of Mesmeric somnambulism usually entailed a dramatic diminution or augmentation of perceptual and cognitive abilities. Some entranced subjects experienced varying degrees of anesthesia; others found their senses and intellectual powers heightened. According to Mesmerists, this acute sharpening of the senses could sometimes rise to the level of "lucidity," a pitch so high as to appear superhuman. Many lucid somnambulists became considerably more articulate and quick-witted than they were in the normal, waking state. A few could perform "autoscopy," diagnosing their own illnesses by looking directly into their bodies. Mesmerists also presented accounts of exceptional somnambulists who were capable of reading letters inside sealed envelopes, determining the contents of locked desk drawers in distant rooms, and even divining another person's thoughts.[7]

Among students of Mesmerism, this distinctive state of altered perception nearly always emerged in the context of a particular kind of relationship.

[5] For biographical information on Puységur, see Bertrand Méheust, *Somnambulisme et médiumnité,* (Le Plessis-Robinson: Synthélabo, 1999), 1:352–356.

[6] Léon Chertok and Isabel Stengers, *Le Cœur et la raison, l'hypnose en question de Lavoisier à Lacan* (Paris: Payot, 1989); Adam Crabtree, *From Mesmer to Freud: Magnetic Sleep and the Roots of Psychological Healing* (New Haven: Yale University Press, 1993); Henri Ellenberger, *The Discovery of the Unconscious: The History and Evolution of Dynamic Psychiatry* (New York: Basic Books, 1970); Alan Gauld, *A History of Hypnotism* (Cambridge: Cambridge University Press, 1995).

[7] Méheust, 1:153, 156–216.

The *magnétiseur* served as manipulator of fluid, objective observer, poser of questions, and documenter of answers, while the *somnambule* became the instrument on which the *magnétiseur* acted. This relationship often involved social inequality; indeed, as Alison Winter argues, it usually seemed to grow out of the tensions such inequality created.[8] *Somnambules* were frequently women, or men of lower class status and lesser education than their *magnétiseurs*. The Mesmerist literature presented somnambulists as by nature shy, delicate, and self-effacing. *Magnétiseurs*, in contrast, were usually men, and were often educated *bourgeois* or aristocrats; those of more humble origins tended to be self-confident autodidacts. Descriptions of prominent *magnétiseurs* stressed their physical vitality—often seen as a reflection of the large quantities of fluid they could command—their assertiveness, their prodigious willpower, and their general air of assured mastery. In the trance state itself, the inequality between *magnétiseur* and *somnambule* seemed to disappear or even be reversed: The distinguished Mesmerist became the channel that allowed the humble *somnambule* to transcend her everyday terrestrial limitations and achieve a state of superhuman lucidity. Puységur's early experiments provided an influential template for this relationship. His first somnambulist was a twenty-three-year-old peasant named Victor Race. In his normal state, Race was a shy, unlettered farmer who spoke in the local patois. Once Mesmerized, he manifested not only heightened powers of mind but also a new social personality: He seemed able to read Puységur's thoughts directly; when he did so he abandoned his rustic patois for the standard French of a more learned man.[9]

Under the Restoration, the resurgence of interest in *magnétisme animal* quickened. Several Mesmerist periodicals appeared, and the phenomena of somnambulism became an increasingly frequent object of study in French hospitals. By the mid-1820s, Mesmerism had caught the attention of the Académie de médecine. Two of its members, Etienne Georget and the Baron Rostan, began to argue strongly for the reality and importance of the phenomena somnambulists produced, and in 1825 the Académie appointed the first of several commissions to study the subject.[10] Many of Georget's and Rostan's colleagues, however, were skeptical of the remarkable feats *magnétiseurs* ascribed to their somnambulists; the difficulty of reliably producing the most spectacular forms of lucidity in a laboratory context further undermined the Mesmerist cause. In 1842, after a series of unsuccessful experiments, and over the loud objections of a minority of its

[8] Alison Winter, *Mesmerized: Powers of Mind in Victorian Britain* (Chicago: University of Chicago Press, 1998), 1–13.
[9] Meheust, 1:14–18.
[10] Ibid., 1:364–367.

members, the Académie de médecine resolved to abandon all further experimental study of lucid somnambulism.[11]

This Academic rejection helped reinforce an oppositional tendency that had been present in Mesmerist circles throughout the nineteenth century. Beginning in the 1830s, but to a considerably greater extent in the next decade, Mesmerism became a facet of *la vie de bohème*. Doctors and aristocrats continued to pursue their experiments but increasingly found their ranks swelled with journalists, Romantic Socialists, literary writers, and visionary working-class autodidacts. Charles Fauvety, a follower of the Romantic Socialist Saint-Simon, studied Mesmerism, as did the Fourierist Victor Considerant; as we have already seen, the *Démocratie pacifique* frequently printed reports of unusual phenomena produced by *magnétiseurs*.[12] Writers such as Honoré de Balzac, Théophile Gautier, and Alexandre Dumas used Mesmerism as a dramatic device in their works and attended Mesmerist gatherings.[13] For many in this new audience, Mesmerism seemed a natural step in a broader Romantic quest for transcendence. The lucid somnambulist, her mind liberated from bodily constraints, freed to seek enlightenment in the ethereal world of pure essences, gave a powerfully concrete form to a central Romantic ideal. Wild-eyed, literally magnetic, and endlessly vital, the *magnétiseur* did the same.

As Mesmerism's most prominent audience became more literary, an ideological shift occurred within the movement. In the late 1840s, a new school of spiritualist Mesmerists began to emerge. These thinkers did not construct their position from whole cloth, but instead built on the precedent of a late-eighteenth-century *Lyonnais* school that included such figures as the mystical Freemasons Louis-Claude de Saint-Martin and Martinès de Pasqually.[14] According to exponents of this spiritualist current, Mesmerism could serve not only as a therapy but as a means of receiving direct revelations from the beyond. In the lucid state, these

[11] The story of this dismissal is a mainstay of the nineteenth-century Mesmerist literature. See, e.g., Ernest Bersot, *Mesmer et le magnétisme animal* (Paris: Hachette, 1854), 41–70. Méheust presents a detailed account of these controversies as well. See Méheust, 1:372–454.

[12] For Fauvety's opinions on Mesmerism, see the 1854 letter reprinted in Goupy, *L'Ether, l'électricité et la matière*, 279–301. For Considerant, see Beecher, *Victor Considerant*, 41.

[13] Honoré de Balzac, *Séraphîta* (Paris: L'Harmattan, 1995 [1835]); Alexandre Dumas, *Mémoires d'un médecin, Joseph Balsamo*, 2 vols. (Paris: Legrand et Crouzet, n.d.); for more on Gautier's interest in Mesmerism, see Anne-Marie Lefebvre, *Spirite de Théophile Gautier, étude historique et littéraire*, unpublished doctoral thesis, Sorbonne troisième cycle, 1978, 15–23.

[14] For more on this current of French thought, see Christine Bergé, *L'au-delà et les Lyonnais, mages, médiums et Francs-Maçons du XVIIIe au XXe siècle* (Lyon: LUGD, 1995), 13–56; David Allen Harvey, *Beyond Enlightenment: Occultism and Politics in Modern France* (DeKalb: Northern Illinois University Press, 2005); René Le Forestier, *La Franc-maçonnerie occultiste au XVIIIe siècle et l'Ordre des Elus Coens* (Paris : La Table d'Emeraude, 1987); Auguste Viatte, *Les Sources occultes du romantisme*, vol. 1, *Le Préromantisme* (Paris: Champion, 1928).

thinkers argued, somnambulists' souls detached from their bodies, acquiring the ability both to see and to converse with spirits, saints, and angels.

The two leading exponents of this approach in the 1840s were Henri Delaage, an aristocratic—by rumor if not by birth—*homme de lettres*, and Louis-Alphonse Cahagnet, a sometime cabinet-maker and cutter of collars.[15] While Delaage and Cahagnet shared the fundamental assumption that somnambulists could contact the spirit world, the conclusions they drew from this belief differed as dramatically as their social stations. Delaage was an ardent Catholic, whose elaborately wrought prose testified to the extent of his formal education. He argued that Mesmerism and the ecstasy of somnambulists confirmed Church teachings.[16] Cahagnet, on the other hand, was an inveterate freethinker and passionate autodidact. He cited Swedenborg as his primary inspiration but readily diverged from the master's text when somnambulists provided him with new information.[17] Both, however, were prolific writers who established a new sense of the possibilities of somnambulism, while simultaneously disconcerting those Mesmerists who preferred to avoid metaphysical speculation in favor of therapeutic practice.

Delaage and Cahagnet also both contributed to a further development in the social meaning of the somnambulist's role. In their spiritualist Mesmerism, the relationship between *magnétiseur* and *somnambule* was still founded on inequality: The *somnambule* Delaage described at greatest length, Alexis Didier, was the shy product of a humble family of artisans; Cahagnet's most remarkable subject, Adèle Maginot, was a woman of undetermined occupation from what one commentator called "the illiterate class."[18] The transcendence these somnambulists achieved in their states of lucidity, however, far outstripped anything Puységur would have imagined.

[15] Both Delaage and Cahagnet are intriguing figures. For a typically imprecise biography of Delaage, who remained calculatedly vague on the subject of his origins, see *L'Union magnétique* 4, no. 53 (Mar. 10, 1857): 1–2; for a more detailed biography of Cahagnet, see ibid. 4, no. 61 (Jul. 10, 1857): 1–2; and 4, no. 62 (Jul. 25, 1857): 1–2.

[16] Delaage's books all make similar arguments. For a typical example, see Henri Delaage, *L'Eternité dévoilée, ou vie future des âmes après la mort* (Paris: Dentu, 1854), especially pages 43–96, which convey his view of the relationship between Mesmerism and Catholicism.

[17] For a comparatively concise account of Cahagnet's views, see the "conférences" on God and Swedenborg in Louis-Alphonse Cahagnet, *Sanctuaire du spiritualisme, étude de l'âme humaine et de ses rapports avec l'univers, d'après le somnambulisme et l'ecstase* (Paris: Germer Baillière, 1850), 4–116, 234–256.

[18] Louis-Alphonse Cahagnet, *Magnétisme. Arcanes de la vie future dévoilés, où l'existence, la forme, les occupations de l'âme après sa separation du corps prouvés par plusieurs années d'expériences au moyen de huit somnambules extatiques*, 3 vols. (Paris: Germer-Baillière, 1848–1854); Henri Delaage, *Le Monde prophétique, ou moyen de connaître l'avenir, suivi de la biographie du somnambule Alexis* (Paris: Dentu, 1853); for "illiterate class," see Alexandre André Jacob [Alexandre Erdan, pseud.]. *La France Mystique, tableau des excentricités religieuses de ce temps, 2e édition* (Amsterdam: R.C. Meijer, 1858), 1:43.

They became heroic visionaries, seers capable either of astonishing feats of clairvoyance or of dialogue with angels. Didier, sitting in Paris, could describe the furnishings in London apartments he had never visited. Maginot had extensive conversations with supernatural beings and the souls of the illustrious dead. In Didier's case, his distinction as a clairvoyant also made him a celebrity, eagerly sought after in the highest literary and social circles.[19] Dumas, for example, praised Didier's uncanny abilities in a series of letters to *La Presse* in 1847, describing several experiments performed before such notables as Ferdinand de Lesseps.[20] This new conception of the somnambulist as celebrity and vehicle of transcendent inspiration served as a crucial precedent for the social role that spirit mediums would occupy, and to a certain extent create for themselves, in subsequent decades.

In the early years of the Second Empire, the Academic rejection of Mesmerism and its ever-closer association with the Romantic, left-wing Bohemianism of 1848 pushed it into a decidedly marginal position. Napoleon III's government proved quite willing to use legal sanctions against Mesmerists and somnambulists who exercised their powers for money. In 1852, for example, Alexis Didier was tried and convicted along with nine other somnambulists under articles 479 and 480 of the *Code pénal*, which forbade the practices of divination and dream interpretation.[21] In a mere fifteen years, Mesmerism seemed to have been transformed from a legitimate, if controversial, scientific pursuit into a form of superstitious trickery no better than fortune-telling.

The *Tables Tournantes* as Saving Grace

For many Mesmerists, the 1853 séance vogue initially seemed to presage a dramatic reversal of this decline. While the new phenomena had no precedents in the literature of *magnétisme animal*, they nevertheless seemed to bear out some of its key concepts. Published commentary in pamphlets and newspapers, much of it written by journalists familiar with the Bohemian world of *magnétiseurs* and *somnambules*, frequently used terms such as "fluid" and "magnetism," which made it easy for Mesmerists to claim the *tables tournantes* as their own. In May 1853 the *Journal du magnétisme*—at the time France's only Mesmerist periodical—began to print extensive quotations from various press reports of the new phenomena. The Baron Jules

[19] Bertrand Méheust, *Un Voyant prodigieux: Alexis Didier, 1826–1886* (Le Plessis-Robinson: Institut Synthélabo, 2003).

[20] Quoted in Méheust, *Somnambulisme et médiumnité*, 1:603–620.

[21] For a partial transcript of this trial, see *Journal du magnétisme* 11 (1852): 550.

du Potet de Sennevoy, the journal's editor and a veteran of the Academic struggles of the 1820s, triumphantly announced that the *tables tournantes* had at last given "Mesmer's discovery" the "universal sanction" it had lacked for so long. Scientists would no longer be able to dismiss the Mesmerist belief that the mind could exert a force of its own by means of the universal fluid; the *tables tournantes* irrefutably proved the reality of that force and its power. As a result, du Potet wrote, "one can say with certainty that today's events are crucially important, that they mark the beginning of a new era the likes of which have never been seen before."[22] The narrow-mindedness of Academic doctors and scientists had long prevented them from perceiving the vast potential of the human mind. After the advent of the *tables tournantes*, this blind disregard of mankind's transcendent nature was no longer possible.

For *magnétiseurs*, the fact that the universal fluid appeared to transform the séance table into what one writer called an "extra-human galvanic battery" added a considerable allure to the new phenomena.[23] Not only did the *tables tournantes* prove the fluid's reality, as du Potet observed, they also seemed to reconcile the therapeutic and spiritualist schools of Mesmerism. Previously, the spiritualists' theories had relied entirely on the verbal assertions of somnambulists who claimed to see otherworldly entities; now, these theories had an objective component anyone could perceive. Moving tables—especially those that tapped out messages—rendered otherworldly forces tangible. By transforming the spiritualist enterprise into a viable empirical project, the *tables* seemed to eliminate the primary reason for therapeutic Mesmerists' dissatisfaction with the spiritualist approach.

In the preface to a pamphlet on the *tables tournantes*, Delaage gave voice to this hope by emphasizing the power of the new phenomena to reconcile idealism and materialism, both within Mesmerism and in society at large. This reconciliation, he believed, would have profound implications for the continued progress of mankind:

> The mission of the first half of the nineteenth century seems to have been to study the properties of matter and its constituent elements. An infinitely more marvelous glory awaits the second, upon which we, sons of the future, have just embarked, our hearts full and our faces aglow with the celestial luster of divine hopes; this new half-century will transform all the sciences into paths leading to the infinite, which is to say, to God.[24]

[22] *Journal du magnétisme* 12 (1853): 221.
[23] Ibid., 189.
[24] Henri Delaage, intr. to Ferdinand Silas, "Instruction explicative des tables tournantes, d'après les publications allemandes, américaines, et les extraits des journaux allemands, français et américains" (Paris: Dentu, 1853), 1.

The *tables tournantes* had begun a new stage in mankind's triumphant march toward total understanding, Delaage wrote, by making the soul and the universal fluid palpable in uniquely accessible ways. Soon, these undeniable facts would lead both therapeutic Mesmerists and conventional scientists to embrace the spiritualist conception of the soul as an autonomous and immortal entity. Once science and spiritualism had fused in this manner, Delaage believed, a new era of human possibility would dawn. Human beings would no longer be alienated from their divine nature; progress would involve not only an ever-increasing mastery of the physical world but also an ever-increasing knowledge of the spiritual one. Moral improvement, Delaage believed, would inevitably accompany this process of discovery.

The extent of Mesmerists' enthusiasm for these new phenomena was evident at the banquet du Potet's Société du Mesmérisme held on May 23, 1853, in honor of Mesmer's birthday. Before an audience of 250 guests, an array of speakers eagerly claimed the *tables tournantes* for Mesmerism. In a long and ardent address, for example, the former Fourierist and Second Republic politician André-Saturnin Morin presented the *tables tournantes* as the latest addition to the "crowd of marvelous phenomena" that *magnétisme animal* produced.[25] It was perfectly logical, Morin argued, for the universal fluid to allow man to "control inanimate objects as he controlled his own organs," even if this ability had appeared only recently.[26]

Like du Potet and Delaage, Morin believed that the séance vogue heralded a new phase in Mesmerism's development, which would eventually have a dramatic effect on French society as a whole. In Morin's account, these phenomena even promised a form of compensation for the failure of the Second Republic. The widespread popular interest in *tables tournantes* and *parlantes*, he declared, marked nothing less than the "dawn of a humanitarian revolution."[27] Like the journalists at the *Démocratie pacifique*, Morin had been forced to abandon his efforts to improve society by political means, but he shared their hope that the transformative vision of Romantic Socialism could still be realized with the help of otherworldly forces.

Where the Fourierist journalists looked to the transcendent power of solidarity, however, Morin looked to Mesmerism and the connection its techniques could forge with the spirit world. Indeed, he observed in a

[25] *Journal du magnétisme* 12 (1853): 251. For a description of Morin's political career under the Second Republic, which included an unsuccessful run for deputy and terms as conseiller municipale of Chartres and conseiller générale of the Eure-et-Loire, see Archives de la Préfecture de Police de Paris, Carton BA 1195, dr. 109274.

[26] *Journal du magnétisme* 12 (1853): 251.

[27] Ibid., 252.

subsequent article on the subject, commerce with disembodied souls promised to benefit humanity in the same way that "the intervention of the superior white race" had benefited "savage peoples." Like the "white race," Morin believed, the spirits would be agents of "progress" and radical transformation. The *tables parlantes,* in his view, were the first beachhead in an impending Divine conquest of mankind; the temporary subjugation communication with spirits would entail promised not to restrict human achievement, but to advance it in untold ways. Man might have proved incapable of inventing a perfect society on his own, but under the stern influence of a colonizing God, improvement would be inevitable.[28]

Even the refusal of the Académie des sciences to consider the renowned engineer Marc Séguin's argument that an unknown, invisible force caused these phenomena initially left Mesmerists unfazed. Morin, like the advocates of scientific open-mindedness in the press, decried the obstinacy of the *académiciens,* but remained hopeful that Mesmerism's more flexible approach would win the day. After all, he asserted,

> Truth does not need a passport from the *Académies,* whose condemnations have never prevented a discovery. Despite the sarcasm of scientific bodies, we examine what has been declared impossible, prove its reality, and will soon popularize it.[29]

Mesmerists had already spent decades advancing their knowledge of the mind in spite of Academic opposition, Morin argued, and it was only natural for them to continue to do so. The wonderful power of democracy—thwarted in the political sphere but still active in the intellectual one—would take care of the rest. Once a sufficiently large number of French people had seen for themselves that tables could spontaneously move and communicate after being "activated" by a mysterious force, Morin believed, Academic opposition would inevitably collapse. Narrow-minded scientific intransigence was no match for empirical truth, particularly when any amateur with a quiet hour, willing guests, and a small table could perform conclusive experiments of his own. Elitist as the Academy might have been, it would prove unable to counteract an otherwise universal consensus. The *tables tournantes* allowed Morin—and many of his equally idealistic fellow Mesmerists—to express republican and revolutionary hopes, even as the prospect of a true French Republic disappeared beneath the authoritarian pomp of the Second Empire.

[28] Ibid., 560.
[29] Ibid., 330.

Spirit Communications and the Problem of Objectivity

The unanimity Morin predicted with such confidence never emerged, even among the small community of *magnétiseurs*. Instead, a gulf widened between advocates and opponents of the new phenomena. Their debates, like discussions of the *tables tournantes* among Academic scientists and their critics, centered on the epistemological ideal of objectivity and the asceticism it seemed to entail. This new moralization of observational detachment posed a vexing challenge to Mesmerists, many of whom still dreamt of a future in which their ideas would win Academic acceptance. For some students of *magnétisme animal*, embracing the new ideal of objectivity seemed like the best way to make the case for the legitimacy of their discipline. In their view, Mesmerism was a science no different from chemistry or astronomy, and therefore needed to follow the same epistemological rules, developing knowledge through the accumulation of objective data in the form of well-documented accounts of Mesmeric cures. For other *magnétiseurs*, however, the stakes were much higher. Mesmerism, in their view, was not simply the equal of all other sciences, it was their queen. Where chemistry or astronomy gathered knowledge of the physical world, they argued, Mesmerism necessarily explored the metaphysical. The therapeutic efficacy of *magnétisme animal*, in their view, was only incidental, an already-proved fact that could safely be taken for granted. A truly objective stance, as they saw it, was one that fully acknowledged Mesmerism's potential to transform the scientific project into something utterly new, a system of total knowledge that would far outstrip the arid discoveries so important to the *académiciens*.

The dispute between these two groups of Mesmerists turned on the question of spirit intervention. At first, it seemed as if the phenomena of American spiritualism, which added a material dimension to the dialogues with the beyond already familiar to spiritualist *magnétiseurs*, would resolve this difficulty by providing empirical evidence for the existence of otherworldly entities. In the end, however, the addition of mysterious raps or spontaneously rotating tables did little to address the fundamental epistemological conundrum that spiritualist Mesmerists had always faced: the lack of a reliable means of proving that the spirits encountered in a séance were autonomous beings and not simply products of the imagination.

The way in which a student of *magnétisme animal* approached this problem depended on his general view of the Mesmerist project. Those who saw Mesmerism as a discipline comparable to the other sciences tended to present advocates of spirit intervention as transgressors of the new morality of objectivity, emotional fanatics whose enthusiasm "push[ed] Mesmerism into the ruts of prejudice and superstition."[30] Those who saw Mesmerism as a primarily metaphysical endeavor tended to give the spirits the benefit of the doubt. These spiritualist *magnétiseurs* made common cause with the journalists and scientists who had argued in favor of an open-minded study of the *tables tournantes*: in their view, objectivity entailed a studied reluctance to declare anything impossible. Dismissing spirit intervention out of hand showed a dangerously unscientific willingness to rely on preconceived ideas. It was exactly this kind of blinkered, subjective refusal to consider revolutionary new data, they argued, that had motivated the Academic rejections of Mesmerism itself.

Even for those who argued in favor of open-mindedness, however, the epistemological difficulties of the spirit hypothesis remained. As Morin investigated the question more closely in the pages of du Potet's journal, for example, his initial enthusiasm gave way to a more nuanced position. Authentic communications from the beyond, he now asserted, seemed to be rare, and often difficult to identify with any certainty. Even when these messages were authentic, he argued, their contents could be of dubious value. In a long review of the Catholic writer Henri Carion's pamphlet on contacts with the beyond, Morin elaborated this idea, critiquing students of the new phenomena who believed that spirit communications could be an infallible source of revelation. All the spirits Carion contacted, Morin observed, whether they were saints describing the joys of paradise, repentant souls in purgatory, or the suffering damned, professed "a perfectly orthodox doctrine." Elsewhere, however, the spirits espoused other points of view:

> In most other circles, where completely different beliefs dominate, the manifestations have another character; spirits there declare that there is no Hell, that natural religion is sufficient, that all men are saved; a large number profess metempsychosis.

Adversaries of the spirit hypothesis, Morin noted, tended to cite this diversity of opinion as a primary reason for denying the otherworldly origin of such communications. For Morin, however, these divergences did nothing to disprove the possibility that spirits might communicate with terrestrial humans. The specific content of the messages spirits dictated was irrelevant, in his view. The crucial fact was that some of these messages, regardless of the specific opinions they espoused, demonstrated "the action of intelligent beings foreign to man," and could not be adequately explained by any other means.[31]

[30] From an anti-spiritualist article originally published in the *Indicateur de Sens*, quoted in the *Journal du magnétisme* 13 (1854): 163.

[31] *Journal du magnétisme* 13 (1854): 184, 220–221.

Many of the communications students of these new phenomena received, however, did not even meet this basic empirical criterion, particularly when automatic writing was involved, as in Carion's case. "The method of writing mediums," Morin wrote, "is one of the most suspect, one that most gives rise to suspicions of fraud or involuntary illusion." When an animated table tapped out a message, the sheer inexplicability of the phenomenon partly confirmed its otherworldly origin. A writing medium gave no equally clear signs of being animated by an external intelligence; as a result, it was considerably more difficult to tell where self-delusion ended and legitimate contact with the beyond began. To prove his point, Morin undertook an experiment of his own, contacting Voltaire as Carion had, by sitting at a desk, placing his hands on a *planchette,* and invoking the spirit in the name of God. Voltaire heeded Morin's call, and provided the following responses to his interlocutor's questions:

Q. Was it you who communicated with M. Carion?
A. No. It was a great-grandfather who took my name: he is a Jesuit.
Q. Are you happy?
A. Yes. . . .
Q. Does Hell exist?
A. No.
Q. And Purgatory?
A. Purgatory is your miserable Earth and the other planets like it.
Q. Do you continue to espouse the same doctrines you did on Earth?
A. More than ever.

Morin's Voltaire then proved he was not a demon by neatly writing the word "God." This communication, Morin observed, was no less valid than Carion's, and it was impossible to prove that either had come from the authentic spirit of Voltaire. Like all "equal and opposite forces," these two communications neutralized one another, demonstrating that the vexed question of Voltaire's status in the afterworld remained unanswered and unanswerable. In the end, Morin argued, Carion's vaunted autograph from the repentant Voltaire was most likely the product of wishful thinking: "it is at least probable that the operator simply obtained this declaration because he wished to; in his presentation, it serves a merely rhetorical purpose." Authentic spirit communications almost certainly occurred, Morin maintained, but human knowledge would have to advance a considerable distance before it could presume to distinguish the true from the false, particularly in the case of phenomena as ambiguous as automatic writing.[32]

[32] Ibid., 220–221, 223.

Questions like the ones Morin raised proved divisive once the initial excitement of the séance vogue had cooled. As early as the fall of 1853, du Potet told the readers of the *Journal du magnétisme* that he was under increasing pressure from two opposing camps of Mesmerists. One group attacked his willingness to discuss the *tables tournantes* and the phenomena of American spiritualism at all, while the other believed he did not discuss them enough.[33] In 1854 and 1855, du Potet addressed this conflict by assuming a stance of editorial impartiality. The *Journal* published articles by critics of Mesmerism's engagement with the new phenomena alongside those of advocates, but it did not completely abandon the subject.

Though du Potet attempted to maintain a degree of critical distance in his editorial stance, he also made no secret of his personal opinion of the new phenomena. He was firmly on the side of those who favored spirit intervention. In du Potet's view, a careful study of spirit communications could provide an important corrective to a disconcerting metaphysical trend among mainstream scientists:

> *MM. les académiciens* are the gravediggers of all moral truths; they have led doubt into the strongest souls—everywhere the sweet hope of an afterlife has ceased to exist. Religion is dead, man a mere animal powered by a type of electricity; our fathers' belief in the hereafter is passed off as a dream; strength today lies in the negation of all that is true and will be so eternally. We walk in darkness—what use are these marvels of industry, these fertile discoveries in the arts or sciences![34]

The phenomena described in American spiritualist newspapers, and increasingly in the publications of French visionaries, du Potet argued, had the power to remedy this situation by providing a new kind of metaphysical certainty based on an empirical understanding of the soul and its fate after death. The purpose of Mesmerism, he believed, was thus not to contribute to an existing edifice of scientific knowledge but to provide the foundations for a new and more satisfactory one.

Perhaps because of du Potet's continued personal support of the spirit hypothesis, the *Journal du magnétisme*'s policy of ideological diversity did not resolve the building tensions within French Mesmerism. In early 1854, a group of *magnétiseurs* that called itself the Société philanthropico-magnétique de Paris established a journal of its own, the *Union magnétique*, to break the monopoly that du Potet's long-lived publication had previously enjoyed. In the first issue, Jules Lovy explained his group's reasons for starting a new periodical. Mesmerism, he wrote, "inevitably seems to

[33] Ibid., 524.
[34] Ibid.

succumb to all the travails of religious sects," and even risked "a schism." This new publication, Lovy maintained, would work to end these conflicts by becoming "the journal for all."[35]

Lovy left the specifics of the "schism" the *Union magnétique* sought to overcome diplomatically vague. The tenor of his rhetoric, however, which presented the foes of unity among Mesmerists as religious fanatics in the grip of a "blind enthusiasm," made the nature of his group's grievances clear.[36] Though the Société philanthropico-magnétique included several Spiritualists among its members, the majority remained skeptical of the Du Potet circle's willingness to accept the reality of otherworldly intervention. Perhaps more important still, the contributors to the *Union magnétique* also regarded the older journal's increasing concern with metaphysical speculation, at the expense of thorough descriptions of clinical experiments, as a diversion from Mesmerism's true mission.

During 1855, the debate intensified. In January, the editors of the *Union magnétique* decided to devote their journal exclusively to "clinical Mesmerism, which is to say, descriptions of cures obtained by this powerful means." The *tables tournantes* and the phenomena of American spiritualism, which had previously been a subject of frequent—though almost always skeptical—commentary in the *Union* now disappeared from its pages. The editors couched this announcement in the rhetoric of objectivity. They would use "the multiplicity and repetition of facts"—of the easily verifiable, indisputable kind Mesmeric cures presented—to prove "that Mesmerism is not the result of a force that works on *certain* individuals, but rather on *everyone* in general."[37] Therapeutic Mesmerism could cure anyone, and its phenomena, at least according to the editors, were easily replicable. The spectacular manifestations described by American spiritualist newspapers and French visionaries, on the other hand, seemed to appear only rarely, for tiny groups of uniquely favored believers and mediums. The best way to ensure Mesmerism's eventual acceptance as a viable scientific pursuit, then, was to focus on its most common and practical applications.

Du Potet, for his part, took the *Journal du magnétisme* in the opposite direction. While he still printed occasional reports of Mesmeric cures, he also devoted an ever-growing amount of space to accounts of experiments with *tables tournantes* and to articles describing dramatic otherworldly

[35] *Union magnétique, journal du Société philanthropico-magnétique de Paris* 1 (1854):1–2, 23.

[36] See, e.g., a sharply critical series of articles by J. A. Gentil arguing against the role of otherworldly intervention in the production of Victor Hennequin's revelations. This series was also published in pamphlet form, as J. A. Gentil, "L'Ame de la terre et les tables parlantes, ou Sauvons le genre humain, ouvrage examiné au point de vue magnétique de l'influence des besoins sur le moral" (Paris: the author, 1854).

[37] *Union magnétique* 2 (1855): 1. Italics in the original.

manifestations, often culled from American spiritualist periodicals.[38] Late in the fall, du Potet made this shift a matter of formal editorial policy. There was a "new movement" underway in Mesmerism, he wrote, and it demanded a radical change in focus. Once, students of *magnétisme animal* had concerned themselves exclusively with the treatment of disease and the physiological study of trance states. Now, the proliferation of new phenomena obliged them to "address the world of the marvelous, to produce new annals."[39]

In subsequent articles, du Potet observed that many of his readers did not seem to share his enthusiasm for this new departure.[40] A long letter from the *magnétiseur* de Guibert de Clelles, published in July 1856, gives a sense of the criticism du Potet received. The *Journal*'s turn toward spiritualism, Clelles argued, constituted nothing less than a complete rejection of the scientific ideals that had previously characterized French Mesmerism. His condemnation was unstinting:

> If, as you have said . . . , your journal no longer needs to record the *common* phenomena of Mesmerism and somnambulism, then *pure* Mesmerists—I use the word *pure* to designate those who approach this science solely as a form of healing—who, quite generally, only have phenomena of this kind to report, will be excluded in favor of those with eyes and ears good enough to *see* and *hear* the *spirits* . . . . In consequence, the fruits of long years of study, observation and documentation will be lost, all because the *Journal du magnétisme*, in a kind of break with its name and its past, will have allowed itself to be completely invaded by the *spirits*.

Mesmerism, for Clelles, was not a metaphysical endeavor but a therapeutic one. To make it metaphysical, in turn, was to deprive it of the scientific rigor most appealing to "pure Mesmerists." When mixed with spiritualism, Clelles argued, Mesmerism became nothing but an excuse for endless, subjective theorizing based on the cryptic utterances of mediums, in which "I have . . . never been able to perceive the faintest *spirit* [*esprit*], nor even genuinely to suspect such a thing." Mesmerists, Clelles argued, would do better to focus on the "phenomena of *pure, natural, human, ordinary*, even down-to-earth Mesmerism."[41]

Critiques like these did nothing to sap du Potet's resolve. These new phenomena might be dismissed by prejudiced skeptics, but they were simply too important to ignore. In the United States, he wrote, Modern Spiritualism

[38] This increasing interest in *tables tournantes* and the phenomena of American spiritualism is obvious in the index to vol. 14 of the journal. See *Journal du magnétisme* 14 (1855): 673–680.

[39] Ibid., 558–559.

[40] Ibid., 643.

[41] *Journal du magnétisme* 15 (1856): 337, 338, 353. Italics in the original.

appeared to be in the process of provoking a widespread social transformation; in France, the number of "serious men" and "sincere scientists" who studied the subject amply justified its importance. Investigating these new phenomena, du Potet insisted, was "a duty for us."[42] His conviction in this regard, he continued, had made him immune to his readers' increasingly emphatic criticism. Indeed, du Potet expected large numbers of his readers to desert him. After all, he wrote, "it is never the masses, but always a few brave men, who open new routes and lead nations."[43] Those readers who elected to join him as part of this far-sighted elite were welcome, but he would not regret the loss of those who chose a less demanding path. Despite such protestations of equanimity, however, his ripostes grew vituperative. One of his articles, for example, denounced those reluctant to attribute the new phenomena to spirit intervention as "timid" and "weak in magnetic power"—among Mesmerists, a cutting insult indeed.[44]

In early 1857, du Potet decided to push this investigation of Spiritualism further. He dismissed his previous *gérant* (managing editor), Hébert, a therapeutic Mesmerist of the old school, and hired a new *rédacteur-en-chef* (editor in chief), a spiritualist named Zéphyre-Joseph Piérart. In a brief essay at the head of the January 1857 issue, du Potet presented this change as an even more dramatic development than the shift he had announced a year before. Under Piérart, the *Journal* would devote the bulk of its attention to "everything related to psychology and even the afterlife."[45] When he began as *rédacteur-en-chef,* Piérart was a comparative newcomer to the higher circles of French Mesmerism. He had performed his first experiments with a *somnambule* in 1854, and had published only a single article on the subject. In his new position, however, he revealed himself to be a voluble writer and pugnacious supporter of his ideas, which differed markedly from those of the contributors who had previously dominated the *Journal.* After a brief period of enthusiasm, writers like Morin had self-consciously avoided engaging in metaphysical speculation about the new phenomena, and had posed difficult questions about the nature of the communications received in séances. Piérart, in contrast, took the reality of spirits for granted, and immediately began to work out the philosophical implications of their existence.

At this point, French Mesmerists had split into two camps, each with its own conception of what kind of "science" the study of *magnétisme animal* should become. One group, represented by the contributors to the *Union magnétique,* continued to approach Mesmerism as an unjustly overlooked

[42] Ibid., 356.
[43] *Journal du magnétisme* 16 (1857): 4–5.
[44] *Journal du magnétisme* 15 (1856): 449.
[45] *Journal du magnétisme* 16 (1857): 4.

field of medicine. These therapeutic Mesmerists focused on the cataloguing of cures which, in their view, constituted a body of empirical evidence crucial to any claim of scientific legitimacy. Another group, represented by the *Journal du magnétisme*, whole-heartedly embraced American spiritualism. From their perspective, a truly objective stance entailed a willingness to acknowledge that spirit phenomena were possible and worthy of study, even if they defied many currently accepted scientific principles. In addition to studying a new range of phenomena and deriving a new set of theoretical conclusions from them, these spiritualist Mesmerists added an innovative array of practices to their repertoire. The old dyad of *magnétiseur* and *somnambule* began to give way to something new—a relationship between seeker and medium. For this group, Mesmerism had ceased to be a medical science and become a metaphysical one.

The Debate over Home and American Spiritualism

As *rédacteur-en-chef*, Piérart filled the *Journal du magnétisme* with translations of articles from the English-language spiritualist press. He gave particular weight to reports from America, which described an ever-widening range of astonishing phenomena. In his accounts, the United States was a nation in the throes of a full-fledged religious transformation: The spirits had made its parlors and concert halls the staging ground for a new form of revelation, which would inevitably travel across the Atlantic, bringing moral improvement and social justice in its wake. For the many spiritualist *magnétiseurs* who were sympathetic to the political left, this turn of events was logical. The United States, after all, was a republic. To them, it therefore seemed only natural that this dramatically superior, utterly modern political arrangement would bring metaphysical benefits as well. While the strict censorship regime Napoleon III put in place after coming to power made it impossible to explicitly discuss the political aspect of this argument in favor of American spiritualism, it nevertheless pervaded the *Journal*'s analyses of the new phenomena—which often seemed to be based more on wishful thinking than on an accurate reading of news from overseas.

In this context, the 1857 French tour of the American medium Daniel Dunglas Home seemed to mark the first step in the spread of this second, metaphysical American Revolution. Home was the most famous of all spiritualist mediums, with a reputation for producing a wide array of spectacular manifestations. When he arrived in France, Parisian aristocrats and notables vied for invitations to his séances, which were exclusive, highly fashionable events. A torrent of publicity in the press—nearly on a par with what the *tables tournantes* vogue had received four years before—increased the medium's *cachet.* The journalists who discussed Home's séances overwhelmingly cast them in a positive light: rather than mocking these events, they burnished their mystique, transforming the American medium into a Romantic hero and saying little that might call the authenticity of his gifts into question. This approach, which probably owed as much to literary fashion as to any real conviction, struck biased observers like Piérart as proof that the final triumph of spiritualism was at hand.

Piérart's engagement with American spiritualism in the months before Home's arrival went beyond a simple enthusiasm for reports of uncanny phenomena. He also explored the new theories that had emerged to explain these manifestations—especially the cosmology elaborated by the visionary Andrew Jackson Davis, which had become an important element of Modern Spiritualism in the United States. This new religion, Piérart noted, seemed ideally suited to the changing reality of the nineteenth century:

> The adepts of this church appear to support a consoling idea practically innate to the human heart: that God did not only speak to the ancients, that revelation did not come at a single moment, but instead comes continually, in forms suited to the progressing needs, capacities, and tendencies of humanity.

The Modern Spiritualist vision of continuous revelation, Piérart argued, was quintessentially American. It owed a great deal to the habits of mind cultivated by the citizens of "this free country, haven for all truths, enemy of all prejudices." The liberty that Americans (or rather, white Americans) enjoyed, Piérart argued, predisposed them to "calm and reasoned observation." By supporting Modern Spiritualism so ardently, they showed a unique ability to slough off the irrational a priori assumptions of the past and to explore a form of religious practice that entailed a complete rupture with old sources of theological authority. This greater flexibility of mind, Piérart argued, made it inevitable that American mediums would be more powerful than their French counterparts. The constraints of French society left aspiring mediums caught in a web of preconceived ideas; as a result, the phenomena they produced remained decidedly unspectacular. American mediums, judging by the accounts in spiritualist newspapers, seemed to be capable of producing considerably more conclusive manifestations.[46]

Though Morin did not share the new *rédacteur-en-chef's* enthusiasm for the specific revelations mediums received, he echoed Piérart's wonder at the American gift for contacting the beyond. The accounts published so

[46] Ibid., 86–87.

voluminously in the spiritualist press, Morin wrote, proved that in the United States "mediums abound [and] marvelous phenomena . . . are popular and accessible to everyone." Spiritualism would never succeed in France unless a similar proliferation of manifestations occurred there. The local supply of mediums, however, seemed incapable of producing sufficiently dramatic results. "In our country," Morin wrote, "subjects are few and very weak, with the majority limiting their powers to the strongly suspect use of the writing-basket or *planchette*; truly spiritualist phenomena are very rare and lack authenticity." The only means of remedying this state of affairs, he concluded, was to have American spiritualists send mediums to France, to act as missionaries for the new cause:

> Humanity needs to determine the reality of these phenomena, which, if authentic, would furnish palpable proof of communication between the living and the dead. France, deprived thus far, is thirsty for knowledge and, in its poverty, begs for help from better-favored countries. Let us hope that our transatlantic brothers will heed this appeal.

While Morin had never witnessed a manifestation that appeared to provide irrefutable proof of spirit intervention, he remained hopeful that powerful American mediums would have the ability to convince him. The evidence local mediums provided in support of the spirit hypothesis might have been slim, but the translations from American newspapers that filled the *Journal du magnétisme* indicated that an entirely different order of phenomena existed. Americans, with their superior medianimic gifts, had an obligation to share their transformative powers with nations whose subjects had not been able to enjoy a full measure of religious and political freedom.[47]

Daniel Dunglas Home was the first of these prodigiously gifted Americans to tour France. A slim, dark-haired young man of Scottish ancestry raised in the United States, Home arrived in Paris in June 1856 and stayed until the spring of 1857.[48] He had little money of his own, but proved adept at attracting patronage from wealthy aristocrats. By the time of his arrival in France, thanks in large part to a spectacularly successful visit to Great Britain—during which he held the séance that inspired Robert Browning's unflattering poem "Sludge the Medium"—Home was a full-fledged celebrity. While in Paris, he frequented numerous aristocratic *salons* and held several séances for the Emperor and Empress at the Tuileries Palace.[49]

[47] Ibid., 391–392.

[48] For an account of this visit, see Daniel Dunglas Home, J. W. Edmonds, intr., *Incidents in My Life* (New York: Carleton, 1863), 1:138–148.

[49] For a Second Empire courtier's descriptions of experiments with Home, see Théobald Walsh, *Dunglas Home et le spiritualisme américain, souvenirs contemporains* (Paris: J. Claye, 1858).

The phenomena that occurred in Home's séances were more dramatic than anything French mediums had succeeded in producing. Edmond Texier, writing in the *Siècle* early in 1857, enumerated some of these spectacular manifestations:

> tables tilt without being touched, and the objects on them remain immobile, contradicting all the laws of physics. The walls tremble, the furniture stamps its feet, candelabra float, unknown voices cry from nowhere—all the phantasmagoria of the invisible populate the real world.

While he noted that, according to court rumor, Home had caused a sensation in his first séance before the Emperor, Texier nevertheless remained skeptical of these phenomena. The exclusive social circles in which Home moved, Texier observed, did not include humble journalists, and hence all the accounts of séances he had heard came to him secondhand. In the absence of direct evidence, he concluded that it would be better for observers to maintain the stance of "philosophical doubt expressed by that free-thinker, Saint Thomas."[50]

Texier's advice went unheeded. As press accounts of the marvels Home produced became increasingly common, journalists grew far more willing to take their aristocratic sources at their word. Rather than expressing skepticism about Home, journalists proclaimed the reality of his gifts. Some newspapers built on this growing interest by supplementing accounts of Home's séances with other articles on spirit manifestations. The *Courrier de Paris*, for example, printed a series of articles by the spiritualist pamphleteer P. F. Mathieu, in which he described a variety of uncanny phenomena he had witnessed in séances with Honorine Huet and other mediums.[51] Not since the heady days of the *tables tournantes* had these mysterious manifestations commanded so much attention.

This remarkably widespread journalistic credulity—not an altogether unexpected development given the enthusiasm many writers had shown in the initial stages of the *tables tournantes* vogue—was probably a response to social and literary imperatives, not metaphysical ones. Young, dark, and slim, Home cut a picturesque and mysterious figure. The séances in which he produced his uncanny phenomena were always exclusive, private gatherings. By telling stories intended to make these events seem as remarkable as possible, aristocrats, *femmes du monde*, and men about town augmented their social *cachet*: a glamorous party was appealing; a small gathering for the production of uncanny manifestations was irresistible. Accepting the aristocrats' stories at face value, therefore, would have provided journalists

[50] Quoted in *Journal du magnétisme* 16 (1857):90.
[51] See ibid., 354.

an effective means of flattering their primary sources of social gossip. More important still, like the *tables tournantes* before them, accounts of Home's feats provided politically innocuous, eye-catching material for *feuilletons*—the diverting reviews, essays, and serial novels that by 1857 wooed readers from the bottom third of every major newspaper's front page.

In addition, a willingness to embrace the possibility that such fantastic events might be real seems to have struck many journalists, in this era of popular Romanticism, as an attractive literary pose, much as advocacy of Mesmerism had been ten years before.[52] The self-consciously poetic tone of many journalistic accounts reinforces this interpretation. In a *feuilleton* that appeared in *L'Estafette* on March 24, for example, Paul d'Ivoi wrote:

> At the Tuileries, they say, [Home] asked the Emperor if he wished to see Providence. After the Emperor gave an affirmative response, M. Hume [sic] placed a sheet of white paper on the table. Immediately, the lights dimmed; in the half-darkened salon a hand appeared, indistinct, radiant as a luminous cloud, and traced some sentences on the paper. The paper, they say, was given to the Emperor, who keeps these mysterious lines as a talisman.[53]

D'Ivoi's language, in this passage, is closer to the *conte fantastique* than to reportage. Belief in the extraordinary powers of mysterious strangers, here, became a means of expressing the openness to the transcendent and uncanny that French Romantics considered so important to the creative temperament. Expressing belief in Home during these weeks in 1857, therefore, provided journalists with an opportunity to display their connections to aristocratic social circles while showing off their up-to-date literary sensibilities.

In his commentaries on Home, Piérart advanced a different view of this sudden journalistic change of heart, one in which his conviction seems to have overwhelmed his capacity to detect irony. These writers' insistence on the authenticity of Home's powers, Piérart argued, proved that "the phenomena described in these articles, extraordinary as they might seem, have nevertheless been recognized as possible and very accurate by a crowd of trustworthy people." This widespread endorsement, he continued, heralded an impending sea-change in French society. Home's triumphant reception indicated that "at this moment, the human spirit is making a great advance, and we are perhaps on the verge of seeing immense truths arise—the most sublime of moral transformations." What began as stories of strange goings-on in aristocratic *salons*, Piérart

[52] See James Smith Allen, *Popular French Romanticism: Authors, Readers and Books in the 19th Century* (Syracuse, NY: Syracuse University Press, 1981).
[53] Quoted in *Journal du magnétisme* 16 (1857): 204.

believed, would soon spread, much as Modern Spiritualism had spread in the United States. Once the French, following the example of their aristocrats and society journalists, had fully embraced the phenomena of spiritualism, a new era of religious understanding and social justice would necessarily dawn.[54]

Du Potet shared Piérart's enthusiasm, particularly after attending one of Home's séances himself. The gathering, held in the home of a General D——, involved only five people: the general, his wife, his daughter, Home, and du Potet. The group sat around a large, heavy table covered with books. At the beginning of the séance, Home touched it. In response to the contact, du Potet wrote, "the table moved by itself; it sometimes leaned to the left, sometimes to the right." The books, which included numerous heavy folios and quartos, did not move at all while these oscillations occurred. After a series of verbal spirit communications that left du Potet "singularly moved," the table, books and all, rose from the floor and "stayed suspended, obeying laws other than those of gravity." These manifestations, du Potet wrote, convinced him of the genuineness of Home's gifts. In addition, the baron concluded, the nature of the phenomena proved Home's power to be "entirely psychic, coming from intelligent forces that exist in space." The exact nature of the "unknown assistance" Home employed remained a mystery, but du Potet was confident that "time will soon reveal . . . its true essence."[55]

After Du Potet's experience with Home, his editorial notes in favor of the spirit hypothesis took an increasingly visionary turn. In the last issue of 1857, for example, du Potet declared that his goal was not simply to shed light on a new class of phenomenon but to bring about a complete transformation of human knowledge by destroying the assumptions on which Academic science relied. To make his case, he alluded to the ambitious project to reshape the urban fabric of Paris that Napoleon III and his Prefect of the Seine, Georges-Eugène Haussmann, had recently begun:

> Yes, we want an act of demolition: down with all these systems, all these false doctrines that scientists impose, and that impede the spirit's growth instead of invigorating it. Ah! I admire what the government is doing here in old Paris, demolishing to bring air and light, paying no heed to the cries and gossip of the senseless crowd. But we wish to see these old dens of materialism—a gangrene that chews at the hearts of all men—destroyed without *indemnity*.[56]

[54] Ibid., 210–211.
[55] Ibid., 593–594.
[56] Ibid., 645. Italics in the original.

By destroying the accretions of prejudice and undermining scientists' arrogant presumptions, du Potet declared, American spiritualism would "Haussmannize" mankind's understanding of the soul and of the nature of God. Medieval structures and haphazard networks of streets would give way to a new conceptual space in which religion and science converged in harmony. With the advent of American spiritualism, du Potet believed, Mesmerism had been transformed from a mere scientific discovery—like the steam engine or telegraph—into a force that would provoke a sweeping process of social, intellectual, and metaphysical renewal.

As arguments in favor of American spiritualism came to occupy an ever-increasing amount of space in the *Journal du magnétisme*, the *Union magnétique* abandoned its clinical focus and resumed printing articles attacking this turn in Mesmerist thought. Home was at the center of this renewed polemic. The *Union* did not share the *Journal du magnétisme*'s enthusiasm for the American medium. On the contrary, Alexis Dureau, the writer in charge of the *Union*'s press summaries, treated Home with contempt, alluding to widespread rumors about his homosexuality and decrying his elitism. In particular, Dureau took Home to task for the way in which he produced his phenomena independently, without the surveillance and control of trained experts. Home, Dureau observed, was hardly the disinterested saint Piérart and du Potet made him out to be. On the contrary, he was generously, if indirectly, paid for his services:

> M. Hume [sic] is a hired *medium* with 40,000 francs of benefits (lodging, maintenance, food and carriage not included), furnished by M. the count B——, a noble and rich Pole who lives near the Madelaine. Some have a passion for opera girls, others for horses; M. the count B——, who, they say, has a pretty million in income, wants to keep his own *spirit*; he is rich enough to be his master, and generous enough to pay him well.[57]

In this attack, by likening Home to a kept woman, Dureau made a rhetorical move typical among the medium's opponents, who tended to question his masculinity and sexual orientation.[58] Excessive effeminacy, however, was not Dureau's primary reason for deploring the American medium's success. Home's refusal to leave the cosseted, exclusive circles in which he moved, Dureau argued, called the authenticity of his gifts into question. Manifestations that would convince a circle of frivolous

[57] *Union magnétique* 4 (Mar. 25, 1857): 4. The variant spelling "Hume" often appeared in French newspapers. Italics in the original.

[58] For a highly speculative discussion of Home's sexuality, in addition to some useful information about his trip to France and previous visit to England, see Eric John Dingwall, *Some Human Oddities: Studies in the Queer, the Uncanny, and the Fanatical* (London: Home and Van Thal, 1947), 91–128.

aristocrats could well have had a very different effect on trained Mesmerists, who were well versed in the experimental study of such phenomena. As long as Home insisted on holding séances exclusively in aristocratic *salons*, Dureau argued, it would be impossible to treat his alleged abilities as anything more than impressive conjuring tricks. "A man is not serious," Dureau wrote, "when his knowledge serves only to amuse guests between quadrilles, to make little boys laugh and horrify little girls. When one is serious, one approaches committees whose members are serious as well."[59] Until Home undertook experiments in the presence of trained Mesmerists, Dureau argued, his powers had to be regarded with suspicion.

Other writers in the *Union magnétique* used these specific critiques of Home as points of departure for renewed attacks on spiritualist Mesmerism as a whole. The journal's *rédacteur-en-chef*, Guillot, for example, decried the willingness of Home's French supporters to embrace exotic theories and phenomena. The peculiar, far-fetched ideas that American spiritualists espoused, Guillot argued, would "lead to an intellectual, moral, and, consequently, social catastrophe." Indeed,

> Imagined by enthusiasts, exploited by the sly, spread among the credulous, these mystical ideas will soon assume enough coherence to reduce science and good sense to silence, before turning them into objects of persecution. This fear is the only motive for our attack on the ideas of our beloved colleagues.[60]

The disconcerting suspension of critical judgment spiritualist enthusiasm provoked, Guillot believed, could bring about a new Dark Age if it became sufficiently widespread. All advocates of reason and progress, therefore, needed to make it their duty to oppose this tendency among *magnétiseurs*, even if, as in Guillot's case, they counted many of its most prominent supporters among their friends. Mesmerists, he concluded, would do well not to underestimate the power of the imagination or the damage it could do if left unchecked by objective self-restraint.

Honorine Huet and the New Role of the Medium

As these dramatically differing opinions indicate, the tension between spiritualist and therapeutic Mesmerists escalated during the late 1850s. One faction of *magnétiseurs* embraced the *tables tournantes* and the phenomena of American spiritualism, speculating freely on their metaphysical

[59] *Union magnétique* 4 (Jun. 25, 1857): 4.
[60] Ibid. (Oct. 25, 1857): 1.

significance, looking to them as sources of revelation, and searching for French mediums with powers that could equal their American counterparts. The other faction approached the new phenomena more cautiously, arguing against the notion that they were caused by otherworldly intervention and voicing concern that an excessive engagement with the subject might jeopardize the intellectual legitimacy of the Mesmerist enterprise. The fact that neither one of Paris's two major Mesmeric societies was unanimous in its views—despite the increasingly polarized stances of their respective journals—made this conflict still more divisive. In the *Journal du magnétisme* and at the meetings of its affiliated Société du Mesmérisme, Morin had voiced increasing skepticism about the otherworldly intervention Piérart so emphatically claimed to see in the new phenomena.[61] Similarly, as Guillot's remark about his friends' beliefs indicates, the Société philanthropico-magnétique, which ran the *Union*, counted many spiritualists among its members, including Henri Delaage.[62] Both groups, then, faced internal conflicts about the proper interpretation of these new phenomena that did not surface explicitly in the pages of their respective periodicals.

In 1858, the rift between spiritualist and therapeutic Mesmerists finally grew too deep for the existing institutional structures to contain. In January, Piérart left the *Journal du magnétisme* and started a periodical of his own, the *Revue spiritualiste*, exclusively devoted to the study of the new phenomena. In it, Piérart took the role of otherworldly intervention for granted. After the departure of its *rédacteur-en-chef*, the *Journal du magnétisme* began to assume a more moderate stance. Morin, who had written infrequently during Piérart's tenure, contributed numerous skeptical articles, and du Potet tempered his own enthusiasm. The *Union magnétique*, for its part, inaugurated a new policy of openness to spiritualist ideas. Readers interested in the new phenomena and related metaphysical speculation could now find articles arguing in favor of spirit intervention in the *Union*'s pages, though they were always prefaced with an editorial note from Guillot reaffirming his fundamental dissatisfaction with such theories.[63]

The Société philanthropico-magnétique also took a more decisive step in its effort to resolve the conflict between spiritualist and therapeutic Mesmerists. In late December 1858, the *Union magnétique* announced to readers that it had begun to receive a great deal of correspondence containing accounts of alleged spirit phenomena. To address this growing interest, Guillot wrote, the Société philanthropico-magnétique had decided

[61] *Journal du magnétisme* 16 (1857): 30.
[62] See, e.g., *Union magnétique* 4 (May 25, 1857): 3.
[63] See, e.g., *Union magnétique* 5 (Apr. 25, 1858): 1.

to establish a "commission of inquiry" to investigate these manifestations and to pronounce definitively on their authenticity. The commission included eminent Mesmerists from both the spiritualist and therapeutic camps. In addition to comparative neutrals such as Morin and P. Tharin, it included the therapeutic Mesmerists Dureau, Guillot, and Millet, and the spiritualists Piérart and Delaage.[64] Morin, in the *Journal du magnétisme*, enthusiastically supported the endeavor. "When it is a matter of phenomena as prodigious as these," he wrote, "sober observation is essential."[65] After a long period of negotiation, the commission succeeded in arranging a series of sessions with the talented French medium Honorine Huet.

If not for her medianimic gifts, Huet's prospects would have been poor. The eldest daughter of a middle-class Jewish family from Marseille, she moved to Paris in the late 1850s with her mother and younger sister, Virginie. The family had undertaken the move to find a suitor for Virginie, who was beautiful and a moderately gifted pianist. Honorine, however, did not take part in Parisian social life. Instead, she worked as a governess for the daughters of the writer Théophile Gautier, whose wife was a friend of Honorine's mother. The elder Mlle Huet's unprepossessing physical appearance probably condemned her to this spinster's role. In the cruel account of her former charge, Judith Gautier, Huet's most remarkable physical trait was a "vast corpulence;" her most remarkable intellectual one, an intense, idiosyncratic, and sanctimonious Christian piety. A polyp in her "Bourbon nose" rendered her already pronounced Provençal accent absurdly nasal.[66]

The séances with Huet provide a revealing example of how much the American spiritualist example had changed the practice of Mesmerism in the years since the séance vogue of 1853. While the most obvious change was the introduction of new phenomena such as table-moving and spirit raps, spiritualism also introduced an even more pervasive shift in social dynamics. Home's tour revealed this transformation clearly: The old relationship between *magnétiseur* and *somnambule* had given way to a new arrangement, which allowed the entranced subject a considerably greater degree of independence. In some ways, Mlle Huet would probably have struck the *magnétiseurs* on this committee as a classic *somnambule*—an intelligent, visibly eccentric, and highly strung woman. Like a *somnambule*, Huet also had a close relationship with an older man of some education, the pharmacist P. F. Mathieu, who publicized her feats and often served as her questioner in séances. Unlike her predecessors, however, Huet did not require the fluid of a *magnétiseur* to enter into communication with the

[64] Ibid. (Dec. 25, 1858): 4.

[65] *Journal du magnétisme* 17 (1858): 665.

[66] Judith Gautier, *Le Collier des jours, le second rang du collier* (Paris: Renaissance du livre, 1909), 15, 5, 36, 7.

beyond: Mathieu was a disciple, not an active collaborator in Huet's clairvoyance. While he admired her ability to produce such manifestations as "direct writing," he paid even greater heed to the content of the edifying communications she received and reacted with equanimity when the spirits did not cooperate. The phenomena that appeared in Huet's séances did so exclusively on *her* terms.

This new situation generated considerable tensions among those seated around Huet's table. The two spiritualists in the group, Delaage and Piérart, adapted readily to the requirements of the situation; the others were considerably more reluctant to concede full authority to this woman who acted as both *somnambule* and *magnétiseur.* Instead of passively viewing and accepting the manifestations she produced, the committee members sought to conduct an array of experiments; in response the spirits frequently refused to oblige. After the most spectacular of the many failures that marked these séances—the experiment with the hidden spirit alphabet—Mathieu placed the lettered piece of cardboard flat on the table, in full view of everyone present. Huet's entity tapped twice, and then spelled out the word "skeptics."[67] In her séances, the negative attitudes of the *magnétiseurs,* not the shortcomings of the medium, were responsible for the spirit's inability to produce conclusive manifestations.

All eight of the men on the committee agreed that these two séances were unconvincing as proof of spirit intervention. Despite the opposition of the spiritualists Piérart and Delaage, six of the committee members voted to publish their conclusion, which appeared in August 1859. After this publication, what had been an intense intellectual debate erupted into ad hominem polemic, which precipitated a schism between therapeutic and spiritualist Mesmerists. Morin, following the example of Academic scientists by pathologizing those who claimed to have witnessed the new phenomena, insinuated that Piérart suffered from "spirito-madness."[68] Piérart called Morin a "born skeptic," who had exacerbated his chronic lack of faith with long service as a lawyer and excessive study of the works of Voltaire. "He has lived his life as a quibbler," Piérart concluded, "and as a quibbler he will die."[69] In 1859 and 1860, both the *Journal du magnétisme* and the *Union magnétique* shifted their attention back to accounts of Mesmeric cures. The spiritualists, in turn, lent their support to Piérart's journal and to a competing periodical—the *Revue spirite*—founded by a former mathematics teacher who called himself Allan Kardec.

[67] *L'Union magnétique* 6:112 (Aug. 25, 1859): 3.
[68] *Journal du magnétisme* 18 (1859): 637.
[69] *Revue spiritualiste* 2 (1859): 269.

After this turn away from spiritualism, the Mesmerist movement was gradually eclipsed, not to reemerge until the early twentieth century. The *Journal du magnétisme* stopped appearing in 1861; that same year, the Société du Mesmérisme and the Société philanthropico-magnétique—both diminished by the loss of their spiritualist members—fused into a single group. The circulation of the *Union magnétique*, France's only remaining Mesmerist periodical for much of the 1860s, dwindled steadily. By 1866, it had shrunk to a mere 300, while that of Kardec's new journal had reached 1,800.[70] This transition, as we will see in Chapter 3, reflected one of the most remarkable developments in Second Empire religious life—the rise of Spiritism.

[70] Archives nationales, F/18/294.

CHAPTER THREE

The Invention and Development of Spiritism, 1857–1869

Contacting a spirit in France in 1867 was a markedly different experience from what it had been in 1859. Where séances had once taken place in darkness, they were now often held in lighted rooms. Physical contact between the séance participants was no longer considered an essential part of the ritual, and in its absence, the atmosphere of playful flirtatiousness that had once characterized so many experiments with *tables tournantes* disappeared. When communicating for the spirits, mediums no longer relied on moving tables, mysterious raps, or a cumbersome *planchette*; instead, they sat quietly, holding pencils in the conventional manner, writing when the otherworldly forces inspired them. While séances still often occurred in family circles, they also took place in formal meetings, where attendees and their otherworldly guests obeyed strict rules of order, like the members of any other discussion society. These rules involved a partial return to the old paradigm of Mesmerist and *somnambule*: mediums managed their own trance states, but instead of determining the direction of communication themselves, they often responded to questions posed by an educated male society president.

The messages spirits sent and the language the living used to discuss them also changed. In published works, the fanciful, oracular communications many visionary mediums had produced in the 1850s gave way to plainer, clearer messages. Where there had previously been a pronounced diversity of opinion among inhabitants of the beyond, a consensus emerged. The spirits most mediums contacted now elaborated a moral system based on charity and the importance of fellow-feeling.

Fig. 8. Hippolyte Léon Dénizard Rivail, better known as Allan Kardec, in the 1860s. (Collection of the author.)

They often espoused a cosmological vision in which the human soul purified itself through reincarnation, expiating its sins and correcting its faults in a series of progressively more elevated lives. The seekers who took these messages to heart increasingly looked to a unified body of authoritative texts for guidance, and new words emerged to describe their ideas and practices. When these believers referred to their conviction that the living could enter into direct contact with the souls of the dead, they called it *spiritisme*; those who accepted this idea were *spirites*.

These dramatic changes owed a considerable amount to the influence of a former teacher named Hippolyte Léon Dénizard Rivail (fig. 8). His career in heterodoxy began in 1857, with a book called *Le Livre des Esprits*, which he published under the pseudonym Allan Kardec. The *Livre des Esprits* was a Second Empire bestseller, making Kardec one of the most widely read philosophers of the period. By 1874, the book had been through twenty-two editions of a minimum of 2,200 copies each—which means at least 48,000 were in circulation at the beginning of the Third Republic.[1] In comparison, the 1853 version of Victor Cousin's *Du Vrai, du beau, du bien*, a widely read work of philosophy

[1] Pierre-Gaëtan Leymarie, afterword to "Actualité, William Crookes, ses notes sur des recherches faites dans le domaine des phénomènes appelés spirites, pendant les années 1870–1873, publiées par le *Quarterly*," by William Crookes, trans. Samuel Chinnery and Jane Jaick (Paris: Librairie Spirite, 1874), 30.

published by the same firm as Kardec's book, had gone through twenty-three editions by 1881.[2] Though Kardec's subsequent books also enjoyed considerable success, he was not content to be a mere writer. Instead, he became a tireless propagandist for his ideas, editing a successful journal and founding an influential society devoted to the holding of séances. By 1862, the movement he called Spiritism had adherents throughout France, particularly in the cities of Paris, Lyon, and Bordeaux.

Kardec's success in what remains a difficult vocation—that of professional heterodox spiritual leader—stemmed from a variety of factors. First, there were the ideas themselves, which exerted a strong attraction on their own terms. Spiritist philosophy, so accessibly presented in Kardec's books, struck many educated middle- and lower middle-class people as rational, consoling, and reassuringly familiar. Rooted in Romantic Socialism and Positivism, it fit the expectations of many seekers of alternative cosmologies—especially those sympathetic to the visionary discourse of the mid-nineteenth-century left. Kardec intensified the appeal of Spiritist ideas by linking them to a set of practices: society meetings, the holding of séances, and the performance of automatic writing. This practical dimension brought individual believers together and gave them opportunities to have a personal, in many ways novel, experience of the sacred. Finally, dramatic as his innovations were, Kardec avoided any appearance of extremism. He did not present himself as a visionary or prophet but instead adopted the role of the rational, even stolid, expert. Communication with the beyond, as he saw it, was a serious endeavor worthy of the attention of "the enlightened classes." He therefore strove to ensure that the ideas and practices of Spiritism met what he perceived as the intellectual demands of educated, level-headed observers—especially male professionals.

For believers, Spiritism seemed to resolve the crisis of factuality in a very direct way, providing a distinctively modern source of metaphysical consolation that did not push too far into the wilds of visionary excess. This chapter will explain how Kardec and his followers accomplished this delicate balancing act by analyzing five aspects of Second Empire Spiritism: (1) the life and writings of its founder, (2) the organizational structures he created, (3) the conversion experiences of believers, (4) the complexities of séance practice, and (5) the new doctrine's ambivalent relation with Catholicism. The Spiritist promise of innovation without rupture provides a revealing glimpse of the unsettled world of the Second Empire, with its urgent need for new certainties to replace old ones rapidly coming to seem obsolete.

[2] Victor Cousin, *Du Vrai, du beau, du bien* (Paris: Didier, 1881).

The Invention of Spiritism

Before he became *chef du spiritisme*, Kardec's life and interests were those of a typical midcentury intellectual with moderate leftist inclinations. He studied Mesmerism in the 1820s, for example, and devoted himself to popular education when the idealism of the Second Republic was in full bloom. Kardec's values were similarly typical: he had a middle-class respect for dignified and educated men, a strong faith in the authority conferred by objective inquiry, a Positivist's suspicion of metaphysical speculation, and a republican's belief in the inevitability of progress. He also shared the existential doubts common among the period's freethinkers, many of whom regretted the loss of their childhood Catholic faith even as they acknowledged its inability to provide them with intellectual satisfaction.

This typicality served him well when he began his studies of spirit communications. His doubts led him to see the potential for consolation the novel phenomena of American spiritualism afforded, and the conventionality of his outlook gave his ideas a uniquely effective foundation. Kardec's experience as a teacher also helped him in his task: even as he appealed to the professional bourgeoisie, he designed his texts with a shrewd sense of both the limitations and the aspirations of moderately educated or self-taught readers such as middle-class women and urban tradesmen.

Becoming Allan Kardec

Hippolyte Léon Dénizard Rivail was born to a comfortably middle-class Lyon family on October 3, 1804. His father, Jean Baptiste Antoine Rivail, was a magistrate, as Rivails had been for generations. Hippolyte was baptized and raised Catholic. After several years of primary education in Lyon, his parents sent him to Yverdun, in Switzerland, where he enrolled in Johann Heinrich Pestalozzi's innovative academy.[3] There, the young Rivail would have been trained in the Pestalozzian manner, which drew from the educational writings of Rousseau and emphasized the importance of learning through direct observation of nature. Pestalozzi also used Rousseau as the inspiration for the religious aspect of his educational program, which encouraged an intuitive, emotional faith over the rote, catechistic memorization of dogma.[4]

Pestalozzi's example appears to have made a deep impression on the young Rivail. In 1832, after several years of military service, he married a

[3] Henri Sausse, *Biographie d'Allan Kardec* (Paris: Jean Meyer, 1927), 18–20.
[4] Gerald Lee Gutek, *Pestalozzi and Education* (New York: Ranodm House, 1968), 137–139.

schoolteacher named Amélie Boudet and founded a Pestalozzi-inspired technical school in Paris. Despite his zealous efforts, the *Institut technique* proved to be an ill-starred venture. Rivail's uncle and primary financial backer was an inveterate gambler, and his losses, which eventually amounted to 45,000 francs, ruined the young teacher, forcing him to close the school. After this setback, Rivail held his creditors at bay by working as an independent bookkeeper for a variety of different enterprises, including a theater and the Catholic newspaper *L'Univers*. This new career proved quite lucrative, earning Rivail about 7,000 francs a year. In his spare time, he wrote pedagogical works on grammar, mathematics, and other topics. Rivail also participated in the Second Republic flowering of popular education: throughout 1849, he lectured on physiology, astronomy, chemistry, and physics at the *Lycée Polymathique* and offered free courses at his home.[5]

Rivail, like many progressive intellectuals of his time, was a casual student of Mesmerism, and had been since the 1820s.[6] Late in 1853, a Mesmerist friend, M. Fortier, told him about uncanny events that had occurred in his experimental séances. Initially, Fortier had succeeded in producing only basic physical phenomena—causing the table to lift off the ground, for example, or to rotate spontaneously under its own power. After several months of continued experimentation, however, Fortier came to Rivail with startling news: his séance table had begun communicating clear messages by means of mysterious tapping noises. Initially, Rivail wrote, he was skeptical. The notion of *tables parlantes* seemed to him nothing more than "a tall tale."[7]

Rivail, of course, would later change his opinion dramatically. The decisive encounter that led to his eventual conversion occurred in May 1855, when he called on Fortier in the apartment of a *somnambule*, Mme Roger. Conversation turned to the *tables parlantes* and to the theory that these strange phenomena were the work of disembodied human souls. One of Fortier's other guests, M. Pâtier, argued for the reality of the "spirit hypothesis," citing examples gleaned from regular séances with a medium named Mme de Plainemaison. Rivail found Pâtier's arguments in favor of the new phenomena convincing, largely because of his interlocutor's unimpeachable respectability: "M. Pâtier was a public official, of a certain age, a very well-educated man, with a cool, grave character; his poised language, untouched by all enthusiasm, impressed me strongly." Pâtier was no wild-eyed visionary or half-cocked enthusiast; he was a serious man who considered

[5] Sausse, *Biographie*, 20–23.

[6] Ibid., 24.

[7] Allan Kardec, *Œuvres posthumes, ed.* Pierre-Gaëtan Leymarie (Paris: Dervy, 1978), 239, 240. The passages quoted here come from a memoir Kardec wrote in the late 1860s.

his judgments with bureaucratic thoroughness. Any phenomenon that could attract the sustained interest of such a personage, Rivail reasoned, was worth further study. Pâtier ended the conversation by inviting Rivail to attend one of his weekly séances at Mme de Plainemaison's. The younger man accepted "eagerly."[8]

Attending Pâtier's séances introduced Rivail to a small but intensely active group of Parisians engaged in the study of these phenomena. At one of Pâtier's gatherings, Rivail met a man named Baudin, who held weekly experimental sessions of his own, at which his wife and two adolescent daughters served as mediums. Unlike Pâtier's séances, which were primarily devoted to the study of "of turning, jumping, and running tables," Baudin's centered on the production of "medianimic writing."[9] At these gatherings, the mediums—usually the two Mlles Baudin—would place their hands on a basket or *planchette* with a pencil attached. The pencil would then write, either spontaneously or in response to questions from other séance participants.

Initially, the séances Baudin hosted were playful affairs. The spirit who manifested most frequently was called Zéphyr and was as likely to provide "humorous quips" as serious moral advice. While charmed by this tone of lightness, Rivail wished to do more in these séances than simply collect the occasional "mordant and witty epigram" from the beyond. Drawing on his Pestalozzian training in the "experimental method," he decided to make the Baudin family séances more rigorous. To this end, he brought a set of questions to ask the spirits each week. Once he began to use these regular contacts with the spirits as a means of systematic inquiry into "the nature of the invisible world," Rivail wrote, the Baudin séances assumed a new gravity and sense of purpose. "Serious people" started attending the meetings, and "trivial questions lost their allure for the majority." The demeanor of the spirits who appeared at the séances also changed to reflect the newfound gravity of their audience. Zéphyr, in particular, dropped his previous mischievousness and became a dignified teacher who—with Rivail's prompting—began the sober revelation of a powerful new solution to the "controversial problem of humanity's past and future."[10]

Rivail's motivation for changing the tenor of the Baudin meetings probably stemmed from something more than simple intellectual curiosity. Like many others during this period, he appears to have found that regular contact with the beyond soothed feelings of grief. A posthumously published conversation between Rivail and the newly serious Zéphyr,

[8] Ibid.
[9] Ibid.
[10] Ibid., 242, 244, 243.

dated December 1855, sheds some light on this more personal aspect of the former teacher's growing interest in the beyond:

> *Q.* Does my mother's spirit come to visit me sometimes?
> *A.* Yes, and she protects you as much as it is possible for her to do.
> *Q.* I often see her in my dreams; is this a memory and a figment of my imagination?
> *A.* No; it is in fact her spirit appearing to you; you must be able to tell by the emotion you feel.
> *Remark.*—This is perfectly correct; when my mother appeared to me in dreams, I felt an indescribable emotion, which the medium could not have known.[11]

In addition to answering his questions about psychology and cosmology, then, the spirits who appeared at the Baudins provided Rivail with a novel form of consolation. Zéphyr's testimony proved that Rivail's deceased mother had not simply evaporated into nothingness; she led an independent life in the spirit world and continued to offer her son palpable comfort by visiting him in his dreams. Even more powerfully, Rivail's mother spoke to both the sentimentalist and the Positivist freethinker in her son: by mentioning the private, intense feeling he experienced upon seeing her, the spirit seemed to give empirical proof of her own existence. Emotionally fraught revelations like this one, of course, could occur only in an appropriately grave and calm atmosphere. Zéphyr's reassurance and Rivail's self-exposure would have seemed inappropriate in the Baudin circle's light-hearted early days.

Gradually, Rivail realized that the information he received from the spirits at the Baudin séances "formed a whole and took on the proportions of a doctrine."[12] The time had come, he decided, to begin collating these spirit communications with an eye to future publication. The Baudin circle responded enthusiastically when Rivail informed them of his intention to produce a book of spirit teachings. Other men who frequented the Baudin séances, including the playwright Victorien Sardou, his father, the writer René Taillandier, and the publisher Alfred Didier, provided Rivail with notebooks of spirit communications they had collected from different mediums, in hopes that the additional data would help him in his endeavor.

The spirit Zéphyr, for his part, communicated his approval of the project through the Mlles Baudin. He also suggested a pseudonym for Rivail to use:

[11] Ibid., 246. Italics in the original.
[12] Ibid., 244.

> You will take the name Allan Kardec, which we give to you. Have no fear, it is yours, you have borne it with distinction in a previous incarnation, when you lived in old Armorica.[13]

Zéphyr's choice of a Celtic-sounding name and his allusion to Rivail's former existence in Western Gaul resonated with a long-standing Romantic Socialist interest in Druidism. Over the previous decade, philosophers such as Jean Reynaud had imagined the religion of pre-Christian France as a rational, indigenous alternative to Catholicism—one that echoed Rivail's spirits by placing particular emphasis on reincarnation.[14] The former teacher and bookkeeper embraced the new identity Zéphyr gave him. Allan Kardec's life, which began when his book appeared in 1857, would be very different from Hippolyte Rivail's. The ordinary, earnest yet doubt-ridden idealist gradually became the magisterial and authoritative leader of a full-fledged religious movement.

In 1856, to accelerate the process of information gathering, Kardec began to frequent the somnambulist Mlle Célina Japhet and her *magnétiseur*, M. Roustan. The Mlles Baudin were both on the verge of marriage, which would end their careers as family mediums. Japhet, in contrast, was a professional *somnambule*, free to devote considerably more time to answering Kardec's questions. In addition, her communications had a consistency of voice and logic that gratified Kardec's *esprit de système*. The automatic writings he used as raw material in the drafting of his work had come from a variety of different hands, and hence sometimes expressed contradictory ideas. Japhet's spirits resolved these contradictions, clarified imprecisions in the spirits' language, and supplied missing logical connections.[15] As his consultations with Japhet continued, and her role in the composition of the manuscript grew, the spirits she consulted began to proclaim the world-historical significance of Kardec's project in increasingly emphatic terms.

Kardec completed his manuscript in 1857. Later that year, it appeared as a quarto entitled *Le Livre des Esprits, contenant les principes de la doctrine spirite,*

[13] Quoted in *Le Spiritisme* 5 (1888): 233. This communication was published in the first edition of the *Livre des Esprits*, but excised from the subsequent ones. The story of Rivail's pseudonym is controversial. The account in the main text is the most common, and the earliest. For an alternative, see an article by the Russian psychical researcher Alexander Aksakof, first published in English in 1875, arguing that Rivail received the name "Allan" from the medium Célina Japhet, and the name "Kardec" from the medium J. Roze. The French translation of Aksakof's article appeared in *La Lumière* 10 (Jan. 1899—Dec. 1900): 38–40.

[14] See Lynn L. Sharp, *Secular Spirituality*, 24–28.

[15] See *La Revue spirite, journal d'études psychologiques* 1 (1858): 36, where Kardec explained Japhet's role in the production of the *Livre des Esprits. La Lumière* 10 (Jan. 1899—Dec. 1900): 38–40 provides further information about Japhet's previous career and her influence on Kardec.

published by Edouard Dentu.[16] From a marketing perspective, Kardec's timing was propitious: as discussed in chapter 2, the celebrity medium Daniel Dunglas Home's visit to France triggered a flurry of renewed interest in spirit phenomena.[17] Dentu took full advantage of this development, much as he had during the *tables tournantes* vogue of 1853. In addition to publishing Kardec's book, he also published similar texts from the heterodox Catholic mystic Girard de Caudemberg and the spiritualist Mesmerist Paul Auguez.[18] Despite this competition, the *Livre des Esprits* enjoyed a remarkable success, which grew as the decade continued. Its first edition sold out quickly. In 1858, Kardec followed it with a revised and augmented edition in the more accessible octavo format. He published this version of the book with Alfred Didier, his friend from the Baudin séances, whose prestigious list included distinguished philosophers such as Victor Cousin.

Though superficially similar to other texts on spirit phenomena, Kardec's book constituted a dramatic innovation in its genre. This novelty helped account for the work's unusual popularity. Ordinarily, when spirits spoke in books like these, they tended to do so in a florid, oracular style. Auguez, for example, quoted an exchange with the spirit of Balzac he had witnessed in a séance:

> — Do you intend, one of the guests asked, to pass all your works in review? There are so many, after all!
>
> — No! Only one more thing for this evening: a trifle! A single petal from each flower you have loved, a single drop among the thousand drops of perfumed dew, pearls of the dawn, that adorn such flowers at break of day! A trifle, a breath to awake the insect asleep in his blossom; a sigh and all it can do to silence and forget the song of a bird; a faint murmur in the foliage—joyous and hidden frolics of happy sylphs—a shepherd's song heard and understood in a poetic scene.

[16] For more on Dentu, who was a believer in spirit communications, though skeptical about Kardec's ideas, see the collection of obituaries published under the title *E. Dentu* (Paris: Dentu, 1884), especially Olympe Audouard's article from *Le Papillon,* reprinted on pages 53–60.

[17] For examples of newspaper coverage of Home, see *Le Courrier de Paris,* Dec. 8, 1857, 2; *L'Estafette,* Dec. 12, 1857, 2; *La Patrie,* Nov. 29, 1857, 2. For books and pamphlets, see Louis Goupy, "Phénomènes de spiritualisme à expliquer" (Argenteuil: Worms et cie., 1857); Baron L. de Guldenstubbé, *Pneumatologie positive et expérimentale, la réalité des esprits et le phénomène merveilleux de leur écriture directe* (Paris: A. Franck, 1857); Lambert-Elisabeth d'Aubert, Comte de Résie, *Histoire et traité des sciences occultes, ou examen des croyances populaires sur les êtres surnaturels, la magie, la sorcellerie, la divination, etc., depuis le commencement du monde jusqu'à nos jours,* 2 vols. (Paris: Vivès, 1857).

[18] Paul Auguez, *Les Manifestations des esprits, réponse à M. Viennet* (Paris: Dentu, 1857); Girard de Caudemberg, *Le Monde spirituel ou science chrétienne de communiquer intimement avec les puissances célestes et les âmes heureuses* (Paris: Dentu, 1857).

For Auguez, the perceived literary élan of this communication amply justified its publication and proved its otherworldly authorship. After all, Auguez argued, its "poetry" far exceeded the capacities both of the medium, Mlle Octavia, and of the other guests at the séance, all of whom were "men of the positive sciences."[19] Whatever the communication's literary value, however, it would have proved frustratingly enigmatic for the reader eager to unlock the secrets of the beyond. Sibylline effulgence of this kind generated a visionary literary atmosphere but failed to provide useful information, moral or cosmological. The spirits in Kardec's book expressed themselves quite differently. In place of rhetorical flights often disconcertingly free of specific content, Kardec's spirits spoke about clearly defined subjects in simple language.

The organization of Kardec's book was also a significant innovation. Where texts like Auguez's and Caudemberg's were dense and repetitive, Kardec's was divided into short segments, set off with convenient headings, each addressing a specific cosmological or moral question, from "The Origin and Nature of Spirits" to "Self-Knowledge."[20] In a review of the *Livre des Esprits* published in 1860, the journalist Louis Jourdan singled out this aspect of the book as its most remarkable characteristic:

> The book we are discussing . . . can serve as a kind of *vade mecum*; it can be picked up, put down, opened to any page, and it immediately piques the curiosity. The Spirits respond to questions that preoccupy us all; their answers are sometimes quite weak, but sometimes they concisely address the most difficult problems, and they always offer interesting information or helpful instructions. I know of no course in moral thought more attractive, more consoling, more charming than this.

Its easily digestible quality, so conducive to browsing, Jourdan wrote, made Kardec's book uniquely appealing to "the great mass of readers, particularly women." Readers not educated enough to maintain the "concentration" necessary for the absorption of a more recondite philosophical work could navigate the *Livre des Esprits* quite comfortably. Kardec's book, then, was not simply a text to be read once and put aside; it was a collection of short, freestanding moral essays to be referred to repeatedly in times of need. In this respect, it stood between the new literature of self-help and an older tradition of devotional texts.[21]

[19] Auguez, *Manifestations*, 135–136, 137.

[20] Allan Kardec, *Le Livre des Esprits, contenant les principes de la doctrine spirite* (Paris: Dervy, 1996 [1860]), 498, 502.

[21] Quoted in *La Revue spirite* 4 (1861): 104.

Kardec also set himself apart from other authors in his chosen genre by inventing a new word for his subject. Since the *tables tournantes* vogue of the early 1850s, the beliefs and practices of people who conversed with the souls of the dead in séances had been called *spiritualisme,* a term French adherents had borrowed directly from their Anglo-American counterparts. Kardec argued that this simple borrowing was, in fact, misleading. The term *spiritualisme* already had a well-defined philosophical meaning in French; by adopting it, Francophone believers had created a source of confusion. "New things require new words," Kardec proclaimed. "Spiritualism" was simply "the opposite of materialism," and hence applied to any person "who believes he has something in himself other than matter." "Spiritism," on the other hand, was much more precise. It explicitly designated a "doctrine" based on "relations between the material world and Spirits, or beings from the invisible world."[22]

Kardec's terminological choice was shrewd. By inventing a specific name to describe both his doctrine *and* the practices that went along with it, he inextricably associated the two. Anyone who held séances was a Spiritist—or *spirite,* in French—and all Spiritists accepted the metaphysical system Kardec outlined in his writings. Coining the word *spiritisme,* therefore, allowed Kardec to emphasize the distinctiveness and specificity of his ideas, while simultaneously creating the impression that everyone who contacted spirits shared them. The benefits of this strategy became clear in the early 1860s, when the word *spiritisme* entered general French usage as the generic term for belief in spirit contacts, a position it still occupies today. Every time a journalist or priest used the word—even in a critical article—he or she indirectly bolstered Kardec's status as the leading authority among these heterodox thinkers by accepting his terminology. As the term *spiritisme* became increasingly common, Kardec's competitors—many of whom persisted in calling themselves *spiritualistes*—were pushed further toward the margins. Because they engaged in the same practices as Kardec's followers, *spiritualistes* became *spirites* in the eyes of the larger public, even though they reviled Kardec's ideas.

The Philosophy of Spiritism

The spirits in Kardec's book said little that was new. For many Second Empire readers, the vision of the beyond Kardec elaborated probably derived much of its authority from its sheer predictability. By combining the "spirit hypothesis" with conventional elements of the period's socialist and progressive thought, Kardec gave his book a broader appeal than earlier, more idiosyncratic compilations of spirit communications enjoyed. At the

[22] Kardec, *Livre des Esprits,* I.

same time, however, Kardec eliminated the revolutionary aspect these ideas had acquired during the 1840s. He accomplished this change of direction by adapting one of the key elements of Charles Fourier's cosmology—the idea of reincarnation—and bolstering it with an epistemology drawn from Comtean Positivism.

According to the *Livre des Esprits,* before God created the material universe, there existed an immaterial *monde spirite* peopled by primitive, disembodied souls. God created the material world to provide these souls with a means of perfecting themselves. In order to reach an ultimate goal of transcendent understanding and moral purity, each soul had to endure a long string of incarnations on various planets. These incarnations served an expiatory function—every life lived entailed a set of tests that, if successfully passed, brought the soul closer to perfection. As souls perfected themselves, they earned the right to inhabit ever-more-hospitable planets. Earth was near the bottom of this planetary hierarchy. On more elevated planets, like Jupiter, social organization was more just, individuals were more enlightened, and the physical aspects of existence became less important. For Spiritists, a soul could only advance through this expiatory cycle of reincarnations, since an ironclad "*loi du progrès*" (law of progress) governed the universe, prohibiting retrograde movement.

Kardec and the spirits he quoted used the Golden Rule as the basis for a fundamentally social conception of morality. Both good and evil, they argued, expressed themselves primarily through an individual's relations with others. Charity and selfishness, therefore, became the two poles of the Spiritist moral compass. In his explanatory commentary, Kardec defined both terms broadly. Charity, he argued, was the essential virtue. It was not "limited to alms," but instead had far larger ramifications: "it encompasses all relations we have with others, whether they be our inferiors, our equals, or our superiors." Selfishness, in turn, was any act that put personal comfort and gratification before the greater good of society—even if this gratification involved devoting one's life to the solitary contemplation of God. An egotistical desire for personal satisfaction and sensual pleasure, Kardec's spirits maintained, lay at the root of every evil act, and was the product of the coarse and unclean physical body's animal impulses.[23]

This conception of a morality based on charity served as the justification for a meliorist social vision. Behavior that tended to stabilize society while offering comfort to the suffering was charitable; behavior that destabilized society, even in the name of greater justice, was selfish. Though all souls were equal in the eyes of God, Kardec wrote, social inequality was a necessary condition of life on Earth. Since the tests each soul endured in

[23] Ibid., 395, 342–343, 407–409.

a given incarnation were related to its past transgressions, some inevitably suffered more than others. Material inequality in human society, therefore, served a cosmic purpose: The poor suffered more because they had more errors from previous lives to atone for.

Here, Fourier's idea of reincarnation ceased being an incentive to create paradise on Earth through social reorganization and instead served to justify the suffering that existing economic arrangements caused. Since the experience of poverty was a necessary form of expiation, those who sought to end social inequality on Earth with a revolutionary, material redistribution of wealth, in Kardec's view, were misguided. As one of the spirits declared, such radicals "are systematizers or jealous and ambitious men; they do not understand that the equality they dream of will soon be overturned by the force of events."[24] Equal distribution of wealth might create a momentary equality, the spirit allowed, but humanity's innate greed would inevitably reassert itself. Indeed, the "systematizers" and "ambitious men" who thought otherwise often based their political schemes on their own "selfishness," arrogantly putting ideological purity and a desire for power ahead of human feeling. The only way to resolve the problems an unequal distribution of wealth created, Kardec asserted, was to create a moral atmosphere in which feelings of solidarity inspired a voluntary abandonment of natural human selfishness. In a Spiritist world, he argued, the rich would feel an obligation to be charitable, while the poor, strengthened by the expectation of a better life to come, would accept gifts with a resigned gratitude.

A similar blend of egalitarianism and acceptance of inequality characterized the Spiritist view of gender. The spirits Kardec quoted maintained that the soul had no sex, but also asserted that male and female bodies were suited for different social functions. According to the *Livre des Esprits,* the roles men and women played in society were a biological inevitability—a man's "physical organization" rendered him incapable of dispensing the kind of love a mother could, just as a woman's rendered her incapable of inhabiting the public worlds of science or politics. At the same time, however, the Spiritist notion of the sexless soul admitted a larger vision of equality, one that did not pose dramatic threats to contemporary social categories. Since the soul had no sex, men and women deserved to be treated equally before the law; because their bodies differed, men and women had different "natural" missions to fulfill on Earth. An ideal society, according to Kardec, acknowledged and accepted this immutable difference, granting women equality "of rights," but not "of functions."[25]

[24] Ibid., 362.
[25] Ibid., 365–366.

Spirit phenomena provided the underpinnings of this eschatological, moral, and social vision. Between incarnations, every soul existed for a period of time as a disembodied "wandering spirit." These spirits filled the universe: though humans ordinarily could not perceive them, they formed an omnipresent throng surrounding the living. When people contacted the beyond in séances, these "wandering spirits" were the beings that appeared. All such spirits had distinct personalities, Kardec maintained, and differed from one another as dramatically as would a randomly assorted crowd of human beings. Some had advanced rapidly through the spirit hierarchy and showed a saintly concern for human welfare; others had progressed slowly and exhibited a mischievous eagerness to lead people astray.

While the invisible world of spirits had always existed, the recent discovery that certain people could act as mediums made this impalpable crowd accessible to human beings in a way it had never been before. The well-trained medium, Kardec believed, was an innovative religious instrument to be placed alongside the innovative scientific instruments that characterized the modern laboratory. Both, through the objectivity they conferred, allowed access to a new realm of truth. Scientists discovered organisms in a drop of pond water by using a microscope, Kardec wrote; Spiritists discovered the mysteries of the beyond by using mediums to contact the souls of the dead.[26]

To explain these new phenomena, Kardec advanced a distinctive conception of human physiology. An incarnated human, he maintained, was a tripartite entity, composed of a physical body, an immaterial soul, and a "semi-material" link between the two, called a *périsprit.* In the living, the *périsprit* manifested itself as the "vital fluid" that animated the body; in the dead, it served as a physical envelope for the wandering soul. According to Kardec, the *périsprit* gave disembodied spirits the ability to produce tangible phenomena. By channeling their "vital fluid" through a receiver—the medium—they could act in the material world. In Kardec's view, therefore, spirit phenomena were not supernatural because they did not involve a divine suspension of the laws of nature. Instead, these manifestations were direct consequences of human physiology, and as such were no more miraculous than breathing. Spirit phenomena appeared to transgress the laws of nature only because human beings did not yet understand the manner in which the body and soul—with the assistance of the *périsprit*—collaborated in their production.[27]

While Kardec's presentation of his ideas was innovative, the ideas themselves were not. Indeed, the doctrine of Spiritism was for the most part a

[26] Ibid., 491.
[27] Ibid., 40–41.

selective compendium of ideas from mid-nineteenth-century French Romantic Socialist thinkers.[28] Kardec's spirits appeared to have borrowed their notion of reincarnation and their critique of eternal damnation from the works of Fourier and Jean Reynaud.[29] Their moral vision, with its emphasis on charity, owed a great deal to the thought of Etienne Cabet, Pierre Leroux, and Henri de Saint-Simon.[30] The Spiritist conception of a universe driven to constant improvement by a "law of progress" reflected the republican optimism of thinkers like Eugène Pelletan.[31] Even Kardec's notion of the *périsprit,* which might strike the modern reader as peculiar, had its antecedents in Fourier's notion of the "aromal body" and in the theories of the Mesmerists. To moderately educated readers, therefore, the philosophical system outlined in the *Livre des Esprits* would have seemed familiar. For readers with progressive inclinations, it would also have seemed to create an unimpeachably modern context for religious experience. As a *spirite,* it was possible to embrace what so many at the time saw as the ideas of the future without renouncing the older-seeming consolations of belief.

Kardec was forthright about the derivative nature of the philosophy outlined in his text. He readily acknowledged his terrestrial precursors, but also insisted that the doctrine of Spiritism was unique in one crucial respect: instead of relying on an individual thinker's speculations, Spiritism rested on a foundation of unassailable fact. As a result, his new doctrine could lay claim to an authority more complete than any previous metaphysical system had enjoyed. Kardec's discussion of this point revealed its roots in the epistemology of Comtean Positivism, another system of thought that many of his readers would have perceived as both forward-looking and familiar.[32] Spiritism, Kardec wrote,

[28] Paul Bénichou, *Le Temps des prophètes* (Paris: Gallimard, 1977); Charlton, *Secular Religions*; Albert Leon Guérard, *French Prophets of Yesterday: A Study of Religious Thought under the Second Empire* (London: T. Fisher Unwin, 1913), esp. 159–179; Frank Manuel, *The Prophets of Paris* (Cambridge, MA: Harvard University Press, 1962).

[29] Jonathan Beecher, *Charles Fourier: The Visionary and His World* (Berkeley: University of California Press, 1986); Michel Nathan, *Le Ciel des Fouriéristes, habitants des étoiles et réincarnation de l'âme* (Lyon: Presses Universitaires de Lyon, 1981); Jean Reynaud, *Philosophie Religieuse, Terre et Ciel* (Paris: Furne, 1864 [1854]); Sharp, *Secular Spirituality*, 1–47.

[30] Georges Brunet, *Le Mysticisme social de Saint-Simon* (Paris: Les Presses Françaises, 1922); Robert B. Carlisle, *The Proffered Crown: Saint-Simonianism and the Doctrine of Hope* (Baltimore: Johns Hopkins University Press, 1987); Christopher H. Johnson, *Utopian Communism in France: Cabet and the Icarians, 1939–1851* (Ithaca: Cornell University Press, 1974); Armelle Le Bras-Chopard, *De L'Egalite dans la différence, le socialisme de Pierre Leroux* (Paris: Presses de la Fondation Nationale des Sciences Politiques, 1986); Frank Manuel, *The New World of Henri Saint-Simon* (Cambridge, MA: Harvard University Press, 1956).

[31] Paul Baquiast, *Une Dynastie de la bourgeoisie républicaine, les Pelletan* (Paris: L'Harmattan, 1996), 19–176.

[32] D. G. Charlton, *Positivist Thought in France During the Second Empire, 1852–1870* (Westport, CT: Greenwood Press, 1959).

> is not at all a theory, a system invented to provide a first cause; it has its source in the facts of nature itself, in positive facts that frequently appear before our eyes, but that until now had unknown origins. It is thus a result of observation, a science in a word: the science of relations between the visible and invisible worlds; a science still incomplete, but growing each day with new studies. It will take its place, you can be sure, in the ranks of the *positive* sciences. I say *positive* because every science based on facts is a positive science, not a purely speculative one.

Other philosophies, like those of Fourier or Reynaud, were at root "products of the imagination." These earlier thinkers, in keeping with the epistemological approach Comte identified with the "metaphysical" phase of human history, had begun with abstract suppositions, which they then developed under the sole guidance of individual reason. Kardec had constructed his philosophical system in a very different way, which made it eminently suited to what Comte identified as the dawning "positive" age. Spirit communications, Kardec argued, were objective documents, not mere bursts of personal insight. Spiritism, therefore, was a science as positive—as much a matter of empirical induction—as chemistry.[33]

The "positive" metaphysics of Spiritism, Kardec believed, definitively solved a crucial intellectual problem of the period: the religious crisis of factuality created by the growing prestige of science, and the concurrent rise of philosophical materialism. Conventional theology, as Kardec saw it, had no empirical component; its notions of God, the cosmos, and the afterlife were based on philosophical speculation and a priori assumptions. Spiritism furnished this missing empirical dimension by making the immortality of the soul, and all the metaphysical propositions it entailed, *real* in a quintessentially modern way. Where orthodox religion—which for Kardec was above all Catholicism—provided a series of "allegorical images that mislead us," Spiritism provided empirically grounded, transparent statements of fact. Spirit communications were eyewitness accounts of the beyond: They gave clear, unambiguous descriptions of what every human being could expect to experience after death. Where the airy philosophizing of orthodox theology did nothing but "generate doubt," the rigorously supported conclusions of Spiritism created a "renewal of fervor and confidence."[34]

Despite these claims of empiricism, in practice Kardec's exposition of his philosophy often seemed closer to old-fashioned metaphysical deduction than to modern scientific induction. This contradiction appeared most clearly when the time came for him to choose the otherworldly

[33] *La Revue spirite* 7 (1864): 325, 323. Italics in the original.
[34] Kardec, *Livre des Esprits*, 71, 70, 71.

communications he would accept as true. Inevitably, the spirits contradicted one another. Some, for example, proclaimed the reality of reincarnation, while others denied it. Faced with such a conundrum, it was impossible for Kardec to provide purely empirical reasons why one view of the afterlife should prevail over the other: Both had strong support from mediums. Numerically speaking, in fact, opponents of reincarnation outnumbered advocates—if only because nearly all Anglo-American mediums accepted the notion that each soul had a single fleshly existence. For a skeptical observer, Kardec's decision to take the word of one group of spirits, and discount that of another, might seem all too subjective.

Kardec addressed this objection by appealing to the truth-determining power of rigorous logic. The rational and the true, he argued, were identical. Hence, in his view, the unparalleled rationality of Spiritist doctrine gave it a greater claim to truth than any other philosophy could command. Indeed, Kardec argued that the ideas his book outlined were so transcendently logical they could handily convince new believers on purely intellectual terms. Witnessing spirit phenomena, therefore, was by no means a prerequisite for Spiritist conversion. No human being unaided by higher powers, Kardec believed, could have invented the system outlined in his books, which he called a "philosophy that explains what NONE other has explained."[35] In this context, the very familiarity—even banality—of Spiritist philosophy conferred an additional authority, making its tenets seem to be transparent expressions of progressive "common sense." For Kardec, simplicity and ordinariness were marks of intellectual perfection, and hence provided evidence of Spiritism's divine origins as concrete as any spectacular levitation or apparition.

Kardec presented the philosophy expounded in the *Livre des Esprits* as a watershed in human history. All humans were fundamentally rational beings, gradually moving toward an ever more rigorous empiricism, so in his view it logically followed that mankind as a whole would inevitably embrace the new doctrine. Spiritism's worldwide triumph would engender powerful feelings of solidarity and fraternal love, as the irrational prejudices that had previously separated one people from another dissolved into irrelevance. Once Spiritism became "a popular belief"—as the *loi du progrès* guaranteed it would—and once the specific doctrine it posited acquired the status of immutable fact, material comforts would finally be recognized as meretricious illusions.[36] In a world where the hereafter seemed to exist with the same irrefutable concreteness as the herebelow, people would think first and foremost of their souls, not their bodies. This new perspective would bring about an age of social harmony, spurring the

[35] Ibid., 483. Emphasis in the original.
[36] Ibid., 356.

rich to perform ever-greater acts of charity, inspiring a virtuous resignation in the poor, and eliminating the sense of cosmic doubt that had seemed to be an inevitable condition of modernity.

Organizing Spiritism, 1858–1869

Kardec's grandiose vision of his doctrine's significance prevented him from being content with literary success. To realize its world-historical mission, Spiritism needed a social component: The individual readers who adopted its ideas had to be forged into a coordinated movement. Kardec accomplished this task by founding a journal, the *Revue spirite*, devoted to the study of communications with the beyond. He then linked it to a spirit society, which he patterned after the more conventional learned discussion societies—*sociétés savantes*—that had become fixtures of French middle-class life. By the late 1860s, a network of spirit societies spanned the nation, and a vast literature had emerged to elaborate on the basic principles that Kardec had established.

Despite the popularity of the *Livre des Esprits*, Kardec's larger organizational project initially seemed ill-fated. His efforts to establish a new periodical met with two early setbacks. First, the wealthy Dutch spiritualist J. N. Tiedeman, who had initially expressed an interest in funding Kardec's venture, withdrew his support. Second, in a manifesto that appeared at the end of 1857, the spiritualist Zéphyre-Joseph Piérart, newly dismissed as editor of the *Journal du magnétisme*, announced his own plans to found a journal and harshly criticized Kardec's approach to the study of spirit phenomena.

Disconcerted by these developments, Kardec decided to consult the spirits for advice. The medium he chose to visit, the young Mlle Ermance Dufaux, was well known among students of spirit phenomena as a virtuoso automatic writer.[37] The anonymous spirit she contacted urged Kardec to bring out his periodical quickly, even if he had to finance it himself, because otherwise his competitors would have an advantage difficult to overcome. In addition, Dufaux's spirit offered Kardec some editorial advice that revealed an acute sense of what made publications on spirit phenomena appealing:

> Especially in the beginning, [the journal] must appeal to curiosity; it must contain both the serious and the entertaining; the serious will attract men of

[37] Dufaux specialized in biographies of famous people, dictated directly by their spirits. Her best-known work in this genre was *La Vie de Jeanne d'Arc, dictée par elle-même* (Paris: Ledoyen, 1858).

> science, and the entertaining will amuse the ordinary reader; the latter part is essential, but the former is the most important, because without it, the journal will lack a solid foundation. In a word, you must keep monotony at bay with variety; if you combine solid instruction and attractive subjects, this will be a powerful auxiliary for your subsequent work.[38]

Kardec heeded the advice he received through Dufaux. At his own expense, in January 1858, he published the first issue of his new journal. He called it the *Revue spirite.* In its pages, he struck a balance between long articles on philosophical questions and shorter, more entertaining pieces.

These shorter articles tended to be either automatic writings or transcribed dialogues between a spirit and a human interlocutor. Both stressed the emotional aspects of Spiritism and the tangible comforts it could provide. In the first issue, for example, Kardec published a dialogue between a bereaved mother and the spirit of Julie, her deceased daughter, who communicated through a medium by automatic writing:

> *Julie*: I no longer have the body that made me suffer so, but I have the same appearance. Aren't you happy that I no longer suffer, since I can speak with you?
> The Mother: If I saw you I'd recognize you, then!
> *Julie*: Yes, certainly, and you have already seen me often in your dreams.
> The Mother: I *have* seen you in my dreams, but I thought it was a figment of my imagination, a memory.
> *Julie*: No, it's me; I am always with you and searching for ways to console you; I even inspired you to summon me here. I have many things to tell you.[39]

Exchanges like this gave the new doctrine the emotional immediacy of sentimental fiction. Kardec did not simply make abstract declarations about the consoling power of Spiritism, he provided concrete examples of how it worked in practice, which allowed readers to share vicariously in the improving pleasure contacts with the beyond afforded. The "spirit phenomenon" this medium described was also accessible. Where an earlier journal might have focused exclusively on rare, spectacular manifestations few could ever hope to witness, Kardec's *Revue* included phenomena its readers could easily relate to their own experience. Many readers, for example, would have been able to remember dream-visits from loved ones of their own and find consolation in the prospect that those mental encounters were contacts from the other world. Kardec himself, as we saw earlier, discovered the consoling power of spirit communications through a similar experience.

[38] Kardec, *Œuvres posthumes,* 269.
[39] *La Revue spirite* 1 (1858): 18. Italics in the original.

The first issue of Piérart's competing journal, the *Revue spiritualiste*, appeared shortly after Kardec's in January 1858. Initially, Piérart seemed better positioned for success. Unlike Kardec, Piérart already had experience managing a periodical. He moved in the most glamorous spiritualist circles: He was among the guests at an 1858 banquet honoring the medium Home, for example; Kardec was not.[40] In addition, Piérart knew most of the writers who had published books and pamphlets on the new phenomena in the wake of the 1853 séance vogue. Many of them—including Paul Auguez, P. F. Mathieu, and Paul Louisy—became regular contributors to his new journal. Finally, and perhaps most important, Piérart's friendships with wealthy *bourgeois* and aristocratic students of spiritualism, like J. N. Tiedeman—Kardec's erstwhile patron—the Comte d'Ourches, and the Baron de Guldenstubbé, helped him secure the financial backing he needed to establish his publication.[41]

The *Revue spiritualiste*, however, proved considerably less attractive to subscribers than Kardec's journal. Where Kardec wrote in simple and authoritative terms, conveying a clear picture of the other world and its inhabitants, Piérart preferred to range more widely in his speculations, avoiding what he saw as the pitfall of a prematurely imposed "orthodoxy" beyond which "there will be only error, heresy."[42] The image Kardec projected was one of methodical rigor and probity; Piérart, on the other hand, both in person and in his prose, was a wild-eyed visionary, prone to rhetorical flamboyance, "imperiousness," and polemical excess.[43] Even Piérart's abortive effort to form an Académie du spiritualisme and his endorsement of a *Livre des Esprits spiritualistes* failed to win him many adherents.[44] By early 1866, the *Revue spirite* counted 1,800 subscribers, the *Revue spiritualiste* a mere 500.[45] Though Piérart continued publishing his journal until 1869, its contents became increasingly eccentric, particularly after 1867, when he began devoting the bulk of it to transcriptions of interviews with his "personal genius."[46]

Heartened by the success of the *Revue spirite*, Kardec decided to found a Spiritist association in the spring of 1858. On April 13, the Ministry of the Interior officially authorized the Société Parisienne des études spirites. According to its first announcement in the *Revue spirite*, the Society was

[40] *La Patrie*, May 2, 1858, quoted in *La Revue spiritualiste* 1 (1858): 72.

[41] For biographical information about Guldenstubbé, see *Le Concile de la libre pensée* 23–24 (1872–1873): 182–188.

[42] *La Revue spiritualiste* 1 (1858): 5.

[43] For a description of Piérart's appearance and "imperious" behavior in séances, see the memoir by Alexandre Delanne published in *Le Spiritisme* 8 (1890): 100.

[44] For a preliminary announcement of the Académie, see ibid., 6; M. Nordmann, *Le Livre des Esprits spiritualistes* (Paris: Patissier, 1863).

[45] Archives nationales, carton F/18/294, "Etat du tirage des journaux (politiques et non politiques) 1er semestre 1866."

[46] See, e.g., *La Revue spiritualiste* 10 (1867): 65–72.

composed "exclusively of serious people," including several "men made eminent by their knowledge and social position."[47] Kardec's society quickly proved to be an inspiring example for other like-minded believers. In the late 1850s and early 1860s, small Spiritist groups began to form throughout France, especially in Paris, Lyon, and Bordeaux. The majority of these groups established corresponding relationships with the Société Parisienne, which the *Revue spirite*'s success had made into the most visible French organization of its kind.

As the influence of the Société Parisienne grew, Kardec worked to codify the practice of Spiritism as he had codified its philosophy. He began this new project by publishing a second book, the 1861 *Livre des médiums*, which established norms for the evaluation of spirit communications and set out rules for proper séance conduct. At the same time, in the journal, Kardec coupled the implementation of these norms with an emphasis on philosophical conformity: To this end, he published an article suggesting two new rules for spirit societies eager to guarantee the necessary "uniformity of doctrine." First, each society was to require its members to make a "categorical declaration of loyalty, and a formal statement of adhesion to the doctrine of the *Livre des Esprits*." Second, societies were to reaffirm this initial commitment by starting each meeting with a reading from either the *Livre des médiums* or the *Livre des Esprits*. The Société Parisienne would sever all ties with any group that refused to accept these new rules and make Kardec's philosophy its primary object of study.[48]

During this period, Kardec also advanced a new scheme for the management of the increasingly numerous spirit societies that had sprung up since 1858. Previously, these groups had been connected to the Société Parisienne only by informal correspondence. Now, Kardec proposed a more structured organization. To give small groups a node around which to congregate, Kardec recommended the creation of a "directing group" in every French city with a large Spiritist population. These groups would serve as coordinating bodies, gathering communications from their regions and corresponding with the Société Parisienne. The Société Parisienne, in turn, would function as the coordinating body for all of France. It would also be in charge of designating local "directing groups." Though it stood at the top of this organizational hierarchy, Kardec emphasized the Société Parisienne's relative lack of power. It could "establish purely scientific relations" with other societies but exerted no other "control," leaving them free to "organize as they see fit." To make the connection even less formal, the Société Parisienne gave no financial support to allied societies

[47] *La Revue spirite* 1 (1858): 148.

[48] Allan Kardec, *Le Livre des médiums, ou guide des médiums et des évocateurs* (Boucherville, QC: Editions de Mortagne, 1986 [1861]); *La Revue spirite* 4 (1861): 373, 378.

and required no membership dues from them.[49] It also remained small, at least on paper: It had only eighty-seven full members in 1862, but it also included what was probably a much larger body of casual, non–dues-paying "free associate members."[50]

Officially, then, Kardec's society did not act as the leader of French Spiritists, but as the first among equals. His description of the Société Parisienne's role, however, dramatically understated the moral influence it exerted. This influence stemmed from Kardec's charisma, the authority of his tremendously successful books, and his position as editor of the most widely circulated spiritualist periodical in France. Any group that strayed too far from Kardec's principles would find its "scientific relations" with the central society abruptly cut off—a rupture that, in the eyes of many Spiritists, would make the dissident society appear illegitimate. Hence, even if membership dues were not required from allied groups, ideological allegiance was.

As Kardec continued this process of codification and centralization, his power as a source of religious authority steadily increased. In 1861 alone, for example, Kardec claimed to have received between 1,200 and 1,500 visitors eager to discuss his new doctrine.[51] By 1863, "not counting a certain number of more or less voluminous manuscripts," he had amassed a backlog of 3,600 spirit communications from various mediums or societies, submitted for his consideration and approval.[52] Other Spiritists also began publishing their own collections of spirit writings, and by 1865, an entire body of literature had emerged that took the philosophy of the *Livre des Esprits* and the practices of the Société Parisienne as points of departure.[53] The growing influence of Kardec's ideas proved even more important than his publications. The basic principles set forth in Kardec's books provided the vast majority of French people who believed in spirit contacts with their fundamental assumptions. Among those in the milieu, Kardec's doctrine ceased being one in an array of competing systems and acquired the status of orthodoxy. The growing authority of Spiritist philosophy influenced mediums as well: increasingly the spirits they contacted spoke the ideas in Kardec's books.

Kardec further augmented his importance by proselytizing on behalf of his doctrine. Beginning in 1860, he made a series of lecture tours through the South, where Spiritism appeared to be growing most rapidly. The largest

[49] *La Revue spirite* 4 (1861): 379, 383.
[50] *La Revue spirite* 5 (1862): 164.
[51] Ibid.
[52] *La Revue spirite* 6 (1863): 156.
[53] See Un Capitaine [pseud.], *Spiritisme élémentaire, théorique et pratique. Faits spirites et entretiens familiers d'outre-tombe contenant la théorie de l'évocation des esprits ou des âmes des morts (d'après les écrits d'Allan Kardec)* (Paris: Ledoyen, 1862); Henri Dozon, "Révélations d'outre-tombe. Espoir et résignation. Semez. Règle de société" (Paris: Ledoyen, 1862); L. T. Houat, *Etudes et séances spirites, morale, philosophie, médecine, psychologie* (Paris: Ledoyen, 1863).

of these tours took place in 1862, when Kardec visited Lyon, Bordeaux, and about twenty other cities, including Avignon, Montpellier, and Toulouse. In Lyon, he spoke to a gathering of six hundred delegates from the city's spirit societies, but most of the groups he addressed were considerably smaller.[54] Kardec's goal in meeting with such groups was not to convince skeptics of the reality of spirit phenomena, but rather to persuade the convinced of the rightness of his doctrine and its superiority to competing alternatives.[55]

The 1862 tour's success inspired Kardec to embark on a more ambitious program of publicity. He wrote and printed a pamphlet titled *Le Spiritisme à sa plus simple expression* (Spiritism Made Simple), priced at 25 centimes, which he advised allied societies to distribute as widely as possible. He also encouraged Spiritists to proselytize among "people exhausted by doubt and horrified by the materialist void."[56] The next two years saw a surge in Spiritist publications. Kardec produced two new compilations of spirit communications. The first, *L'Evangile selon le spiritisme,* appeared in April 1864 and contained a series of meditations on the moral implications of Christ's teachings. The second, *Le Ciel et l'Enfer ou la justice divine selon le spiritisme,* which appeared in August 1865, further elaborated the idea of expiatory reincarnation. A steady growth in the number of spirit societies, especially in the provinces, inspired a flurry of new periodicals, all of them devoted to Kardec's doctrine. Spiritist journals inaugurated during this period included *L'Avenir* (Paris, 1864); *La Vérité* (Lyon, 1863); the *Revue spirite d'Anvers* (1864); the *Médium évangélique* (Toulouse, 1864); the *Echo d'outre-tombe* (Marseille, 1865); and the *Ruche spirite,* the *Sauveur des peuples,* the *Voix d'outre-tombe,* and the *Union Spirite,* all in Bordeaux (1863–1865). A wide variety of other books and pamphlets appeared as well, including a spate of popular novels and newspaper stories.[57] The vast majority of them, whether sardonic and unflattering, like Alfred de Caston's *Tartuffe spirite,* or eerie and romantic, like Théophile Gautier's best-selling *Spirite,* drew heavily on Kardec's writings, which had come to provide the most widely accepted heterodox description of the soul's

[54] Kardec, *Voyage spirite en 1862* (Paris: Vermet, n.d. [1862]), 3–4, 6. For a description of a meeting of fifteen "adepts" in Toulouse, see Archives nationales, carton F/18/10927, letter dated Sept. 27, 1862.

[55] Kardec, *Voyage spirite,* 20.

[56] *La Revue spirite* 6 (1863): 70, 73–74.

[57] For examples not cited elsewhere in this chapter, see Auguste Bez, *Miracles de nos jours* (Bordeaux: l'auteur, 1864); J. B. Bourreau, "Comment et pourquoi je suis devenu spirite" (Niort: Favre, 1864); Emilie Collignon, "L'Education Maternelle, conseils aux mères de famille, le corps et l'esprit" (Paris: Ledoyen, 1864); J. Condat [pseud. J. Chapelot], *Réflexions sur le spiritisme, les spirites et leurs contradicteurs* (Paris: Didier, 1863); Evariste Edoux, *Spiritisme pratique, appel des vivants aux esprits des Morts* (Lyon: Librairie moderne, 1863); Dr. Guyomar, *Etude de la vie intérieure ou spirituel chez l'homme* (Paris: Delahaye, 1865); B. J. B. Lussan, *Quelques pages sur le spiritisme* (Toulouse: Chauvin, 1865); Camille de Montplaisir, *Qu'est-ce que le spiritisme?* (Lyon: Girand et Josserand, 1863).

experience in the beyond.[58] Journalists and *hommes de lettres* now spoke of *spirites*, not *spiritualistes*.

Kardec's fervor grew as his doctrine became steadily more influential. In the late 1860s, the communications he published assumed a millenarian tone. His final book, the 1868 *Genèse selon le spiritisme*, closed with a series of communications and commentaries declaring that "the time chosen by God has come" and that a new generation of highly evolved souls was in the process of being incarnated on Earth. These more elevated, intelligent humans would transform the planet's social organization, introducing a golden age of charity and fraternity. By the dawn of the twentieth century, Spiritism would become "the pivot on which the human race will turn," the basis for an unshakable new faith in the immortality of the soul and the profoundly moral nature of the universe.[59] Kardec did not live to assess the validity of this prophecy. After several years of faltering health, he died of a heart attack on March 31, 1869. He was sixty-four years old.

Who Were the Spiritists?

Spiritism's success was due in part to the organization-based model of authority Kardec developed. Formal *sociétés spirites* fostered feelings of solidarity among believers and, even more important, served as a regulatory framework for communication with the other world. Conversation with spirits could be unruly: Disembodied souls expressed a bewildering array of opinions that tended to vary from medium to medium. The model of séance practice Kardec invented, which made dialogue with the beyond into a rigorously controlled, "serious" undertaking, seemed to tame the wild landscape that this new form of democratic revelation created. By following Kardec's example, Spiritists sought to preserve the power of the distinctive experience of the sacred their beliefs made possible while simultaneously imposing a measure of doctrinal coherence and ritual structure that had been missing from the French spiritualism of the late 1850s.

This emphasis on coherence and structure distinguished French Spiritism from its Anglo-American counterparts. British and American spiritualists tended to be considerably more tolerant of philosophical or

[58] Alfred de Caston, *Tartuffe spirite, roman de mœurs contemporaine* (Paris: Librairie Centrale, 1865); Théophile Gautier, "Spirite, nouvelle fantastique," in *L'Œuvre fantastique II—Romans* (Paris: Bordas, 1992 [1866]), 202–323; Georges de Parsevel-Deschênes, *Gardeneur, histoire d'un spirite* (Paris: Librairie du Petit Journal, 1866).

[59] Allan Kardec, *La Genèse selon le spiritisme* (Montréal: Editions Sélect, 1980 [1868]), 225, 232. The term *pivot* comes directly from Charles Fourier, who used it to refer to a crucial element in a series of attributes or phenomena.

ritual differences among mediums and groups. In large part, the French preference for codification had to do with religious context: The majority of Spiritists, even if they had abandoned orthodox faith, still conceived of religious authority in Catholic terms. Spiritism's centralization and coherence, therefore, was a collaborative creation, the product of an interaction between Kardec's ideas and his audience's expectations. Believers and mediums often embraced the philosophical and organizational structure Kardec provided because doctrinal uniformity and a solid institutional foundation struck them as crucial elements of religious legitimacy. For these believers, Spiritism derived a key aspect of its "unique realism" from the coherence Kardec's ideas imposed on spirit communications and séance practice—a coherence that imparted what observers using a Catholic religious grammar would perceive as a crucial mark of truth.

The process of creating this coherence, however, also generated tensions and contradictions. Not all mediums readily surrendered the autonomy they had enjoyed in the early days of French spiritualism by accepting the judgments of those who presented themselves as authorities; not all believers willingly adopted the sometimes strict discipline "serious" Spiritists prescribed. To understand these conflicts, and the way formally elaborated ideas and practices worked for believers, we need to look more closely at the seekers Kardec's texts inspired.

Spiritist Demographics: Class and Gender

Unfortunately, the evidence for a rigorous demographic study of Second Empire Spiritism does not exist. The archival sources are fragmentary, the correspondence of societies has been either lost or scattered, and the remaining material—mostly books, pamphlets, and periodicals—is impressionistic and biased.[60] At best, the published texts indicate how Spiritists *wanted* their movement to appear, while leaving the question of its actual composition unanswered and unanswerable. The personal nature of much Spiritist practice also obscures the subject. For every medium who received messages from the beyond at spirit society meetings or sent automatic writings to journals for publication, there could well have been many more who practiced alone or in family séances, holding intimate conversations with deceased loved ones. Similarly, the evidence does not help us count non-mediums who may have read Spiritist tracts, but did not choose to join spirit societies. Any account of the

[60] Kardec's correspondence and the papers associated with the early years of the Société Parisienne appear to have been destroyed in the early 1880s. See Berthe Fropo, *Beaucoup de lumière* (Paris: Imprimerie polyglotte, 1884), 29–30.

movement's overall composition, therefore, is necessarily a matter of speculation.

Despite these shortcomings, the available sources give a sense of the movement's public face. The most striking piece of evidence of this kind is a long article published in the January 1869 *Revue spirite.* In it, Kardec, drawing on a decade of correspondence, sought to present an empirically based description of Spiritism's adherents. This account, while problematic, nevertheless conveys a sense of the kinds of people Kardec hoped to attract with his ideas. Organized Spiritism, as he constructed it, was a respectable endeavor—one that preferred moderation in religious matters and was dominated by educated men. Society members, in Kardec's account, generally considered themselves Catholic but "not attached to dogma." Though Spiritism drew the majority of its believers from the urban "*petite bourgeoisie* and the working class"—especially artisans, clerks, and shopkeepers—its most influential members came from "the enlightened classes."[61] As Kardec never tired of affirming, some spirit societies were "almost exclusively comprised of members of the bar, magistrates, and government officials."[62]

For Kardec, Spiritism's respectability and seriousness were fundamentally masculine. Men were always in the majority in spirit society meetings, he asserted; when couples disagreed about Spiritism, it was usually the wife who refused to allow her husband to explore the new doctrine. Critics were therefore mistaken when they alleged that Spiritism "has found most of its recruits among women, because of their penchant for the marvelous." Indeed, Kardec argued,

> it is precisely this penchant for the marvelous and for mysticism that makes women, in general, more resistant than men. [Women's] predisposition lets them embrace uncritical, blind faith more easily; Spiritism, on the other hand, only permits a reasoned faith based on reflection and philosophical deduction—modes of thought to which women are less suited, because of the narrow education they receive.[63]

The Spiritist enterprise was scientific, not mystical; its adherents prized objective discovery over intuitive insight. This rationalism, Kardec asserted, made Spiritism the business of men.

Journals and accounts of society meetings indicate that the kinds of people Kardec portrayed as typical Spiritists were indeed involved in the movement, though probably not in the numbers he suggested. Spirit societies

[61] *La Revue spirite* 12 (1869): 5, 6.
[62] Kardec, *Voyage spirite,* 5.
[63] *La Revue spirite* 12 (1869): 7.

were unusually open to female members, who in this period were often excluded from more conventional *sociétés savantes.* Still, male lawyers, army officers, doctors, and other professionals figured prominently in Spiritist circles.[64] Men also made up the majority of the mediums who published automatic writings in Spiritist periodicals during the period. Of the 116 mediums who contributed to the *Revue Spirite* between 1858 and 1869, 46 were women. The spirit society president Henri Dozon's *Révélations d'outre-tombe* (Revelations from Beyond the Grave) had a larger proportion of female contributors, with 7 out of 15. In the three Bordeaux Spiritist journals of the period, *La Ruche spirite, Le Sauveur des peuples,* and *L'Union spirite,* 62 mediums published, of whom 20 were women. *La Vérite* of Lyon published communications from 35 mediums; of those, 9 were women. The Marseille *Echo d'outre-tombe* counted 6 women among the 16 mediums who contributed.

The gender imbalance among mediums who published messages from the beyond raises interesting questions. Most strikingly, it points to another key difference between French Spiritism and its Anglo-American counterpart. British and American mediums who published communications in this period were much more likely to be women. The comparatively smaller proportion of published female mediums in France again stems in part from a difference in religious context. For men and women in Britain and the United States, spiritualism, though perceived as unconventional, nevertheless fit into a continuum of Protestant sectarian pluralism. In France, on the other hand, the Catholic Church dominated the religious landscape, and in the eyes of the clergy and many of the devout, Spiritism was a heresy far beyond the pale of acceptable religious diversity. At the same time, Catholic devotion in this period—particularly the exuberant, innovative forms that centered on the Virgin Mary—was seen as a distinctively feminine concern. For a French woman, embracing Spiritism required a willingness to abandon this religious mainstream and the social network that went along with it.[65] Among French men, on the other hand—particularly those with republican sympathies—anticlerical skepticism was

[64] Dozon, for example, was a retired army officer, Houat was a doctor, and Timoléon Jaubert was a lawyer. Provincial journals also featured a few contributors with "respectable" occupations, which were occasionally indicated beneath the contributor's name. Writers published in *La Ruche spirite* of Bordeaux included M. Auzanneau, an *inspecteur d'assurances*; Jean Bardet, printer and typographer; A. Bonnet, architect; J. L. Jean, lawyer; Dr. A. Chaigneau; and C. Guérin, "vérificateur de poids et mesures." See *La Ruche spirite* 1 (Jun. 1863–May 1864): 175, 188, 169, 55, 168. Contributors to *La Vérité* in Lyon included Tibulle Lacy, an alumnus of the *Ecole Polytechnique*; A. R., a primary school teacher in Grenoble, and B. Repos, a lawyer. See *La Vérité,* Mar. 8, 1863, 4; Aug. 2, 1863, 3; and Oct. 11, 1863, 7.

[65] For a useful discussion of the importance of this network, see Bonnie G. Smith, *Ladies of the Leisure Class: The Bourgeoises of Northern France in the Nineteenth Century* (Princeton: Princeton University Press, 1981).

a defining characteristic of masculinity, and sociability revolved around secular organizations such as Masonic Lodges or *sociétés savantes.* For male seekers operating in this more open environment, the social penalty for experimenting with Spiritism would not have been as strong. In addition, many male seekers seem to have found Kardec's doctrine attractive because of the way in which it allowed them to enjoy the consolations of religion while simultaneously retaining the intellectual perspective of progressive free-thought.

The preponderance of male automatic writers in Spiritist publications, however, should not be taken as simple proof that men were more likely to be mediums. The gender imbalance in published messages from the beyond also probably reflected a bias toward male mediums on the part of those selecting communications. As we will see, the criteria used to distinguish worthy automatic writings from unworthy ones tended to exclude women. According to Kardec, elevated spirits wrote in the language of highly educated men, a feat that would have been difficult for women with more limited training to achieve. Women, who usually received Catholic educations, were also less likely to be familiar with the Romantic Socialist ideas that Kardec considered most "rational" and "true." They also may have been less inclined to submit their communications for publication in the first place, preferring instead to keep their dialogues with the beyond private. Since the majority of spirit communications were intimate messages from loved ones, not visionary or philosophical essays, a preference for discretion would be logical.

Kardec's assertions about the class status of Spiritists appear somewhat more accurate. Judging by the *Revue,* Parisian Spiritists from the lower ranks of the working class seem to have been comparatively rare. In 1863, for example, the members of the Société Parisienne treated the humble burial of M. Costeau, a "simple worker" and stalwart member of their group, as an exceptional occasion. One dignitary from the society, along with two mediums, spoke at the gravesite, and a description of the event—the only one of its kind to appear during the whole decade—occupied six pages.[66] The sheer exceptionality of this account, and the amount of attention it received, indicates that most Parisian Spiritists were likely better off than Costeau, coming either from the upper reaches of the working class or from various strata of the *bourgeoisie.* Anecdotal evidence about Bordeaux and Lyon, on the other hand, suggests a greater number of working-class Spiritists.[67]

[66] See *La Revue spirite* 6 (1863): 297–303.

[67] For such evidence, along with Kardec's article of 1869, see the description of spirit society membership in Lyon and Bordeaux presented in *La Revue spirite* 4 (1861): 290, 327–328; and the account of a subscription for *Lyonnais* workers, ibid., 5 (1862): 55–57.

Conversion Experiences

In every socioeconomic class, Spiritists were very much a minority, and conversion to the new doctrine was at root a matter of individual temperament. Adopting Kardec's ideas and joining a spirit society required a streak of independent-mindedness, bolstered by a willingness to act in visibly unconventional ways. A lack of concern with received ideas, however, was only a prerequisite for conversion. Those who adopted Spiritism did so above all because it soothed their grief and provided what they perceived as an intellectually compelling alternative to religious doubt.

Stories of individual conversions convey a sense of the kinds of comfort Spiritism provided. Contacts with deceased loved ones appear to have been the most common spur to conversion.[68] The possibility of speaking directly to a deceased parent, child, or spouse provided a powerful source of consolation for many, particularly those who had come to regard Catholic teachings—and hence the very notion of an afterlife—with skepticism. A letter from a M. Georges, printed in an early issue of the *Revue spirite*, relates a typical conversion of this kind. Before his first visit to medium Ermance Dufaux, Georges wrote, he had been a staunch materialist and "doubted it all: God, the soul, and the afterlife." In the séance Georges attended, Dufaux contacted his recently deceased father. The spirit conversation that ensued, Georges wrote, was enough to eliminate his nagging doubts:

> This lady spontaneously mentioned many precise details concerning my father, my mother, my children, my health; she described all the circumstances of my life with remarkable exactitude, even recalling facts that I had long forgotten; she gave me, in other words, proof of the marvelous faculty of somnambulic lucidity so clear that it changed my views immediately. In the session where my father revealed his presence to me, I witnessed, if you will, the extra-corporeal life of the soul.

After this experience, according to Kardec, Georges quickly became "one of the most fervent and zealous adepts of Spiritism."[69]

Some years after Georges' letter appeared, Kardec observed that many of those who became Spiritists did so after having had a convincing spirit

[68] This was also true in Anglo-American spiritualism. Janet Oppenheim provides several short case studies that illustrate this point, including those of Samuel Carter Hall and Florence Maryatt. See Oppenheim, *The Other World: Spiritualism and Psychical Research in England, 1850–1914* (Cambridge: Cambridge University Press, 1985), 34–35, 38–39. Turner describes the case of F. W. H. Myers and his 1899 contact with Annie Marshall. See Frank M. Turner, *Between Science and Religion: The Reaction to Scientific Naturalism in Late Victorian England* (New Haven: Yale University Press, 1974), 114.

[69] *La Revue spirite* 1 (1858), 21.

conversation with a deceased loved one. In these exchanges, the spirit frequently proved its identity by revealing what Kardec called "intimate facts," personal details that appeared impossible for the medium to have known under ordinary circumstances.[70] In séances like these, the mundane often opened the door to the transcendent: an apparently trivial detail, like a description of the food eaten at a picnic held several summers before, could entail a complete change of metaphysical perspective. These seemingly ordinary revelations possessed such power because they seemed to demonstrate not only the immortality of the soul but also its continuing connection with life on Earth.

"Familiar interviews" of this kind became a central element of many spirit society meetings. Alexandre Delanne, for example, described the séances he held in the early 1860s in the cramped upper-floor Paris apartment he shared with his wife, a prolific writing medium. Sometimes, as many as forty people would pack into the small rooms to witness the proceedings. In these meetings, Delanne actively courted new adherents. For at least an hour, he allowed first-time attendees the opportunity to sit at the séance table with a medium and perform "personal invocations." These dialogues with the beyond usually involved contact with a deceased relative. The spirit evoked would often reveal telling domestic details that many saw as "authentic proofs" of its identity. Contacts of this kind proved to be emotionally powerful spurs to conversion, Delanne wrote:

> How many sweet tears of tenderness flowed before us. How many mothers, sons, fathers, found hope by recognizing beings they believed to have been lost forever! How many souls gnawed by doubt finally discovered their road to Damascus.

These emotional conversations occurred in full view of the gathered society members. The reassuring and sentimental spectacle they provided, Delanne noted, made them the part of Society meetings most popular among regular members. Spirit communications, in this context, provided the opportunity not only for newcomers to become converted but also for the other society members to partake of an uplifting bit of ritual theater, reaffirming their convictions and justifying their involvement with the Spiritist cause.[71]

Converts were also drawn to Spiritism for more cerebral reasons. Many saw these new ideas and practices in much the same way Kardec had in his

[70] *La Revue spirite* 8 (1865), 48.

[71] *Le Spiritisme* 7 (1888): 8–9. The meetings Delanne described in these reminiscences took place in the early 1860s.

early days—as a remarkable solution to the metaphysical conundrums of the age. This was certainly the case for the young Camille Flammarion, who would become one of the nineteenth century's most successful popular science writers. While Flammarion's later fame makes him exceptional, his correspondence and memoirs provide a uniquely well-documented, though in many respects typical, account of an intellectual conversion to Spiritism.

In 1860, the eighteen-year-old Flammarion was a student astronomer at the Paris Observatory under Urbain-Jean-Joseph Le Verrier. He was also in the midst of a spiritual crisis. When he arrived at the Observatory two years before, he had a strong, literal-minded faith, the product of a Catholic education and the example of a devout mother. He was convinced "of the divinity of Jesus and His real presence in the Eucharist," attended Mass every Sunday, and confessed his sins regularly. By 1860, this certainty had eroded. Regarded with a scientifically informed eye, he wrote, the words of the Bible came to appear "quite novelistic." The text he had once assumed to be unassailably true now seemed to offer nothing more than "pure, naïve, unverifiable and even contradictory fictions." By the end of his eighteenth year, Flammarion had ceased to believe "in the divinity of Jesus, in the sacraments, and in all the teachings of the Church."[72]

In the midst of this crisis, Flammarion discovered Spiritism. One evening in the spring of 1861, browsing in the bookstalls near the place de l'Odéon, he came across a copy of the *Livre des Esprits*.[73] The text impressed the young man deeply. Kardec's book appeared not only to resolve the logical inconsistencies of Catholic dogma but also to suggest a new, empirical way of answering metaphysical questions. After visiting Kardec, and being pleasantly surprised by the older man's reasonable, calm demeanor, Flammarion attended a meeting of the Société Parisienne.[74] The spirit contacts that occurred in the meeting intrigued him, but they were all carried out through automatic writing, which he did not find entirely convincing as proof of otherworldly intervention.

Flammarion received the definitive proof he craved in a private séance. A female "enthusiastic believer" who was also at the society meeting, probably Honorine Huet, invited Flammarion to this gathering.[75] While

[72] Camille Flammarion, *Mémoires biographiques et philosophiques d'un astronome* (Paris: Flammarion, 1911), 168, 182, 181, 184.

[73] Patrick Fuentès and Philippe de la Cortadière, *Camille Flammarion* (Paris: Flammarion, 1994), 68–72.

[74] Fonds Camille Flammarion de l'Observatoire de Jusvisy-sur-Orge [FCF], ms. Copybook marked *Miscellanées 1861*, letter to the Abbé Collin dated Oct. 7, 1861.

[75] Ibid., letter to Charles Burdy dated Nov. 1, 1861.

Kardec eschewed the production of spectacular phenomena, she told the young astronomer, this family group excelled in such matters. The séance far exceeded Flammarion's expectations. A spirit named Balthazar appeared, as he often did in séances where Huet was present. Balthazar was an unrepentant *gourmand* who refused to accept the reality of his death because it would mean abandoning the pleasures of the table; this assertive attachment to things corporeal made him quite willing to produce spectacular physical phenomena.[76] In the séance Flammarion attended, the spirit caused an impressive array of raps, then lifted a table off the floor, holding it suspended in midair. Flammarion wrote his friend Charles Burdy that he had been able to turn the table's rollers freely and had felt it tilt gently when he pressed its surface. These manifestations settled the question in his mind. "In the presence of such phenomena," he wrote his friend, "it is impossible to deny the existence of invisible agents."[77]

On November 2, 1861, only two days after his séance with the spirit Balthazar, Flammarion wrote a letter to Kardec asking to be admitted as an associate member of the Société Parisienne des etudes spirites.[78] Kardec accepted his application on November 15.[79] By late December, Flammarion had become an enthusiastic adherent of Kardec's doctrine. He described his new convictions in a letter to the Abbé Berillon of Langres, who had been his confessor during his days at the cathedral school there:

> Have you heard of Spiritism? It is a new science that has just appeared on the horizon, and emanates from God himself, through the ministry of His spirits. . . . This religion surprises at first, but is rational, and will be the culmination of Christianity; it explains all the dogmatic truths of the future life that have previously been so mysterious.
>
> I do not ask you to reflect on this new doctrine, my dear Superior, since I know you always reflect. If you would like, I could discuss it with you at greater length, and, if you will, *ex professo*, since I am in intimate relations with spirits who have already lived on Earth, particularly Galileo and Fénélon; they have taught me the same truths that other spirits have dictated throughout the world. I should warn you in advance that I am not in the presence or under the influence of any evil spirit: I study Spiritism as I study mathematics.[80]

[76] See Camille Flammarion, *Les Habitants de l'autre monde, révélations d'outre-tombe* (Paris: Ledoyen: 1862), 2 :79–94.
[77] FCF, *Miscellanées 1861*, letter to Charles Burdy dated Nov. 1, 1861.
[78] Ibid., letter to Allan Kardec dated Nov. 2, 1861.
[79] Ibid., note dated Nov. 15, 1861.
[80] Ibid., letter to the Abbé Berillon dated Dec. 31, 1861.

Spiritism's empirical basis made the immortality of the soul an incontrovertible fact. It also allowed adepts reassuringly immediate contact with the beyond: In his times of doubt, Flammarion could pose his direct questions to Galileo and the theologian François Fénelon, who would provide him with revelations suited to his personal circumstances. In fact, Galileo inspired the young astronomy student to write a long essay on the origins of the universe, which Kardec canonized by including in his 1868 *Genèse selon le Spiritisme*.[81]

After his conversion, Flammarion quickly became one of the most visible apologists for Kardec's doctrine. He published three books on the subject between 1862 and 1865, and contributed a long series of articles on Kardec's ideas to the literary journal *La Revue Française*.[82] By the latter part of the decade, as his youthful ardor cooled and his reputation in the scientific community grew, however, Flammarion began to distance himself from Spiritism. He remained interested in supernormal phenomena, but became increasingly skeptical of the way in which Kardec used spirit communications as the basis for philosophical speculation.

Despite its relatively short duration, Flammarion's conversion provides a revealing example of the way in which believers in Spiritism became convinced. For some, Kardec's doctrine appealed primarily for emotional reasons because it provided a novel and intense form of consolation, a way to continue cherished relationships with deceased loved ones. For others, the appeal was primarily intellectual. Unlike the abstract, philosophical spiritualism of Kant or Cousin, Spiritism was more than a theory: it had an empirical basis, which any curious newcomer could experience palpably in spirit society meetings. Balthazar's ability to suspend a table in midair, or Galileo's power to communicate by means of automatic writing, seemed to offer concrete proof that the human soul could exist independently of the body. This reassuring idea laid the groundwork for a new kind of religion, one in which metaphysical truths would cease being matters of intuition and speculation, and become simple matters of fact. For many, like Flammarion, who cherished the consolation and moral certainty religion could provide, but who also believed in the ultimate truth-determining power of experimental inquiry, Spiritism appeared to be a definitive solution to a deeply disturbing philosophical problem.

[81] See Kardec, *Genèse*, 58–78.

[82] Flammarion, *Les Habitants de l'autre monde ;* Flammarion, *Des Forces naturelles inconnues*; *La Revue française*, ed. Adolphe Amat, 4 (Jan.–April 1863): 215–237, 365–388, 493–515.

Spiritism in Practice

The stories of Spiritist converts reveal the crucial part mediums played in the doctrine's propagation. By speaking for deceased relatives, contacting errant spirits, and producing strange phenomena, mediums single-handedly made the beyond tangible for other believers. They also served as valuable sources of authority, becoming conduits for spirits who could provide answers to all manner of doctrinal and moral questions. Kardec himself placed these gifted intermediaries at the center of Spiritist practice. A spirit society, he wrote, could not fully carry out its mission without the direct access to the spirit world a medium provided.[83] The commentary of an elevated spirit, gently correcting flaws, suggesting lines of inquiry and offering advice, gave the society's discussions their power to edify. Where purely terrestrial analyses of Spiritist texts might raise questions and suggest diverse interpretations, only a spirit, writing through a medium, could explain the unambiguous truth.

Important as mediums were to the Spiritist project, they nevertheless occupied an ambivalent place in spirit societies. An eyewitness account of a séance by the journalist Jules Claretie suggests this complexity:

> Old women with avid eyes, skinny and tired young people, a mix of stations and ages, of neighborhood doormen and great ladies, calico and satin, poetesses who have happened by and prophetesses chanced upon in the street, tailors and members of the *Institut*; in Spiritism, they fraternize. They wait, they make tables turn, they levitate them, they declaim the jottings Homer or Dante have dictated to the seated mediums. These mediums are immobile, hands on sheets of paper, dreaming. Suddenly their hands fidget, run, thrash about, cover sheet after sheet, move, move still more, brusquely stop. Someone then breaks the silence, names the Spirit who has just dictated, and reads.[84]

While probably exaggerated for color, Claretie's description remains revealing because it is a rare firsthand account by a nonbelieving observer. Where convinced Spiritists presented séances as they wanted them to appear, Claretie did so with somewhat more evenhandedness. The scene he described sat oddly with the self-conscious rigor and empiricism of Spiritist ideas. It also conflicted with Kardec's demographic presentation of the movement as a fundamentally intellectual enterprise for educated men. Here, instead, Spiritism appeared to accord a prominent place to women and to stress the subjective experience of inspiration rather than the objective process of experimental study.

[83] *La Revue spirite* 4 (1861): 46.
[84] *L'Evénement*, Aug. 26, 1866, 1.

The behavior of the mediums Claretie described is also telling. In his account, they seemed less like mechanical instruments than like oracles, or even shamans. This aspect of séance practice proved one of the most contentious and difficult for Spiritists to address. The very notion of an individual communicating for a disembodied spirit engendered a set of disturbing and powerful associations, most of which resonated more with old-fashioned "superstition" than with contemporary scientism. Like demonic possession or mystical ecstasy, the medium's trance was a state observers perceived as liminal. Its strangeness could elicit a decidedly nonrational sense of awe, mystery, and danger, which mediums appear to have heightened by using unusual behavior to signal spirit possession. The writing mediums Claretie described, for example, *performed*, dramatizing their contact with the beyond by seizing the pencil abruptly, staring blankly into space, and moving in strange ways. Such behavior was necessary to the séance's success because it made the invisible spirit's presence seem real.

At the same time, the uncanny aspects of this ritual served to underline the vast distance between the séance room and the laboratory. In the name of seriousness and experimental rigor, Kardec and his followers sought to impose an ever-growing number of limits on the medium's behavior in séances: they turned away from the raps Huet had been so famous for producing, for example, and abandoned the old practice of table-moving. It proved impossible, however, to banish all traces of the uncanny. Contacting the beyond, after all, was not simply an objective experimental procedure; it was an act with powerful emotional implications. In large part because of the sense of liminality it created, the séance could induce strong feelings of fear, awe, consolation, and love. Even for believers like Kardec and Flammarion, who saw themselves as rationalists, this intensity was an important part of Spiritism's appeal. As Claretie's account indicates, even in the regulated atmosphere spirit societies sought to create, the liminal position of mediums gave them a unique, multifaceted, and potentially disruptive form of power.

The content of the spirit communications mediums produced could also seem to complicate Spiritism's aspirations to objectivity and empirical rigor. As we have seen, Kardec presented his philosophy as the first irrefutably true metaphysics, proved not only by rational analysis but also by clear-cut physical evidence, which took the form of spirit communications. The evidentiary value of these communications, however, could often be problematic. When he acted as a compiler of revelations from different mediums, Kardec sought to give the impression that the spirits espoused a uniform doctrine with a uniform voice, but in practice the souls of the dead spoke with as many voices as there were mediums. Often, the spirits agreed—for example, they reliably sang the praises of the Golden Rule.

But on other matters, they could differ in disturbing ways. Finding methods by which to explain or minimize such differences became a central concern of Spiritist groups.

Kardec attempted to resolve the numerous tensions mediums introduced by codifying Spiritist practice. With the creation of a set of rules intended to make spirit society séances serious and objective, he attempted to domesticate mediumism's unruly aspects, while simultaneously furthering his project to bring Spiritism in line with contemporary standards of respectability. The central document of this codification was the 1861 *Livre des médiums.* In it, Kardec presented a detailed collection of instructions—both his own and spirit-authored—for the conduct of séances, the behavior of mediums, and the evaluation of spirit communications. This work proved influential, establishing the dominant paradigm for French séance practice in the 1860s and creating an enduring set of norms. Kardec pursued two strategies in his effort to transform the Spiritist séance into a respectable endeavor: first, he carefully differentiated it from other, "superstitious" forms of contact with the beyond; second, he established rules to control mediums' behavior and the ideas they expressed. In the course of the séance itself, however, the medium proved a difficult creature to tame. For Kardec, philosophical coherence would come only at the price of a constant struggle against the unstinting inventiveness of the spirits.

The Serious Séance in Theory and in Practice

Those who attended Spiritist séances, Kardec wrote, needed to "remain serious in every sense of the word." Superior spirits did not waste their time attempting to communicate eternal wisdom to people in search of mere amusement. Spiritist séances, then, were not diversions for the casual thrill-seeker. The discarnate souls who appeared in them did so calmly, in decidedly unspectacular fashion; their main purpose was to instruct, not to entertain. Any hint of humor, lightness, play, or irony in a séance jeopardized its legitimacy as a vehicle for the accumulation of spiritual knowledge, Kardec believed, because such foolishness repelled all truly wise and evolved spirits.[85]

This atmosphere of seriousness required the rigorous exclusion of all practices that might evoke more popular forms of communication with the spirit world. "The medium," Kardec wrote, "must avoid everything that might turn him into a consultant, which, in the eyes of many people, is synonymous with a fortune teller."[86] Serious mediums rendered their services

[85] Kardec, *Livre des médiums,* 436.
[86] Ibid., 356.

for free, and never did so in the context of a theatrical presentation or a carnival show. They also did not appropriate practices from the fortune-teller's repertoire. For many groups, this rule proved difficult to impose. Some converts seem to have attempted to synthesize mediumism with other practices of divination and folk-healing. Parisian fortune-tellers were certainly eager to meet this new demand, judging by the confiscated handbills preserved in police archives: During the 1860s, a growing number began to list mediumism alongside the other divinatory services they provided.[87]

Kardec criticized this trend sharply in the *Revue spirite.* In his account, the "zealous apostles" who embraced fortune-tellers distributed not only Spiritist texts but also "books of *magic* and *sorcery*, or unorthodox political writings," and they often strayed from the rules of decorum in their séances. "There are some," Kardec warned his readers,

> who organize or ask others to organize meetings where they choose to study exactly what Spiritism recommends that believers avoid . . . ; there, the sacred and profane are offensively mixed; the most revered names are associated with the most ridiculous practices of black magic, including kabbalistic signs and words, talismans, sibylline three-legged tables and other accessories; some add cartomancy, palmistry, divination by reading coffee grounds, paid somnambulism, etc.—using them either as a supplement, or as lucrative products.

Spiritists, Kardec maintained, needed to avoid any practices or ideas that smacked of extremism or superstition. This new doctrine was entirely rational and hence demanded rational behavior from its adherents. Eccentric visionary pronouncements, strange acts of conjuring, card-reading, and other forms of divination characterized the atavistic approach to the beyond that Spiritism sought to replace. In Kardec's view, these deviations also had a political significance. An excessive enthusiasm for popular trappings, in his account, brought a subversive view of social relations—an interest in "unorthodox political writings"—along with it. This sort of association with the *classes dangereuses*, Kardec believed, did little to advance the cause of Spiritist respectability.[88]

The Spiritist séance as Kardec envisioned it, then, had none of the carnivalesque, sometimes transgressive flash he decried in fortune-tellers and in performing mediums like the American Davenport brothers, who had caused a sensation with glowing disembodied hands and spectral guitars during their 1865 visit to Paris.[89] Instead, in Kardec's conception, legitimate

[87] Archives de la Préfecture de Police de Paris, dr DB 215.

[88] *La Revue spirite* 6 (1863): 77. Italics in the original.

[89] *La Revue spirite* 8 (1865): 319. For a more positive account of the Davenport brothers and their French tour, see Zéphyre-Joseph Piérart, "La Vérité sur les Davenport" (Paris : Dentu, 1865).

contacts with the beyond were resolutely antitheatrical. Serious séances were dry events centered on automatic writing—by far the most ordinary-seeming of all spirit phenomena. In fact, Kardec observed, the spirit manifestations produced in an "experimental séance" could often be so unspectacular that only initiates would be able to perceive them.[90] An outsider, unfamiliar with Spiritist thought and practice, would see only a person writing, in a lighted room, surrounded by others posing predetermined questions. While Claretie's account indicates that this antitheatrical vision proved difficult to realize fully in practice, his description also shows how deeply Kardec's strictures—in particular the emphasis on automatic writing—influenced the conduct of mediums and the societies they served.

In the meetings of the Société Parisienne, Kardec maintained an atmosphere of seriousness by asserting his authority as president. To ensure the "silence and reverence" that elevated spirits required, Kardec forbade all members to speak during séances unless he granted them permission to do so. Every communication submitted to the society had to receive the president's approval before being read to the group. Most important, Kardec chose which spirits would be invoked at each meeting and what questions they would be asked. These policies helped guarantee that suitably grave philosophical issues would be addressed in a systematic way and that the more unruly aspects of the medium's inspiration would not wrest control of the proceedings from the president. Other groups readily followed the procedural model Kardec established: Both Delanne and Dozon, for example, implemented the "rule of silence" in their own meetings, forbidding those in attendance from asking unsolicited questions of the spirits who appeared.[91]

Spirit societies tended to make the impartial questioner, not the medium, the true leader of their séances. The questioner's task, Kardec maintained, was to keep in check the spirits who manifested themselves:

> Beings from the beyond must be treated carefully: [an interlocutor] must know how to use language appropriate to their nature, their moral qualities, the degree of their intelligence, and the rank they occupy; to be either dominant or submissive with them, depending on the circumstances, sympathetic to those who suffer, humble and respectful with superiors, firm with the bad and the stubborn, who only dominate those who listen to them complacently; [the interlocutor] must finally know how to formulate and methodically organize questions, in order to obtain the most explicit responses; and how to

[90] Kardec, *Livre des médiums*, 34.

[91] Kardec published his group's *règlement* in the *Livre des médiums*, pp. 458–467. For other examples of these rules in action, see Alexandre Delanne's descriptions of séances, cited in note 124, the rules outlined in Henri Dozon, "Révélations d'outre-tombe," and Kardec's descriptions of Spiritist meetings in *Voyage spirite*, 3–22.

note characteristic traits in these responses, important revelations that escape the superficial, inexperienced or casual observer.

The spirits did not always volunteer their revelations directly; sometimes an expert needed to be on hand to coax them forth. Distinguishing inferior entities from superior ones, in this situation, was a "true art," which Kardec believed he had mastered.[92] The procedure of using an authoritative, non-entranced questioner—usually a male society president—to guide the medium became common practice in Second Empire Spiritist circles. The spirit society presidents Henri Dozon, Alexandre Delanne, L. T. Houat, Jobard, Pierre Patet, and A. Lefraise, for example, all served as posers of questions but not as mediums themselves.[93]

The rules of Kardec's serious séance also established a set of norms governing the form and content of spirit communications. In keeping with his effort to limit the presence of the uncanny, Kardec asserted that the communications of elevated spirits always appeared rather ordinary at first glance. Messages that flamboyantly announced their otherworldly origin with "signs, figures, useless or childish emblems, [or with] a script bizarre, spasmodic, intentionally contorted, exaggeratedly sized, or assuming ridiculous and anachronistic forms" were almost always the work of inferior spirits and hence unworthy of close attention. The communications of good spirits, Kardec argued, bore the same marks of distinction as texts produced by learned men. Elevated spirits expressed themselves clearly, saying "good things" in "terms that absolutely exclude all triviality." Language that met this requirement was not only free of humor, irony, and vulgarity; it was also straightforward and concise. Elevated spirits indulged in rhetorical pyrotechnics and poeticism only when such techniques served to advance a suitably worthy moral message. Good spirits, for Kardec, inspired mediums to speak not with the visionary fire of prophets but with the controlled simplicity of schoolteachers.[94]

Most crucially of all, according to Kardec, good spirits never contradicted one another when addressing points of doctrine.[95] As a result, spirit communications that bore all the stylistic earmarks of elevation—restraint, concision, clarity—but nevertheless contradicted aspects of

[92] Kardec, *Livre des médiums*, 175.

[93] For Dozon, see "Révélations d'outre-tombe," and the journal that followed it, also called *Révélations d'outre-tombe* (1862–1864); for Delanne, see his memoirs in *Le Spiritisme*, esp. vol. 7 (1889): 8–10, and numerous descriptions of his wife's feats as medium (see, e.g., an article in *L'Union spirite* 1, no.2 (Jun. 8, 1865): 44–47); for Houat, see Houat, *Etudes et séances spirites*; for Jobard, see *La Revue spirite* 1 (1858): 197–199; for Patet, see Delanne in *Le Spiritisme* 8 (1890): 148–150; for Lefraise, see *La Lumière pour tous, journal de l'enseignement des esprits* 1 (Apr. 1864–Mar. 1865).

[94] Kardec, *Livre des médiums*, 251, 334, 338.

[95] Ibid., 337.

the Spiritist doctrine already accepted as true, posed the greatest interpretive challenge for discerning students of the beyond. These deceptive heterodox communications were the work of an insidious class of inferior disembodied soul—the *Esprit faux savant* (poseur spirit). These beings were not malicious, Kardec believed. They simply had not yet succeeded in overcoming the intellectual prejudices that had limited their thinking while alive and therefore passed misinformation along to unsuspecting mediums.[96] By positing the existence of this category of spirit, Kardec created an elegant way of discounting communications that met his linguistic criteria but contradicted the established precepts of Spiritist doctrine. The notion of the *Esprit faux savant,* therefore, provided a crucial safety valve—a way to de-legitimize the compelling, logical but awkwardly divergent communications some mediums produced.

In the published minutes of the Société Parisienne, Kardec provided several examples of the method he used to expose *Esprits faux savants.* In October of 1860, for instance, he devoted a general meeting of the society to the discussion of communications produced by a spirit who claimed to be Saul, King of the Jews. For some time before this meeting, the alleged Saul made regular visits to a spirit circle frequented by the medium Mlle B. The cosmology this spirit elaborated differed markedly from the one outlined in Kardec's work:

> In this young lady's circle, the spirit that communicates using [the name of Saul] has propounded an idiosyncratic system with two primary tenets: 1. The earlier a spirit's first terrestrial existence, the more enlightened it is; from which it follows that Saint Louis, for example, is less advanced than [Saul], because he has not been dead for as long a time. 2. That Spirits are only incarnated on Earth, and that these incarnations number only three—never more, never less—which is enough to advance them from the lowest degree to the highest.

Kardec announced that he found this theory to be "irrational and disproved by the facts." To prove his point, he requested that "Saul, King of the Jews" be evoked. The spirit appeared, writing through an unidentified medium, and strongly argued for the reality of his heterodox theory. Eventually, though, the self-proclaimed Saul retreated from Kardec's barrage of probing questions: "once summoned, [the spirit] failed to defend his system, but refused to admit defeat, and requested to be heard in a private séance with his usual medium." Kardec attended Mlle B's next private

[96] Kardec, *Livre des Esprits,* 48–49. See also Kardec, *Livre des médiums,* 408–409.

séance, where he continued to question Saul about his theories. Eventually, the spirit was undone by a series of questions about his notion that reincarnation could occur only on Earth. The spirit maintained that Earth was the only "*solid* globe," and that all other planets were merely "*fluidic* globes." A notion this absurd, which flew in the face of accepted scientific knowledge, Kardec maintained, irrefutably demonstrated that Mlle B's Saul was an *Esprit faux savant.*[97]

These ignorant spirits posed the greatest danger, Kardec maintained, when they felt confident that their listeners would uncritically accept the irrational ideas they espoused. The only way to ensure that good spirits frequented a circle, he wrote, was to subject every communication to the strict methods of "control" he had used on Mlle B's Saul. This process entailed rigorous philosophical analysis; corroboratory evocations of the same spirit by different mediums; and, often, direct confrontations between the expert questioner and the recalcitrant spirit as embodied by its "usual medium." Any communication that failed to withstand this scrutiny—for example, by revealing itself as "irrational" or in conflict with already accepted scientific facts—needed to be rejected out of hand. Groups unwilling to criticize heterodox or bizarre spirit communications, Kardec warned, made themselves dangerously attractive to mischievous spirits. In matters as serious as the exploration of the beyond, there was such a thing as too much tolerance.

By seeking to impose coherence on otherworldly communications in this way, Kardec and other spirit society leaders who followed his example developed a relationship between medium and authoritative male questioner that resembled the old one between *magnétiseur* and *somnambule.* Both of these pairings derived their power from social inequality: mediums, like *somnambules,* were usually either men of lower class status than their questioners or women. In the new paradigm elaborated among Spiritists, however, the relationship between entranced speaker and normally conscious questioner could acquire an adversarial edge. Where the distinguished *magnétiseur* allowed the subordinate *somnambule* to transcend his or her perceived limitations in a context of collaboration, the Spiritist questioner used the power stemming from his superior knowledge to determine whether the spirit speaking through a given medium was "good" or "bad." Here, the medium was free to write whatever communications he or she chose, but these messages became legitimate only when they received the questioner's approval. A good medium, in turn, was to accept these judgments without complaint.

[97] *La Revue spirite* 3 (1860): 331, 332. Italics in the original.

Subduing the Unruly Medium

Despite the strictures Kardec and other like-minded spirit society presidents imposed, the role of medium exerted a strong attraction for many Spiritists. The rules governing séance practice might have limited a medium's possibilities for expression, but the ability to be a vessel for the spirits nevertheless conferred a great deal of authority. Perhaps unsurprisingly, this authority seems to have appealed strongly to those who would have had precious few opportunities to enjoy a similar degree of influence in everyday life, like women and working or lower middle-class men.[98] The role of medium allowed people who would ordinarily have been perceived as "unqualified" to become respected contributors to a public intellectual discourse.

While mediumism could function as a way for the relatively powerless to make their voices heard, it is important not to lose sight of the remarkable diversity of people who appear to have served as vessels for the spirits during this period. In fact, mediums could also be aristocratic men, doctors, or literary writers.[99] In spirit societies, the status of medium was a mark of distinction—Kardec's funeral procession, for example, featured a group of mediums near the head of the assembly, immediately behind the new president, vice-president, and secretary of the Société Parisienne.[100] Mediumism, then, was remarkably democratic, a position available to all people capable of producing communications that satisfied the necessary criteria. Kardec himself asserted that mediumism had nothing to do with characteristics like gender or intelligence. It was a "natural faculty," encountered "in children, women and old men, in the learned and in the ignorant."[101]

At the same time, however, Kardec indirectly acknowledged the tendency for mediums to be women or men of lower economic status when he made his case for the otherworldly origins of the communications they produced. To prove that automatic writings were the work of disembodied intelligences, Kardec often emphasized discrepancies between texts and

[98] Discovering the class background of mediums published in Spiritist journals is more difficult than determining their sex. Important male mediums with humble backgrounds and relatively limited formal education include Pierre-Gaëtan Leymarie, a tailor; Evariste Edoux, a pharmacist (see Archives municipales de Lyon, carton I2 61, document 634); and Jean Hillaire, a *sabotier*. For Hillaire, see Auguste Bez, *Les Miracles de nos jours*.

[99] For examples, see the twelve mediums with titles or *particules* published in *La Revue spirite* between 1858 and 1869, including the prolific Alis d'Ambel; the young Dr. R—, mentioned in the *Union spirite* 1:3 (Jun. 15, 1865): 78; and the famous example of the playwright Victorien Sardou.

[100] See a letter to M. Finet from Muller, dated Apr. 4, 1869, quoted in Sausse, *Biographie d'Allan Kardec*, 88.

[101] *La Revue spirite* 2 (1859): 192.

the perceived capacities of the mediums who produced them.[102] "Much of the time," Kardec asserted, these communications from the beyond,

> particularly when they address abstract or scientific questions, entirely exceed the knowledge, and sometimes the intellectual capacity, of the medium—who is often entirely unaware of what is being written under his influence; who, frequently, does not hear or understand the question posed.[103]

Mediums, Kardec believed, often produced automatic writings at an intellectual level far above that of their conscious personalities. Under the influence of spirits, a humble, uneducated believer with no scholarly inclinations could produce sustained essays on moral philosophy; a medium who spoke only French could write phrases in English or German. The less prepossessing the medium, therefore, the more powerful the automatic writings became as indicators of spirit intervention.

This tendency to make the perceived discrepancy between writers and their communications a proof of authenticity reveals a crucial contradiction in the medium's social position. In the context of Spiritist publications, society meetings, and séances, mediumism provided a way for people to transcend the limitations of gender or class. At the same time, however, the tendency of women and less-educated men to become mediums supported Kardec's insistence that many automatic writings displayed a perspicacity that exceeded the author's personal abilities. In order to give otherworldly communications their full measure of authority, therefore, Spiritist discourse reaffirmed existing ideas about social inequality, even as it created a means by which those ideas could be subverted. A female medium's spirit communications appeared in their full grandeur, for example, only if they could be shown to exhibit an allegedly "masculine" rationality and depth of knowledge.

The case of Honorine Huet exemplifies both the power and the vulnerability of the medium's role. Huet's unsuccessful séances with the Mesmerist *commission d'enquête* did nothing to impede her career as a member of Kardec's society. By 1860, she had become the exclusive voice of Saint Louis, the Société Parisienne's spirit guide. Before this period, the builder of the Sainte Chapelle had emerged as one of the most prolific authors of

[102] Other Spiritist publications often used a similar strategy, especially when the medium was young or poorly educated. See, e.g., a communication by Mlle Dunand, "âgé d'onze ans" in *La Vérité*, May 24, 1863, 4; and one by Mme J. B., who was "complètement illettrée," published in *Le Sauveur des peuples*, Mar. 26, 1865, 3. Alison Winter describes a similar phenomenon in her study of British Mesmerism. Evaluating the veracity of the phenomena somnambulists produced, she notes, was often a matter of "observing and interpreting social characteristics." See Winter, *Mesmerized: Powers of Mind in Victorian Britain* (Chicago: University of Chicago Press, 1998), 66.

[103] Kardec, *Livre des Esprits*, IX.

spirit communications. Many mediums, both independent and affiliated with the society, produced writings with his signature. After he had begun speaking regularly through Huet, however, he made the exclusive nature of his commitment to Kardec's group quite clear:

> In many of the numerous communications that are attributed to me, another spirit has taken my name; I communicate very little outside the Society to which I have given my patronage; I like meeting places devoted primarily to me; it is there alone that I enjoy giving opinions and advice; you should be suspicious of the Spirits who often use my name.

In exchange for this loyal patronage, Kardec extended Saint Louis considerable power over the conduct of the society's meetings. The spirit, always writing through Huet, came to serve as a final arbiter, particularly in contentious or awkward situations. When disputes arose about the conduct of meetings or the resolution of disagreements, Kardec would ask Saint Louis for advice; the distinguished spirit, in turn, seems to have felt no qualms about contradicting the society president.[104]

Huet was not able to sustain this remarkable influence, however. Communications credited to her ceased appearing in the *Revue spirite* after 1861, and by 1867, she had struck out on her own with a short-lived journal called *Le Progrès spiritualiste*—for readers familiar with the French heterodox milieu, this choice of title would have indicated a clear break with Kardec.[105] Unfortunately, the exact reasons for Huet's departure from Spiritism's inner circle are impossible to determine because the minutes of the Société Parisienne ceased being published in 1861. A speech Kardec delivered in January 1862, however, sheds some light on the mystery. A certain number of mediums, he told his audience, had recently withdrawn from his group because "they wished to stand before the Society as exclusive mediums, and as infallible interpreters of the celestial powers."[106] The timing of this statement, coupled with the evidence of the society's previously published minutes, leaves little doubt that Huet was among these presumptuous mediums. While Kardec was willing to allow mediums a certain amount of influence in society meetings, he nevertheless insisted that their power be rigorously controlled, and as president, he did not hesitate to exclude those whose ambitions he deemed threatening.

Huet was not the only medium to run afoul of Kardec. Indeed, the problem of keeping mediums in their places appears to have been among the

[104] *La Revue spirite* 3 (1860): 98; for Saint Louis contradicting Kardec, see 130, 365.

[105] Though she ceased to play an active part in the *Société Parisienne*, Huet enjoyed a successful career as a medium that lasted into the 1890s.

[106] *La Revue spirite* 5 (1862): 166–167.

greatest difficulties Kardec faced as *chef du spiritisme.* Mediums, he often remarked, could be reluctant to hear their communications criticized. This paradoxical authorial pride in the absence of real authorship, in his view, posed a serious threat to the progress of Spiritism and stemmed primarily from the influence of *Esprits faux savants,* who often sought to gain influence by flattering mediums they perceived as being vain. A spirit who led a medium to believe he or she was a chosen prophet, and not merely a conduit for information, was necessarily "of bad quality," Kardec believed.[107] Mediums were passive instruments, not divinely anointed visionaries, and needed to view their communications accordingly, as bits of data for objective analysis.

Despite these warnings, however, all too many mediums, when presenting their communications to Kardec for evaluation, appear to have done so already convinced they had received wisdom from superior beings. These mediums would, of course, have been displeased to hear the *maître* proclaim that deviations from already established points of doctrine called the origin of their revelations into question. From the medium's perspective, after all, a spirit communication was the physical trace of a powerful, deeply personal experience of inspiration and transcendence. By giving a communication his authoritative stamp of approval, Kardec proved the authenticity of that moment of inspiration; if Kardec refused the communication, on the other hand, it meant the medium had mistaken an inferior spirit's fantasies for enlightenment. Kardec's further implication that such inferior communications were consequences of the medium's own "weakness and credulity" would have made his refusal doubly painful.[108]

Despite the resentment it provoked in a few cases, this authoritarianism served Kardec well.[109] By 1864, the overwhelming majority of groups devoted to spirit contacts accorded a central role to Kardec's texts and acknowledged the preeminence of the Société Parisienne. The popularity of Kardec's books, the simplicity of the ideas they contained, and their accessible style made Spiritism the philosophical lens through which the French—believers and critics alike—understood séances and the otherworldly contacts that occurred in them. The communications mediums received reflected this growing consensus by echoing the doctrine Kardec espoused. By the end of the Second Empire, Kardec's ideas had come to assume an important place in the French heterodox imagination, which they would continue to occupy well into the twentieth century.

[107] *La Revue spirite* 2 (1859): 34.

[108] Ibid., 33.

[109] See, e.g., the 1866 letter to Kardec published in Jean-Baptiste Roustaing, "Les Quatre Evangiles de J-B Roustaing, réponse à ses critiques et à ses adversaries" (Bordeaux: Durand, 1882), and the rival spiritual system elaborated in J. Roze, *Révélations du Monde des esprits, dissertations spirites obtenues par J. Roze, médium,* 3 vols. (Paris: Ledoyen et Dentu, 1862).

Spiritism and Catholicism

As his emphasis on doctrinal consistency and organizational structure indicates, the example of Catholicism was never far from Kardec's mind. In fact, in his view, Spiritism derived a crucial aspect of its legitimacy from its power to corroborate the most important teachings of the Church. At first glance this assertion seems strange because Spiritist philosophy often conflicted with crucial tenets of Catholic dogma. To Kardec and many of his followers, however, these philosophical dissonances were mere diversions from a broader religious agreement. To understand why so many advocates of the new doctrine took this position, we need to look more closely at Spiritism's simultaneously affirmative and critical approach to the Church and its teachings.

Unsurprisingly, the French clergy rejected Spiritism and responded to believers' efforts at conciliation with a barrage of polemical books, pamphlets, and sermons. These Catholic critiques, and the Spiritists' sometimes evasive responses to them, reveal the complex place Kardec's ideas occupied in Second Empire religious life. Spiritists presented their new doctrine not as a cause for revolution but as a means of regeneration, a way to reconcile what they saw as the increasingly destructive conflict between "progress" and "tradition" that had come to characterize French society. Kardec's ideas, they believed, would allow France to reclaim a lost certainty and stability while benefiting from the social justice and technological advancement modernity seemed to promise. Maintaining this hope for future regeneration, however, demanded some self-censorship and restraint from Spiritists. The only way to make the grand synthesis to which Spiritism aspired seem possible was to ignore many of the very conflicts the new doctrine claimed to resolve.

Making Spiritism Catholic

When discussing the manner in which his doctrine functioned for spiritual seekers, Kardec went to great pains to assert that it was not a religion in the conventional sense. Instead, he maintained, the new doctrine supplemented religious belief. It provided a simple proof of the immortality of the soul that "fortifies religious sentiments in general, and applies to all religions."[110] Spiritism actually did not have many of the characteristics nineteenth-century French people associated with a conventional religion. It had no liturgy, no formal churches, and no sacraments. Religion in the conventional sense, for Kardec, was essentially a matter of "conscience": Catholics, for example, went to Mass because the experience

[110] *La Revue spirite* 5 (1862): 38.

gratified a personal sense of moral obligation. Spiritism, on the other hand, was a matter of reason: Believers went to séances to discover the objective justification for the subjective dictates of their consciences.

Kardec limited this rationalistic universalism, however, by presenting his new doctrine in fundamentally Christian terms. Christ occupied a central place in the Spiritist moral universe, Kardec wrote:

> For man, Jesus is the epitome of the moral perfection to which humanity can aspire on Earth. God gave him to us as the most perfect model, and the doctrine he taught is the purest expression of God's law, because Jesus—the purest being to appear on Earth—was animated by the divine spirit.[111]

Whatever deviations from Catholic orthodoxy Spiritism might have entailed, Kardec was certainly not ready to renounce the divinity of Christ or his status as moral exemplar. Indeed, Kardec relied on Christ's authority because Christian principles furnished the a priori moral postulates on which the discursive rules of Spiritism depended. Kardec determined whether a communication was "true" or "false" by evaluating it in terms of Christian morality: Any communication that deviated from Christian moral principles was, by definition, the work of an inferior spirit. For Spiritists—who were overwhelmingly Christian—this use of Christ's teachings as an a priori moral standard seemed natural and indeed served as a powerful sign of the new doctrine's metaphysical authority.

Kardec took this Christianization of Spiritism a step further by arguing that his doctrine had a particular affinity with Catholicism. Indeed, he maintained that the teachings of the spirits reinforced the Catholic Church's claim to religious primacy. Spiritism, Kardec wrote,

> is found everywhere, in all religions, but it appears yet more—and with more authority—in the Catholic religion than in all the others. In [Catholicism] we find all the important principles: Spirits of every rank, their occult and visible relations with men, guardian angels, reincarnation, disengagement of the soul from a living body, second sight, visions, manifestations of all kinds, and even tangible apparitions.[112]

This argument for the fundamental identity of Spiritism and Catholicism was premised on several shaky assumptions—the assertion that the Church taught reincarnation was particularly far-fetched. At the same time, however, Kardec made shrewd use of the new openness to tangible forms of religious experience that characterized Catholicism in this period. From the

[111] Kardec, *Livre des Esprits*, 286.
[112] Ibid., 486.

Spiritist perspective, the mystical experiences of the orthodox devout were simply spirit manifestations in a different context; the principles governing both were identical.

This acknowledgment of Catholicism's importance and the consequent effort to present Spiritism as its helpmeet appealed strongly to many believers. The majority of people attracted to Spiritism in France came from Catholic backgrounds—the teachings of the Church, even if renounced, would have influenced their sense of the form a viable religion ought to take. Catholicism, in other words, would have provided the "grammar" to which many believers would have fit the vocabulary of Spiritism. This Catholic context, therefore, played a crucial role in shaping the choices Kardec and others made when constructing their new religious system: to appear legitimate to French seekers, Spiritism had to accommodate the religious ideas those seekers already took for granted.

The communications Spiritist mediums produced indicate how this accommodation worked. Messages from the beyond published and praised as particularly "sublime" and "pure" tended to echo the language of the Gospels and to come from prominent figures in Catholic history, such as Saint Louis, Lamennais, and Saint Augustine, whose posthumous pronouncements in favor of reincarnation figured prominently in the *Livre des Esprits.* The authority of these names would have reassured people approaching the new doctrine from a Catholic perspective, and would have appeared to lend credence to Kardec's sometimes disingenuous assertions that only a "difference in name" separated the teachings of the spirits from those of the Church.[113]

The Catholic Critique of Spiritism

In fact, however, a considerable gulf separated Spiritism from Catholicism, as clerical critics frequently observed. In 1861, when the doctrine emerged as a force in French religious life, a steady stream of books, articles, and pamphlets attacking Spiritism from a Catholic point of view began to appear. At the same time, priests in towns with significant Spiritist populations sermonized against Kardec's ideas, frequently using the new Catholic tracts on the subject as inspiration.[114] The Church did not take official action against these new ideas, however, until 1864, when their popularity had become too widespread to ignore. On April 20, the Vatican issued a decree placing works by Kardec and other Spiritist authors on the

[113] Ibid.

[114] In the anonymous pamphlet "Sermons sur le spiritisme prêchés à la cathédrale de Metz les 27, 28 et 29 Mai 1863 par le R.P. Letierce de la Compagnie de Jésus, réfutés par un spirite de Metz" (Paris: Didier, 1863), 44, for example, the author notes that Père Letierce relied heavily on the writings of the Jesuit Père Nampon.

Index, the Church's list of forbidden books. The texts were condemned *ex regulo IX indicis*, which forbade "all books and writings that discuss superstitious practices."[115]

The burgeoning French Catholic anti-Spiritist literature probably influenced the Vatican's condemnation. These tracts, mostly written by Jesuits, built on the official ecclesiastical critiques of spiritualism issued in the wake of the 1853 séance vogue, but added new arguments specifically intended to refute Kardec's ideas. During this period, Catholic anti-Spiritist authors borrowed liberally from one another, generally making similar arguments but varying their style to suit the intended audiences—which ranged from learned bishops to local *curés* and ordinary laypeople. The critique of Spiritism that emerged from these texts emphasized four points: (1) the problem of authority in the evaluation of spirit communications; (2) the impossibility of assimilating the idea of reincarnation into Christian theology; (3) the essentially diabolical nature of the séance; and (4) the threats the new doctrine allegedly posed to French society as a whole. At the same time, these texts addressed a more delicate issue: the difference between the phenomena of Spiritism and the miracles at the heart of the new surge in Catholic piety.

One of Spiritism's primary flaws, according to Catholic critics, was the way in which it justified its use of spirit communications. Kardec, these writers observed, freely admitted that mischievous, underevolved spirits often interfered in séances.[116] Given the deviousness of some of these inferior spirits, Catholic critics observed, the task of separating revelation from fantasy could prove impossibly difficult. The Jesuit Ambroise Matignon, a liberal and prominent writer for the journal *Etudes théologiques*, made this argument quite forcefully in his attack on Spiritism, which he presented as a dialogue between a Spiritist and a theologian.[117] When verifying the identity of a living being, the theologian argued, "I can study his actions, follow all his conduct." Spirits, on the other hand, made themselves known only by fragmentary "signs," which lent themselves to easy forgery. The insufficiency of this evidence appeared to contradict Spiritism's claims to transcendent certainty. In fact, Matignon asserted, the new doctrine's authority came from the blind credulity of those who trusted the spirits' assertions despite the weakness of the material proof they furnished. This uncritical

[115] *La Science et la Foi, journal religieux, scientifique et littéraire* 1, no. 7 (Feb. 2, 1865): 97.

[116] Ambroise Matignon, *Les Morts et les vivants, entretiens sur les communications d'outre-tombe* (Paris: Adrien Le Clère, 1862), 8.

[117] This dialogue was a more accessible version of the learned refutation he included as part of a previous, much longer work. See Ambroise Matignon, *La Question du surnaturel, ou la grace, le merveilleux, le spiritisme au XIXe siècle* (Paris: Adrien Le Clère, 1863 [1862]), esp. 542–565. For a short biography of Matignon, see Jean-Marie Mayeur and Yves-Marie Hilaire, *Dictionnaire du Monde religieux dans la France contemporaine*, vol. 1, *Les Jésuites* (Paris: Beauchesne, 1985), 194.

trust, in his view, demonstrated that Spiritists were guilty of the very irrationality for which they reproached orthodox Catholic theologians.[118]

In addition to resting on shaky evidentiary foundations, these critics argued, Kardec's notion of expiatory reincarnation dramatically contradicted the fundamental principles of Catholicism. Spiritist eschatology, the Abbé Jean-Baptiste Marouseau observed, transformed Adam and Eve into "a myth" and reduced original sin to "the sum of errors committed in an alleged previous existence." These redefinitions, which might have seemed innocuous to those unfamiliar with theology, Marouseau warned, undermined the basic precepts of Christianity:

> If this is the case, it necessarily results that Jesus Christ is not the Son of God, descended from Heaven to efface original sin and reinstate the lost rights of human nature; clearly there is no longer a reason for his incarnation, and the majestic edifice of religion crumbles entirely; there is nothing left, not even the sublime moral teachings you admire, because now, without the sanction of the word of life, they cease to be obligatory; nothing could be clearer. Thus, on the most fundamental point, Spiritism is the negation of Catholicism, the most complete contradiction of it.[119]

According to Marouseau and other clerical critics, Spiritist eschatology made the status of Christ problematic. If a universal *loi du progrès* ensured that every soul was engaged in a constant process of improvement from the moment of its creation, what became of Christ's role as redeemer? If Christ was not sent to "reinstate the lost rights of human nature," then why did he appear, and what was the purpose of his death? By eliminating the theological principles of original sin and redemption, Catholic critics asserted, the Spiritists left themselves unable to justify the unique authority of Christ's teachings. If Christ's role was simply that of an inspired moralist, as Spiritists often seemed to argue, then there would be no reason to accord him any more importance than Moses, Mohammed, or Buddha. This cavalier approach to the traditional theological underpinnings of Christianity, Matignon ruefully observed, brought Spiritism "quite close to religious indifference, as understood by the false philosophers" of the eighteenth century.[120]

The practices of Spiritism, these critics argued, were as anti-Catholic as its ideas. A séance was an inherently evil activity, they maintained, regardless of the intentions of those in attendance. To support this assertion,

[118] Matignon, *Les Morts*, 16, 17.

[119] Jean-Baptiste Marouseau, *Réfutation de la doctrine spirite au point de vue religieux* (Paris: Raveau-d'Artois, 1865 [1861]), 45–46.

[120] Matignon, *Les Morts*, 35.

they drew on Catholic theology, comparing spirit-summoning to canonical rituals like baptism or the sacrament of Communion. The Church taught that a ritual, if carried out in the approved manner, would be effective regardless of whether the person performing it actually believed in God and the Church. A baptism, for example, was effective even if the priest performing the rite did not believe in its power. For these writers, this concept of the intrinsic efficaciousness of ritual carried over to the séance, which, Matignon argued, could be considered "a type of diabolical sacrament."[121] Even if all those participating in the act did not believe in the Devil, their practices would nevertheless summon his agents. Since, as the Jesuit polemicist Nampon wrote, "no angel, no spirit docile to God could respond to a question illicitly posed without becoming an accomplice to and perpetrator of sin," the only way for a good Catholic to preserve his or her soul was to avoid these dangerous practices.[122]

For Catholic critics, the ideas and practices of Spiritism did more than jeopardize a few misguided souls; they also threatened the very foundations of French society. Nampon, for example, observed that the statutes of the Société Parisienne forbade anyone not in sympathy—or at least open to—Spiritist ideas from attending their meetings. This act of exclusion, he argued, made Spiritism a "secret society," with all the threatening political implications the phrase had come to assume in the wake of the Second Republic, when leftists had used clandestine organizations to coordinate popular revolt. Worse still, Spiritism was not merely a human secret society but a diabolical one that united "all the powers of Earth and Hell against the Church of Christ." Soon, Nampon warned, the demonic intelligences speaking through mediums would use their influence to wreak political as well as spiritual havoc. "Could we have to fear," Nampon asked, "that one night, when the police are not on guard, someone might get the idea to call on the spirit of Brutus to save the republic, or ask the spirit of Orsini to provide the recipe for his infernal bombs?" For these Catholic critics, Spiritism's powerful appeal to the modern mind, and the beachhead it afforded demonic invaders, made it a full-fledged "moral epidemic."[123]

While they agreed about the danger séances posed, Catholic critics of Spiritism stopped well short of forbidding conversation with the dead. Indeed, as many of these writers observed, the lives of the saints were full of visitations from deceased friends and loved ones. More strikingly still, the popular Catholic press used such dialogues as a means of dramatizing the role of purgatory and the continuing relationship between the living and

[121] Ibid., 69.
[122] Nampon, *Du Spiritisme* (Paris: Girard et Josserand, 1863), 22.
[123] Ibid., 46, 37, 40.

the dead; journals devoted to the subject had emerged as an important part of the period's revitalized religious life.[124] The evil of the séance, then, did not have to do with spirit contacts themselves, but with the circumstances under which those contacts took place. In spirit dialogues authorized by the Church, Matignon asserted, "Heaven has the initiative."[125] These visitations occurred spontaneously, as miraculous manifestations of divine grace. Spirit communications took on a completely different significance if human beings used superstitious methods to produce them, because presuming to have ghosts at one's beck and call was a sin of pride. When Spiritists held séances, they arrogantly assumed that their individual efforts could play a role that was properly God's alone; the unfortunate consequences their meetings inevitably entailed were retribution for this derogation of divine power.

The Spiritist Rebuttal

Spiritists generally countered these clerical arguments by asserting that their doctrine was fundamentally more rational than the constructions of Catholic theologians, and hence better suited to the intellectually sophisticated world of the nineteenth century. Spiritism was not the negation of Catholicism, as commentators like Marouseau argued, but a divine attempt to provide new, stronger justifications for the most important principles the Church espoused. For Spiritists, then, the role of communications from the beyond was not to invalidate Scripture but to reveal its meaning more completely. This revelation could have occurred only in the nineteenth century, Kardec believed, because previously mankind had not been ready to approach these truths directly. As a result, the real meaning of the scriptures had been hidden beneath a protective veil of allegory. Spiritism pierced this veil with direct testimony from the beyond, giving "things a clear and precise meaning that cannot be subject to any false interpretation."[126]

For Kardec and other Spiritists, the notion of eternal damnation and related concepts like original sin were the aspects of Catholic theology most in need of otherworldly rectification. In the *Livre des Esprits*, Kardec published critiques of these ideas signed by the spirits of a variety of impressive authorities, including Saint Augustine, Plato, Lamennais, and the

[124] See the popular Catholic journals *L'Echo du purgatoire* (1865–1940) and *Le Liberateur des âmes du purgatoire* (1862–1884). See also Lynn Louise Sharp, "Echoes from the Beyond: Purgatory and Catholic Communication with the Dead" (paper delivered at the thirty-first annual conference of the Western Society for French History, Newport Beach, CA, Oct. 31, 2003).

[125] Matignon, *Les Morts*, 67.

[126] Kardec, *Livre des Esprits*, 467–468.

Apostle Paul. All agreed that scriptural references to the eternal suffering of the damned were allegorical, exaggerations tailored to the needs of a more brutal age. The modern mind found such savagery repellent, and hence was prepared to accept the subtler truth. Saint Paul, for example, asserted that "the idea of Hell, with its fiery furnaces, its boiling cauldrons, could be tolerated, which is to say forgivable, in a century of iron; but in the nineteenth century, it is no more than a vain specter useful at most for frightening little children." Mankind had now attained intellectual maturity, Paul's spirit wrote, and was therefore capable of understanding that the true goal of punishment should be "rehabilitation," not the infliction of suffering for its own sake.[127] Expiatory reincarnation, because it entailed the inevitable improvement of every soul, provided the only truly just system of posthumous recompense.

Spiritists took a similar approach when rebutting Catholic assertions that the entities contacted in séances were agents of the Devil. The whole notion that Satan existed, they argued, revealed the superstitious absurdity of orthodox Catholic theology. For modern, philosophically sophisticated believers, one pamphleteer argued, the Devil was a logical impossibility, a "hypothesis" that contradicted both "the idea of divine omnipotence" and that of God's "infinite goodness."[128] Certainly, malevolent spirits existed, but their evil was not part of their essential nature. Instead, as Kardec asserted, "impure spirits" were simply less evolved; given enough time, they too would inevitably become good. This revision of traditional Catholic theology, Spiritists argued, elegantly allowed for the problem of evil while affirming God's essential justice.

The Spiritists' reforming zeal had its limits, however: Certain crucial aspects of Catholic theology struck adherents of the new doctrine as eminently worthy of preservation. Perhaps the most important of these was the idea of Christ's divinity. In a long rebuttal to Nampon's critique, which attacked Spiritism's treatment of Christ in terms similar to Marouseau's, Kardec devoted particular attention to this issue:

> You claim that Spiritism denies the divinity of Christ; where have you seen this proposition formulated explicitly? It is, you say, the consequence of the entire doctrine. Ah! If you enter the realm of interpretation, we can go further still. If we had said, for example, that Christ did not achieve perfection, that he needed the suffering of a corporeal existence to advance; that he required his passion in order to ascend in glory, you would be right, because we

[127] Ibid., 463, 464.

[128] "Sermons sur le spiritisme," 35. See also John McManners, *Death and the Enlightenment: Changing Attitudes to Death in Eighteenth-Century France* (Oxford: Oxford University Press, 1981), esp. 120–147.

> would have made him a simple mortal who could only advance by suffering—not even a *pure Spirit* sent to Earth with a divine mission. What passage have you found in which we say this? Well, you have said it, but it is something we have never said, and will never say.[129]

Tellingly, Kardec couched this defense in negative terms. He argued that he had never *denied* the divinity of Christ; at the same time, however, his philosophy left the exact nature of that divinity imprecise. To reconcile Christ's role with Spiritist eschatology, Kardec transformed the messiah from a redeemer into an inspired teacher. Jesus, in Kardec's view, was a "moral legislator" with a "Divine mission," but his death had not transformed the fundamental situation of mankind, as the Catholic Church taught. This reconception of Christ's role may have seemed innocuous at first glance—after all, Kardec still accorded him an "exceptional nature"—but it also introduced a series of philosophical inconsistencies that Catholic critics, as we have seen, pointed out with relish.[130] Kardec's strategy, in the face of these attacks, was to avoid specific discussion of this aspect of his doctrine, since any effort to clarify it would have appeared heretical to all but the most freethinking Catholics.

Kardec's strategic use of silence and imprecision was not limited to the vexed question of Christ's true nature, however. In the pages of the *Revue spirite*, he self-consciously avoided engaging in extensive critiques of Catholic teachings and urged other Spiritists to do the same. Adherents of the new doctrine, he wrote, were to focus on their own moral improvement, not on abstract questions of theology. When adepts showed less restraint, Kardec singled them out for criticism. A revealing example of Kardec's approach to those who took their critiques of Catholicism too far appeared in the *Revue spirite* early in 1863. In February of that year, a medium named M., from the town of Tonnay-Charente (Charente Inférieur), submitted a long set of spirit communications dictated by the spirit of "Jesus, son of God." The spirit provided answers to a variety of questions, including such potentially explosive ones as the following:

> 4. What should I think of communion? Are you in the host, my Jesus?
> 5. What do temporal and spiritual power have in common that prevents them from being separated?

M's Jesus urged that his answers to these questions be published. The medium included a preface declaring the epochal importance of the document, and requested that the Société Parisienne convene a "formal meeting"

[129] *La Revue spirite* 6 (1863): 173. Italics in the original.
[130] Ibid., 9.

to discuss it. Kardec's reply was prompt. He printed the questions in the *Revue*, but not their answers, and accompanied them with a ringing condemnation issued in the name of the society. Spiritism's goals, he wrote, were to fight for "the destruction of materialism and the moral improvement of mankind." The societies that propagated these ideas had no business discussing "the dogma of a particular sect." "Progress and time" would eventually "purify" all religions of "controversial dogmas," Kardec concluded, but Spiritists would do well not to hasten the process. Instead, they were to allow the superior rationality of their doctrine to speak for itself—once the truths of Spiritism had earned general approbation, the Church would inevitably accept them, just as it had accepted the Copernican model of the solar system.[131]

As his cautious approach to M's questions indicates, Kardec sought to strike an extremely delicate balance between doctrinal innovation and religious accommodation. He wanted his doctrine to make Catholicism appealing to modern believers, not to serve as the basis for a new Protestant sect. Quixotic though it appears, Kardec's aspiration to reform the Church instead of replacing it probably did much to make his ideas attractive to spiritual seekers in a French Catholic context. As Kardec and other Spiritists presented it, their doctrine served as the Church's "most fervent auxiliary," renewing the faith of those who would have otherwise been led astray by the apparent conflicts between Catholic dogma and modern science.[132] The Catholic critique of Spiritism, however, reveals the ultimate untenability of this vaunted synthesis. The very ideas and practices Kardec singled out as being crucial to Spiritism—reincarnation and the séance—were incompatible with the Church's teachings.

During his lifetime, Kardec used his charisma, and the control it gave him over Spiritist discourse, to avoid the most difficult theological problems Catholic critics raised. After his death, however, this state of discursive suspended animation proved impossible to maintain. Once believers began to address the questions Kardec had deferred or suppressed, their perception of their role in French religious and political life changed dramatically. After Kardec, Spiritist hopes for accommodation with both Catholicism and political conservatism began to give way to a more oppositional approach. This shift and its complex repercussions will be discussed in chapter 4.

[131] Ibid., 84–85.
[132] M. J. B., "Lettres sur le Spiritisme écrites à des ecclésiastiques" (Paris: Ledoyen, 1864), 8.

CHAPTER FOUR

Spiritism on Trial, 1870–1880

On the first anniversary of Allan Kardec's death—March 31, 1870—a small group of Spiritists gathered at a construction site in the Père Lachaise cemetery. The monument they had come to inaugurate was a dolmen made of rough-hewn granite slabs, sheltering a bronze portrait bust of the deceased *chef du spiritisme* and paying tribute to his past life as a Druidic sage (fig. 9). Kardec's widow, Amélie, and Pierre-Gaëtan Leymarie, the new editor of the *Revue spirite*, had conceived this project in ambitious terms. The total weight of the slabs exceeded 30,000 kilograms (33 tons), which meant that the underground chamber holding Kardec's remains had to be specially engineered. Construction encountered a number of delays, and it was not until the morning of the anniversary that a team of masons succeeded in hoisting the 6,000–kilogram slab that formed the dolmen's roof. When the Spiritists arrived, scaffolding still surrounded the monument, which had not yet received the deeply carved inscription it would eventually bear: "Birth, death, rebirth and unceasing progress: that is the law."[1]

Scaffolding dismantled, Kardec's tomb became one of the most celebrated in the cemetery, where it still stands, a stalwart of the guidebooks, commandingly positioned at the crest of a hill. Spiritists made it a place of pilgrimage, gathering there every year on March 31. Initially, these meetings were small. The first, in 1870, attracted only a few devotees; the second, held during the opening weeks of the Paris Commune of 1871, was

[1] "Discours prononcés pour l'anniversaire de la mort d'Allan Kardec, inauguration du monument" (Paris: Librairie Spirite, 1870), 5–12.

Fig. 9. The commemorative dolmen erected for Kardec in Père Lachaise cemetery, as it appeared in 1870. (Collection of the author.)

similarly modest. By the middle of the decade, however, the annual commemoration of Kardec's disincarnation had become the defining ritual of the Spiritist year, attracting adherents from all over France. In 1875, a crowd of eight hundred believers assembled at the monument, adorned it with twenty-two extravagant wreaths, and listened to six formal addresses.

To the Spiritists, this crowd was yet another indication of their movement's growing strength: A new era was beginning in France, and they believed Spiritism would play a crucial role in it. The years following Kardec's death had been tumultuous. In 1870, Napoleon III's authoritarian government collapsed after an ignominious defeat at the hands of the Prussians, who went on to besiege Paris. Then, during the spring of 1871, the city erupted in the last of the century's great popular insurrections, brutally crushed in its turn by the leaders of the newly elected National Assembly at Versailles. The Second Empire had given way to a new republic, albeit one of a decidedly conservative bent. After an overwhelming right-wing victory in the election of 1873, the new prime minister, Marshall Patrice Mac-Mahon, took this conservatism a step further, creating what supporters termed a "Government of Moral Order," closely allied with the Catholic Church and devoted to repressing the various forms of urban radicalism that had emerged so violently in 1871. Though the architects of this conservative regime initially intended to pave the way for a return to monarchy, the intransigence of the fanatical heir to the throne made a restoration impossible. By 1875, the political winds had begun to shift toward secular democracy.

Spiritists responded to this turmoil by reshaping the intellectual and political character of their movement. Kardec had considered his task above all a philosophical one: the creation of a rational, coherent system from the spirit communications mediums received. After his death, the focus of Spiritists shifted. Their goal was no longer to elaborate points of doctrine but instead to provide further empirical evidence for the conception of otherworldly intervention that gave Kardec's philosophy its authority. A growing number of studies by well-known British scientists, which inaugurated the new field of psychical research, lent credence to the hope that rigorous laboratory experiments might one day definitively prove the "spirit hypothesis." While these studies tended to deny the reality of spirits, accounting for mediumistic phenomena in psychological terms, Leymarie and his followers remained confident that an array of novel manifestations—especially spirit photography—would eventually reorient the field, making the material presence of disembodied souls impossible to refute. At the same time, Spiritists assertively linked their doctrine to the democratic left. Under Kardec, these political tendencies had been muted; under Leymarie, they intensified. The rational and scientific character of Spiritism, many of its adherents argued, made it the ideal religious foundation for a new French republic.

This enthusiasm, however, generated problems of its own. In the tense climate of the early 1870s, when France was still reeling from the popular uprisings that had occurred in Paris and other cities, Spiritism seemed dangerously subversive to some observers on both the right and the left. As the influence of the Church grew, the old Catholic concern that séances might serve as incubators of revolution took on new life. At the same time, from the increasingly antireligious perspective of the mainstream left, Spiritism came to seem like a form of destructive, atavistic superstition. Though none of the believers present at Kardec's tomb on that late March afternoon in 1875 would have known, their movement, largely ignored during the Second Empire, had begun to attract attention from the authorities. The size of the crowd had surprised the police officer patrolling the cemetery enough that he warned his superiors by telegraph, instead of filing the usual after-the-fact report. Shortly thereafter, the Paris police decided to intensify their investigation of the movement, which one detective deemed to be a social threat on a par with absinthe.[2]

The consequences of these developments crystallized during a trial that began in June 1875 and came to be known as the *procès des spirites.* In it, Leymarie, a photographer named Edouard Buguet, and an American medium named Alfred Firman were found guilty of making and selling fraudulent spirit photographs. All three of the accused received prison terms and stiff fines, and their cases attracted considerable publicity. As the epithet indicates, the *procès des spirites* was more than a simple case of fraud: In the eyes of the press, Spiritism as a whole—and with it, the idea of faith itself—was on trial. The Spiritists who testified on behalf of the accused considered their metaphysical concerns an integral part of a democratic political program, with roots in the ideas of midcentury Romantic Socialists. This political stance led Catholic journalists to present Spiritism as a telling example of the malign religious chaos a purely secular republic would foster. Left-wing journalists used accounts of the trial as an opportunity to launch ruthless attacks on religious belief in general, thereby distancing themselves from the visionary current that had been so important to midcentury republicanism. As this polemic demonstrated, belief had acquired a new political meaning under the Third Republic—one that thwarted many of the hopes Spiritists held dear.

Spiritism's New Direction

The development of Spiritism after Kardec's death owed a considerable amount to the influence of Pierre-Gaëtan Leymarie. His background

[2] Archives de la Préfecture de Police de Paris, dr. BA 1243, report dated Jun. 14, 1875, 2.

and opinions differed markedly from those of his predecessor. Kardec had been a formally educated political moderate, descended from the professional bourgeoisie; Leymarie, in contrast, was a lower-middle-class autodidact and radical. He was born in 1827 to a large family in the northern industrial city of Tulle. As a teenager he took a position as an apprentice tailor in Paris, where he quickly became active in the clandestine organizations of the democratic left. Like many others who participated in the upheavals of 1848, Leymarie fled the country to avoid prosecution after Louis-Napoleon Bonaparte's coup d'état; he did not return home until 1859, when the emperor declared a general amnesty for former revolutionaries. Once back in Paris, Leymarie married and opened a tailor shop, which never prospered and likely went bankrupt in 1871. Spiritual and political successes, however, compensated for his business difficulties. Soon after his return to France, he joined Kardec's Société Parisienne, where he played an increasingly prominent role as a writing medium. At the same time, he resumed his involvement with leftist causes: most notably, in the mid-1860s, he helped found the *Ligue de l'enseignement*, an organization that would play a crucial role in the development of French republicanism after the fall of the Second Empire.[3]

Leymarie also brought a very different temperament to Spiritism. Where Kardec had been self-consciously restrained, Leymarie was an exuberant, pugnacious activist. He had reached his position as Kardec's successor through hard work in the evenings and on Sundays since the rest of his time was devoted to the thankless business of tailoring; Spiritism had always provided him with a glorious escape from the grinding, obscure life of a simple artisan. Perhaps as a result, he had a fervent conception of the movement's social mission. Leymarie's vision of Spiritism is suggested in an 1869 communication he wrote as a medium, in which a spirit named Sonnette explained why tailoring was the most commonly represented trade among Kardec's disciples. Tailors, the spirit wrote,

> must be organized, frugal, careful, tasteful, they must be artists to some extent, and more important still, they must be patient, know how to wait, listen, smile and greet with a certain elegance; but after all these little conventions, which mean more than one might think, they must still calculate, organize their books by debit and credit, and suffer, suffer continually. In contact with men from all classes, taking note of their complaints, their secrets, their tricks, their false faces, they learn a great deal!

[3] J. Malgras, *Les Pionniers du spiritisme en France* (Paris: Librairie des sciences psychiques, 1906), 102–109.

Where Kardec had seen Spiritism as a justification for the existing social order, Leymarie's communication presented it as a force that revealed the flaws of the status quo. By sharpening awareness of the distinction between the world of the flesh and the world of the soul, Spiritism gave believers a unique ability to transcend the pretensions and paradoxes that made a tailor's "multiple life"—the doomed attempt to be both artist and businessman—such a persistent cause of suffering.[4] For Leymarie, then, Spiritism fostered political consciousness and hence could serve as a powerful tool for bringing about social change. Certainly, the movement Kardec founded had provided a form of deliverance for Leymarie himself: His new position as editor of the *Revue spirite* allowed him to abandon his old trade, close his shop, and devote himself to the life of the mind in a way that would ordinarily have been beyond the means of someone of his class.

Leymarie's rise depended on his willingness to join forces with Kardec's widow, who spent the period immediately after her husband's death working to provide the movement with a more solid fiscal basis. In the last year of his life, Kardec had applied for a *brevet de libraire* (bookseller's permit), which would have made it possible for him to supervise the publication and distribution of his writings personally. He died before finishing the application process, so his widow received the permit in her husband's place.[5] From a commercial point of view, this move was eminently sensible because Kardec's books had proved to be extraordinarily popular. To make efficient use of this legacy, Mme Kardec founded a commercial company, the Caisse générale et centrale du spiritisme, alongside the bookstore; this new organization managed the publication and distribution of Kardec's works, the *Revue spirite*, and a variety of other Spiritist books and pamphlets.

Mme Kardec's commercial initiatives, which were in part a widow's effort to guarantee herself a livelihood, unsettled many Spiritists. She had established the Caisse générale as an institution officially separate from, but organizationally closely connected to, the noncommercial and "purely scientific" Société Parisienne. According to a police report, a faction of dissident Spiritists believed that Mme Kardec's new organization "used science as a pretext, and appeared to have a pronounced commercial character that would lead it, sooner or later, to make use of the Société Parisienne des études spirites—as an instrument, and to subordinate its interests to those of the newer Society."[6] The line between the commercial society and the scientific one appeared disconcertingly blurry for many

[4] *La Revue spirite* 12 (1869): 39, 40.

[5] Archives nationales, carton F/18/1819, dr. Rivail. The permit was granted to Mme Kardec in August of 1869.

[6] Archives de la Préfecture de Police de Paris, dr. BA 1243, report dated June 14, 1875, 4.

who saw Spiritism as an enterprise founded on objective study. By 1871, however, the various factions had reached an uneasy peace, and Leymarie consolidated his position as Kardec's successor.

Leymarie's ability to reconcile these opposing groups probably stemmed from his vision of the movement's future, which reflected a broad consensus among Spiritists. Leymarie and his critics agreed that after Kardec's death, the best way to advance his ideas was to turn away from the philosophical speculation that had previously been at the center of the Spiritist enterprise. Instead, the focus would shift to the study of spectacular phenomena that occurred in séances. The *Revue spirite* announced this change of direction in January 1870. In the previous phase of the movement,

> the spirits provided numerous instructions, because it was a question of establishing a doctrine. Since this phase is now complete, Spiritist meetings will assume a different character. Mediums, having received the elements their instruction requires, are like the pupil who has finished his classes, who has no further need for elementary lessons. The spirits will only repeat themselves.[7]

The age of revelation had passed with Kardec. The new task Spiritists faced was to prove their doctrine's truth, which required the accumulation of a suitably impressive quantity of scientifically controlled, physical evidence. This self-consciously empirical approach became increasingly important to Spiritists as the decade continued. The experimental study of spirit phenomena seemed to be the next step the movement needed to take in its ongoing effort to resolve the crisis of factuality that plagued religious life.

This growing interest in the experimental study of spectacular phenomena accompanied a shift in Spiritist political discourse. During Kardec's lifetime, Spiritist tracts, the *Revue spirite,* and the numerous journals that followed its example discussed politics only in general terms. While they frequently mentioned the utopian future humanity would inevitably enjoy, they left its specific details indistinct. Kardec, for his part, had worked to dissuade his followers from engaging in political projects. The *loi du progrès* made social change inevitable, Kardec argued. Efforts to speed the process of reform through direct political action, however, constituted a "perilous path" that good Spiritists needed to avoid at all costs.[8] The stringent laws governing associations during the Second Empire probably influenced this self-conscious avoidance of politics. Kardec was acutely aware that the

[7] *La Revue spirite* 13 (1870): 5.
[8] *La Revue spirite* 5 (1862): 37.

continued existence of his Society depended on its scrupulous avoidance of "all questions involving controversies of religion, politics, and social economy."[9]

Kardec's reluctance to commit his society to an explicit political agenda did little to curtail the activism of individual Spiritists, however, particularly as the Second Empire liberalized in the late 1860s. During this period, discussions of Spiritism began to appear in venues closely associated with the political left. The Romantic Socialist Charles Fauvety, for example, began to publish articles on Kardec's ideas in his new journal *La Solidarité*, and Maurice Lachâtre, a well-known radical and freethinker, included a comprehensive definition of *spiritisme* in his *Dictionnaire Universel.*[10] Most important, Spiritists became deeply involved in the *Ligue de l'enseignement*, a society devoted to lay education and the founding of popular lending libraries. Kardec expressed reservations about the group, but it held its first meeting in Leymarie's home, and the secretary general of its Paris chapter was Emmanuel Vauchez, a convinced Spiritist.[11]

After Kardec's death, Leymarie allowed these connections between Spiritism and the political left to solidify. Increasingly, the Spiritists expressed their vision of social transformation in terms that directly referred to current events. In the pages of the *Revue*, articles about women's rights, the socialist factory in Guise founded by Jean Baptiste André Godin, and the laic religion envisioned by Charles Fauvety began to appear. French Spiritism was coming to resemble its Anglo-American and German counterparts, which tended to view progressive social reform as a logical complement to their metaphysical concerns. Spiritist writers also began to adopt increasingly anti-Catholic positions: Where Kardec had dreamed of reaching an accommodation with the Church, Leymarie and his followers dreamed of replacing it. Indeed, for many of its adherents, Spiritism appeared to be the ideal religious basis for a new, republican France. Writing in the *Revue spirite*, the republican magistrate Valentin Tournier insisted on the close ties between Spiritism and the left:

> As far as the doctrine [of Spiritism] is concerned, I do not understand how a republican could have the courage to mock it. Is there a more democratic

[9] *La Revue spirite* 6 (1863): 85. Kardec's phrasing of this injunction is the same as that used on the official forms Second Empire nonpolitical organizations filed when announcing their incorporation.

[10] Kardec published Lachâtre's definition in *La Revue spirite* 9 (1866): 31–32. For more on Lachâtre, see Jacqueline Lalouette, *La Libre pensée en France, 1848–1940* (Paris: Albin Michel, 1997).

[11] See Katherine Auspitz, *The Radical Bourgeoisie: The* Ligue de l'enseignement *and the origins of the Third Republic, 1866–1885* (Cambridge: Cambridge University Press, 1982); *La Revue spirite* 10 (1867): 79–80, 110–118. Vauchez's connection with Spiritism is discussed in J. Malgras, *Les Pionniers du spiritisme*, 191–197.

> doctrine? One better devised to encourage men to treat one another as equals and brothers? Is there one that places duty on larger, more solid, more rational foundations? Is the republican not a man of duty? How does one demand sacrifice from a person who did not exist a few days ago, and could cease to be in an instant? How does one interest him in generations past and future, if he feels no connection to them? How, further, does one teach him love of country, of humanity?[12]

Spiritism, with its "rational basis" and its Positivistic insistence on the importance of empirical evidence, was a quintessentially republican belief system, Tournier wrote, uniquely able to provide the non-Catholic foundation for social morality that so many republicans sought. Republicanism was a force for "progress in politics"; Spiritism, Tournier maintained, played a similar role in religion.[13] Each, he believed, would reinforce the other.

Despite these grand ambitions, Spiritists continued to suffer from what Leymarie called "the prejudices of the pulpit and of journalism": Kardec's death did little to stop critiques of his doctrine.[14] In 1874, for example, the archbishop of Toulouse declared that Spiritism "is nothing other than communication with demons and a return to the monstrous superstitions of idolatrous peoples," while the newspaper *La République française* denounced it as a grotesque manifestation of "brain softening."[15] This constant criticism did not discourage the Spiritists, however. They were certain that definitive scientific proof of the reality of spirit phenomena was already being accumulated and would vindicate them in the near future.

The Mixed Blessing of Psychical Research

French Spiritists looked to Great Britain as the most potent source of support for their beliefs. There, the "phenomena of spiritualism" had begun to attract scientific scrutiny from eminent and established figures. In

[12] *La Revue spirite* 18 (1875): 14. Tournier made related arguments in two widely distributed pamphlets: Valentin Tournier, "Le Spiritisme devant la raison," 2nd ed. (Paris: Librairie Spirite 1875 [1869]); and Julien Florien Félix Desprez, Archbishop of Toulouse, and Valentin Tournier, "Instruction pastorale sur le spiritisme par Mgr. l'Archevêque de Toulouse, suivie d'une réfutation par M. V. Tournier" (Paris: Librairie Spirite, 1875). The latter of these, a strongly anticlerical statement, was one of the *Caisse générale*'s biggest post-Kardec publishing ventures. According to a subsequent account of Leymarie's, the Librairie spirite printed 20,000 copies of it, to be distributed to major newspapers and political figures. At the end of his life, Leymarie would argue that the distribution of this pamphlet inspired the Buguet investigation. See J. Malgras, *Les Pionniers du spiritisme*, 95.

[13] *La Revue spirite* 18 (1875): 9.

[14] *La Revue spirite* 17 (1874): 3.

[15] Desprez and Tournier, 6; *La République française*, Oct. 2, 1874, 1.

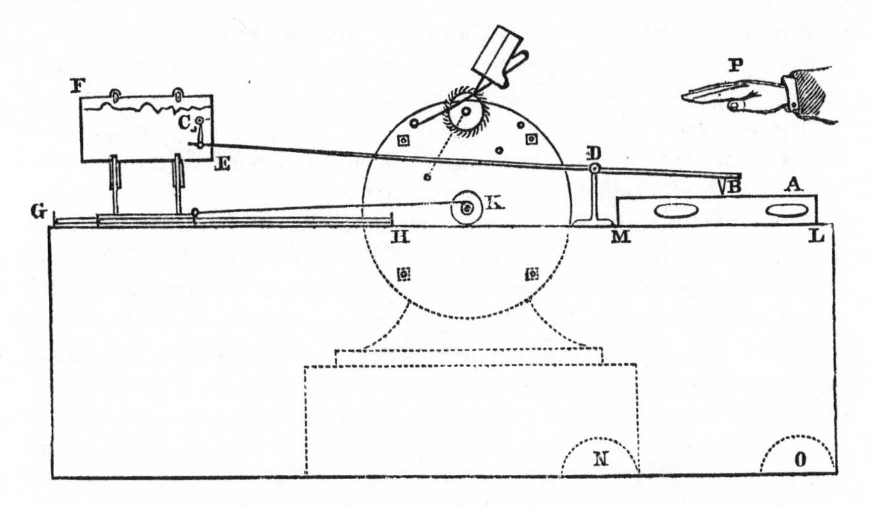

Fig. 10. The device William Crookes constructed to measure the psychic force emitted by the medium Home. When Home placed his hand in position P, Crookes found that the needle (C) steadied where it had previously been agitated. Dr. Puel published this image in an 1874 issue of his short-lived *Revue de psychologie expérimentale.* (Collection of the author.)

1870, the chemist William Crookes, discoverer of thallium and fellow of the Royal Society, published the first in a series of articles in the prestigious *Quarterly Journal of Science* describing his experiments with Daniel Dunglas Home (fig. 10). The data Crookes presented seemed to prove the authenticity of some phenomena the medium produced. A year later, the London Dialectical Society—a prestigious intellectual discussion group—issued a voluminous report on the new phenomena that declared the subject "worthy of more serious attention than it has hitherto received."[16] In 1874, the naturalist Alfred Russell Wallace, coauthor with Charles Darwin of a crucial paper on the theory of natural selection, published an ardent "defense of Spiritualism." At roughly the same time, Crookes took his engagement with the phenomena a step further, publishing several short articles in a spiritualist newspaper attesting to the authenticity of the "full form materializations" produced by the medium Florence Cook. Other major figures in English intellectual life followed the paths of Crookes and Wallace, including the Cambridge moral philosopher Henry Sidgewick, who was among the most important British thinkers of the period.[17]

[16] Committee of the London Dialectical Society, *Report on Spiritualism of the Committee of the London Dialectical Society, together with the evidence, oral and written, and a selection from the correspondence* (London: Longmans, 1871).

[17] Janet Oppenheim, *The Other World: Spiritualism and Psychical Research in England, 1850–1914* (Cambridge: Cambridge University Press, 1985).

Where in 1853 previous observers, such as Michael Faraday and François Arago, had sought to dispatch these phenomena with a single experiment, this new group approached the subject with self-conscious open-mindedness. For these scientists, the mysterious phenomena of spiritualism were complex enough to merit the creation of a new discipline, which would come to be called "psychical research."

These developments captured the imaginations of many French Spiritists, but also complicated their intellectual project. On one hand, the work of these eminent Britons seemed to prove that séance phenomena were authentic; on the other, many of the new publications stopped well short of endorsing spirit intervention and adopted a very different approach. Psychology, rather than metaphysics, provided these thinkers with their explanations. Crookes, for example, suggested that Home's mysterious powers were the product of a "psychic force" originating in the medium's own mind. The manifestations that occurred in séances, psychical researchers argued, revealed that the mind had unsuspected powers but did not necessarily prove that the living were capable of communicating with the souls of the dead.

The first publication to introduce this new approach to France was a small journal, *La Revue de psychologie expérimentale*, edited by Dr. T. Puel. The *Revue* was short-lived—appearing irregularly throughout 1874, then even more sporadically until 1876—but it included extensive translated excerpts from many early classics of psychical research, including the 1871 report of the London Dialectical Society and Crookes's studies of Home. Puel's sober-sided journal, which self-consciously downplayed the more fantastic aspects of the phenomena it described, nevertheless provided the Spiritists with powerful encouragement. The publication had all the earmarks of unimpeachable seriousness and advocated strongly for the further study of phenomena Spiritists had been exploring for over a decade.

At the same time, however, Puel emphasized that empirical proof of spirit intervention—let alone of Kardec's particular philosophy—was far from a fait accompli. Students of these strange phenomena, Puel noted, were subject to a double pressure: on the one hand from "adepts of one or another hypothetical theory that they view as incontrovertible," and on the other from "scientists" who, "misled by an exaggerated positivism," insisted that the phenomena were impossible. Puel argued that the best solution was to follow a middle way. For him, this "path of scientific truth" demanded openness to the evidence and freedom from preconceived notions.[18] Puel echoed his British counterparts by insisting that at this early stage, all a true scientist could do was observe and

[18] *La Revue de psychologie expérimentale* 1, no. 1 (1874): 49–50.

document the phenomena; he also shared the supposition that the most likely explanation would owe more to psychology than to otherworldly forces.

While the Spiritists viewed this growing scientific interest with enthusiasm, they remained ambivalent about the psychical researchers' reluctance to endorse the "spirit hypothesis." Spiritists thirsted for the prestige science could give their ideas but were both frustrated and disappointed by the caution even the most open-minded psychical researchers exhibited when seeking to explain the prodigies they had observed. This ambivalence was evident in the Société Parisienne's first attempt to use psychical research for propaganda purposes, an 1874 pamphlet that presented translated excerpts from Crookes's articles with an explanatory afterword by Leymarie. In the excerpts Leymarie had chosen, Crookes clearly expressed his preference for the hypothesis of "psychic force" rather than spirit intervention.[19]

Despite his fundamental disagreements with the conclusions Crookes reached, Leymarie nevertheless praised the British scientist's work. These rigorous experiments, supported with telling graphs and detailed descriptions of procedure, Leymarie asserted, provided Spiritists with powerful ammunition to use in their battle against skeptics:

> By devoting an hour to Mr. W. Crookes, the unconvinced, journalists, academicians, materialist or positivist doctors, will know the stark and exact value of so-called spirit phenomena. What is at issue here is no longer the doctrine of Allan Kardec and the *madness* of the adepts who accept his philosophical conclusions, but a member of the *Royal Society of London*, a chemist, a scientist of the first order, who, with a certain pride, claims to be a materialist even after these investigations.[20]

With Crookes's experiments, Leymarie believed, Spiritists had finally found a way to answer the pundits and scientists who so peremptorily asserted the falsehood of all séance phenomena. No longer would such people—arrogant victims of a prejudice they mistook for reason—be able to question the sanity of Allan Kardec's followers. Crookes, by using all the methods and tools of modern science, had proved that the phenomenal basis of Spiritism, the foundation on which the entire edifice was constructed, was *real*. These manifestations were no longer the exclusive province of amateurs and visionaries; they were now the business of the

[19] William Crookes, "Actualité, William Crookes, ses notes sur des recherches faites dans le domaine des phénomènes appelés spirites, pendant les années 1870–1873, publiées par le *Quarterly*," trans. Samuel Chinnery and Jane Jaick (Paris: Librairie Spirite, 1874), 27.

[20] Leymarie, afterword to ibid., 29. Italics in the original.

scientific establishment, which would soon be forced to retract its ill-considered refutations.[21]

For Spiritists, then, psychical research was a double-edged sword. It appeared to vindicate the ideas they held dear, but did so only partially and in a way that deprived Spiritism of its previous authority. Kardec's theories now had to compete with those advanced by members of the British Royal Society. While Leymarie sought to bolster Spiritism's intellectual prestige by constructing an image of Kardec as an eminent experimentalist in his own right, he nevertheless found himself in a tenuous position. Most disturbing, psychical research introduced a new way of understanding these phenomena: as products of the mind rather than the soul. For many more sophisticated students of these manifestations, this psychological explanation, self-consciously purged of metaphysical trappings, imparted an air of objectivity, and hence scientific legitimacy, that Kardec's Spiritism lacked.

The Promise of Spirit Photography

As Spiritists saw it, the rise of psychical research presented them with a new challenge: rather than simply proving the authenticity of séance phenomena in a general way, their task was now to establish the objective reality of spirit intervention in particular. When Leymarie discovered spirit photography in the early 1870s, he believed he had found an ideal means of accomplishing this goal. Spirit photographs, which showed spectral figures alongside living sitters, offered a forceful answer to the questions of evidence and method that psychical researchers raised when critiquing the spirit hypothesis. Nothing appeared more objective, after all, than an image captured through the camera's lens. Telekinesis, automatic writing, trance speech, and levitation could be ascribed to the medium's own powers of mind, but the ghostly images in these photographs, which sitters often identified as deceased loved ones, seemed incontrovertible evidence of disembodied souls.

For the Spiritists, this new form of evidence was quite literally a godsend, as the spirit Céphas argued in a communication published in 1873, after news of the British photographer Frederick Hudson's experiments had reached France.[22] The communication declared that these new manifestations marked a turning point in human history:

[21] See also *La Revue spirite* 17 (1874): 167; and 18 (1875): 45.

[22] See John Beattie in *The Photographic News* 17 (1873): 334, 371, 383, 406–407, 443–444; James Coates, *Photographing the Invisible: Practical Studies in Spirit Photography, Spirit Portraiture, and other Rare but Allied Phenomena* (London: L. N. Fowler, 1911); Sir Arthur Conan Doyle, *The Case for Spirit Photography* (London: Hutchinson, n.d.); J. Traill Taylor, *The Veil Lifted: Modern Developments in Spirit Photography* (London: Whittaker, 1894).

> You have been told of a time when the manifestations of Spirits would become more common and, so to speak, palpable; in such a way that skeptics will not be able to deny them, and will be obliged to yield before obvious facts. As a result of these experiments, large numbers will become sympathetic to Spiritism and will rally round it . . . Photography is a means the Spirits have at their disposal for giving irrefutable proof of their existence and their presence among you.

Spirit photography was a gift from the beyond, a new weapon for Spiritists to use against "skeptics." The communication went on to assert that while spirit photography was a "quite rare phenomenon" at present, it would soon prove to be one of the primary forms the new manifestations would take.[23] The impending flood of spirit images, indisputably objective evidence of otherworldly intervention, would silence those who questioned the rationality of Spritists' beliefs, and thereby pave the way for the establishment of the new doctrine in its rightful place as the religious and moral foundation of a future republican order.

Leymarie took the communication from Céphas to heart. As reports of spirit photographs produced in Britain became more widespread, he used the *Revue spirite* to call for similar experiments in France. When confronted with spirit photographs, he wrote, "the most skeptical are forced to surrender before the evidence." Hence, "we engage our brothers in Paris and the *départements* to continue their efforts [to produce spirit photographs]; perseverance is not only a virtue, but a duty."[24] Since no French spirit photographer had appeared on the scene, Leymarie began to sell reproductions of American spirit photographs. These images, he believed, were extraordinarily convincing pieces of evidence for the reality of the spirit world and hence would prove to be effective spurs to conversion. When confronted with a large album of these images, all showing the palpable presence of spirits, even the most tenacious "*négateur*" would have been hard-pressed to find a terrestrial explanation, Leymarie believed.

Edouard Buguet, a dark-haired, impressively bearded man of thirty-two, answered Leymarie's appeal for a local spirit photographer in the final months of 1873. Buguet, a portrait photographer who had recently arrived in Paris with his wife and two daughters, discovered Spiritism through a childhood friend he had met by chance shortly after his arrival in the city. This friend, a comic actor at the Théâtre de la Gaîté who used the stage name Etienne Scipion, introduced Buguet to Puel. The photographer became a regular guest at Puel's private séances, where, Buguet

[23] *La Revue spirite* 16 (1873): 127.
[24] Ibid., 306.

later said, "there were mediums who only did phantasmagorical things."[25] As a member of Puel's circle, Buguet met such prominent Spiritists and students of spirit phenomena as Leymarie, Camille Flammarion, the writer Louis Jacolliot, and the Russo-German psychical researcher Alexander Aksakoff. After hearing about spirit photography, Buguet began to try his own hand at it, with surprisingly successful results (fig. 11).

By the end of the year, Leymarie and Buguet had established a business relationship: The editor loaned the photographer 3,500 francs from the Caisse générale to expand his spirit photography concern. The loan was interest free and to be repaid in kind with spirit photographs. The *Revue spirite* then sold the images to its subscribers at a premium, as it had previously done with the imported American images. Leymarie also began to write extensively on Buguet in the *Revue*, emphasizing the photographer's ability to withstand the "experimental" scrutiny of various experts. Leymarie's first article on Buguet's spirit photography appeared in the January 1874 issue of the *Revue*. It described Buguet as "an artist without pretensions, full of affability, who understands his faculty for what it is—a pure and simple act of mediumism."[26]

That Buguet remained "full of affability" was indeed impressive given the scrutiny Leymarie forced the photographer to endure. After he had discovered Buguet's gift, Leymarie made it the subject of a series of experiments. The first of these, which Leymarie described exhaustively, was typical of his test sittings. Leymarie and a group of "several people" arrived at the photographer's studio; on their way, they had purchased a piece of glass to serve as their photographic negative. They marked the glass by cutting off one of its corners, then presented it to Buguet, who polished and immersed it in the usual collodion bath. Leymarie and his companions supervised this preparation and watched as Buguet took the glass and placed it in his camera, which had been thoroughly scrutinized "internally and externally." After his guests had assumed their poses in front of the camera, Buguet called for "calm and silence," made a "mental invocation," and took the picture. The print he made, again under supervision, showed "the images of spirits . . . with their faces half veiled" alongside those of the human sitters. As a final verification, Leymarie matched the removed corner to the exposed plate. Nowhere in this elaborate process, Leymarie asserted, had he or his guests detected the slightest deception.[27]

For Leymarie, Buguet's ability to withstand this intensive scrutiny proved the authenticity of his gift. After the photographer had successfully produced his images in a controlled, experimental context, they ceased

[25] Marina P. G. Leymarie, ed., *Le Procès des spirites* (Paris: Librairie Spirite, 1875), 2.
[26] *La Revue spirite* 17 (1874): 6.
[27] Ibid., 7.

Fig. 11. Buguet and the spirit of his uncle. (Image courtesy of the Archives de la Préfecture de Police de Paris.)

being mysterious curiosities and acquired the solidity of scientific fact. By describing the experimental procedure in such punctilious detail, Leymarie sought to include the reader in this process of verification: The journal's subscribers could follow the experiment for themselves and, weighing the evidence, appreciate the probity of Leymarie's conclusions. Leymarie, for his part, appears to have been convinced of the authenticity of the images he described, though subsequent events would indicate that his scrutiny was probably not as rigorous as he claimed.

Several months later, when Leymarie began inviting scientifically trained observers such as Flammarion and Alexander Aksakoff to these experimental sittings, Buguet bridled. In a letter dated April 30, 1874, he complained about a recent sitting with Flammarion and emphatically laid out the conditions for subsequent experiments:

> It is clearly understood that these men will only be simple spectators—which is to say I will allow no one to touch my products. If I have the ability to magnetize them, it is my affair, I do not want it to be as it was with Flammarion, I will only hold a single sitting. I will [only] perform with these men under these conditions. I have attended two séances held by Williams and Firman, and they have been enough to let me know what skeptics are.[28]

Buguet liked the credibility that his willingness to submit to Leymarie's constant experiments gave him, but he also seemed to be aware of the risks that such practices entailed. Flammarion, presumably, tried to examine Buguet's procedures more closely than the other experimenters had and therefore marked himself as persona non grata.[29] Buguet, cannily enough, couched his reluctance to submit to excessively vigilant scrutiny in physical terms. Skeptics, he argued, taxed his endurance by draining undue amounts of his "vital fluid," placing his health at risk.[30]

Buguet appeared to have had less difficulty convincing British observers, who studied him during a trip to London in the summer of 1874. In the course of his extended visit, he performed experiments with W. H. Harrison, editor of *The Spiritualist,* and with Stainton Moses, who made the photographer's gifts the subject of a series of articles in the spiritualist magazine *Human Nature.*[31] Buguet even won the endorsement of Crookes

[28] Quoted in Leymarie, ed., *Le Procès des spirites,* 112.

[29] The FCF includes a spirit photograph of Flammarion by Buguet.

[30] See *La Revue spirite* 17 (1874): 123. This insistence on the power of skeptics to prevent phenomena from appearing was common among perpetrators of this type of fraud. For more on fraudulent mediums and their strategies, see Ruth Brandon, *The Spiritualists: the Passion for the Occult in the Nineteenth and Twentieth Centuries* (New York: Knopf, 1983).

[31] For excerpts from Moses' articles on Buguet, see Mrs. Henry Sidgwick, "On Spirit Photographs, a Reply to Mr. A. R. Wallace," in *Proceedings of the Society for Psychical Research* 7 (1891–1892): 282–288.

himself. In a letter to Leymarie, Buguet triumphantly wrote: "Last Saturday I had an experiment with M. Crookes, of the Royal Academy [sic], it was among the most complete. This gentleman gave me all his congratulations, and promised to send me a letter from London when I have returned to Paris; he took the photographic plate with him."[32] Other observers were more skeptical: A writer for the *British Journal of Photography*, for example, could not detect any signs of fraud, but noted that Buguet's spirits always had "an unmistakable French cast of features."[33] Hesitancy of this kind did little to diminish Leymarie's enthusiasm. Buguet's images had received the imprimatur of the most eminent exponent of British psychical research and hence appeared to be the definitive proof of the "spirit hypothesis" that Kardec's followers had anticipated with such firm conviction.

Buguet's scientific success became a commercial one as well, in part because he charged twenty francs per sitting—a moderately stiff fee for the period—and in part because he understood the unspoken rules of Spiritist ritual so well.[34] In his studio, he struck exactly the right balance between the mundane and the mysterious. A shopkeeper named Caillaux described his sitting with the photographer, which was altogether typical, in an 1875 letter. After informing Buguet of the purpose of his visit, Caillaux waited in the reception area until he was called into Buguet's studio:

> [E]ntering the laboratory, I told M. Buguet: "I have come seeking to obtain the portrait of my mother or my father; I have prayed, and I hope." "That is exactly how you must act in order to obtain a result," he told me simply, "those are indeed the conditions under which one must come here; I will do what I can to obtain what you desire." We were alone; his lens was already in place; he had me sit and, indicating the point on which I was to focus my eyes, said: "let us pray." At the same time, he turned his gaze heavenward; then, taking his head in his hands, he went to lean his elbows on a small table; he remained there for several seconds. A second operation was carried out in the same manner.[35]

One of the developed photographs, Caillaux maintained, depicted his mother—he recognized her likeness, as did his brothers and other relatives. The other photograph showed a man's image, but Caillaux was not certain that it was his father, because the older man had died in Caillaux's early childhood.

[32] Leymarie, ed., *Le Procès des spirites*, 114.
[33] *British Journal of Photography* 21 (1874): 346.
[34] In Paris during this period, a set of small portrait photographs cost between twelve and thirty-five francs. The higher price was charged only by the best studios, like Nadar's. See Leymarie, ed., *Le Procès des spirites*, 68.
[35] Quoted in Ibid., 159.

Caillaux's account of his sitting is telling. Perhaps the most striking aspect of the transaction is the lack of theatrical and "mysterious" trappings. Buguet's studio looked ordinary in every respect; in these familiar surroundings, he brought only the subtlest elements of religious ritual to what was otherwise a typical photographic sitting. Having a spirit photograph taken was not a strange, mystical experience but an act in keeping with the ordinary business of life. This matter-of-factness was crucial for Spiritists, who saw it as one of the primary signs of the rationality of their belief. Nevertheless, by adding the element of prayer—along with some well-timed hieratic gestures—Buguet stirred the expectations of his sitters. Emotions helped considerably: Buguet's clients tended to feel the loss of their loved ones acutely and to wish ardently for some sort of consolation. The reassuringly direct, familiar form this consolation took probably made it doubly attractive; it showed that the spirits of the lamented dead were not only palpably present but also readily contacted, even in the most ordinary of situations.

Buguet's photographs struck a similarly shrewd balance between the everyday and the uncanny (figs. 12 and 13). The "terrestrial" component of each image was remarkably ordinary: The human sitters posed conventionally, though they usually left room in the image for a spirit to appear; the background was that of a typical middle-brow photographer's studio; and the images themselves took the familiar form of *cartes de visite.* These prosaic elements highlighted the strangeness of the ethereal spirits. Though their faces were distinct enough to be recognizable, the spirits appeared translucent and possessed only partial bodies; instead of ordinary clothing, they wore flowing robes, which sometimes appeared to envelop the human sitter. Indeed, the spectral presences in these photographs strongly echoed traditional ideas of what ghosts ought to look like. The Spiritists, for all their emphasis on the modernity of their ideas, probably found this element of representational continuity reassuring. Though they appeared in an unambiguously modern context—this kind of photography on paper was a fairly recent invention—Buguet's spirits looked the way spirits had always been supposed to look.[36] In Buguet's photographs, however, what had previously been the stuff of folklore appeared as objective fact, presented in an unmistakably up-to-date manner.

Buguet's expanding, successful practice generated a flood of enthusiastic reader contributions to the *Revue spirite.* These ranged from emotional testimonial letters to theoretical musings on the scientific significance of the new phenomenon. Taken together, the articles help reveal the factors

[36] Paper photographic *cartes de visite* first became popular in the late 1850s. For a history of their development, see Elizabeth McCauley, *A. A. E. Disderi and the Carte de Visite Portrait Photograph* (New Haven: Yale University Press, 1985).

Fig. 12. Pierre-Gaëtan Leymarie, Colonel Carré, and the spirit of M. Poiret, in a Buguet photograph reproduced in the *Revue spirite.* (Image courtesy of the Archives de la Préfecture de Police de Paris.)

Fig. 13. Mme Amélie Rivail and the spirit of her husband, Allan Kardec, as photographed by Buguet. (Image courtesy of the Archives de la Préfecture de Police de Paris.)

that made Buguet's photographs so appealing to Spiritists. His images managed to be objective yet emotionally charged, scientific yet freighted with metaphysical significance. As such, they seemed to be the veritable embodiments of the new rational faith that Spiritism sought to create.

Short letters from customers who recognized spirits in photographs became a staple of the *Revue's* coverage of Buguet. For the most part, these letters were formulaic. A typical example, written by a Parisian named M. Rousset-Guillot, described a sitting by proxy, in which Buguet produced a spirit image by rephotographing the *carte de visite* of an absent client:

> Dear Sir and Brother in Belief,
>
> Upon leaving Mézy, Mme. Bouhey told us: "Try to get the likeness of my daughter." We made an appointment with the medium-photographer for the following Monday. We summon the child every day, and she promised us a good success . . . The plate was developed in our presence, and, alongside the portrait of Mme. Bouhey, we saw a graceful child's face appear. We had two pictures taken, and on the second, there was the likeness of an old woman.
>
> The cards were sent to M. and Mme. Gaberel, our decent and worthy brothers in Mézy (Seine-et-Oise), who quickly took them to M. Bouhey; upon seeing them, he cried: "That's my daughter! That's my mother!" After some explanations, he saw the light, and thanks to spirit photography, we have now gained another believer.
>
> When my wife posed several days later, we obtained the likenesses of her grandmother and great-grandmother. Let us thank M. Buguet, and may our good guides protect him.[37]

Most of the letters printed in the *Revue* follow a similar pattern, describing first anticipation and desire to see the image of a particular relative, then satisfaction with the result. In this case, as in many other letters, there is also a conversion—usually a skeptical relative who becomes a Spiritist after recognizing the image of a deceased loved one. Finally, the letters often end with a small expression of gratitude to the spirits and to Buguet.

These testimonials appeared alongside letters that took a more technical approach, attempting to provide scientific explanations for Buguet's photographs.[38] In a typical letter of this kind, a contributor from Saint-Etienne, Jacques Clapeyron, argued that the key to the mystery of spirit photography was the notion of "fluids." These mysterious forces, which seemed by turns to be electricity or impalpable liquid, could be harnessed by the will. "Fluids," he wrote,

[37] *La Revue spirite* 17 (1874): 310.

[38] One enthusiastic reader, Augustin Boyard, even claimed to have produced spirit photographs of his own, which he submitted to the Académie des sciences. See Ibid., 341–342.

> are forces that act on matter; their specific nature is almost entirely unknown to us. Spirits endowed with will, therefore, can produce a sort of condensation or, more accurately, a modification in the constituent molecules of their ethereal bodies—can make them pass temporarily from a normal fluidic state to a more or less gaseous state capable of making an impression on the sensitized plate in the camera's chamber.

For Clapeyron, spirit photography was neither miraculous nor uncanny. It was a scientifically explicable fact. Indeed, he noted, scientists had already documented that animals were capable of perceiving types of light invisible to the human eye; in its "more or less gaseous state," perhaps the spirit's "ethereal body" was similarly present but imperceptible. Though working gamely with the conceptual tools at hand, Clapeyron admitted that his description remained imprecise and speculative. After all, he wrote, "the discoveries of Spiritist science have only just stammered their first divine lessons." More detailed analyses would have to be the work of future generations.[39]

Buguet also began to attract attention from commentators outside the circle of the *Revue spirite*. Probably the most effective publicity came from the feminist lecturer Olympe Audouard, who had been a believer in spirit phenomena since the late 1860s. Like Flammarion before her, she had been converted by a séance with Honorine Huet. For Audouard, however, the deciding phenomenon had been a spirit message from her deceased son, which included the last words the boy had whispered on his deathbed. Audouard did not discuss her beliefs publicly until 1874, when she wrote a book on the subject called *Les Mondes des esprits* (The Worlds of the Spirits). In addition to a general summary of Spiritist ideas and a collection of spirit communications she had produced, Audouard included an extensive, laudatory discussion of Buguet. For Audouard, his photographs were unquestionably authentic and deserved the close attention of "our scientists, our chemists."[40]

Buguet also began to attract the interest of the *grande presse*; its portrayals of the photographer, while more skeptical than Audouard's, were still surprisingly positive. In May of 1874, for example, the *Petit moniteur* published a front-page article on Buguet. While the reporter's ironic tone betrayed his skepticism, he admitted nevertheless that even qualified observers had failed to discover any fraud: "The most capable photographers of Paris, assisted by the most illustrious chemists, have gone to witness this prodigy. They have seen it! They have seen it; they have examined

[39] Ibid., 179–180.

[40] Olympe Audouard, *Les Mondes des esprits ou la vie après la mort* (Paris: Dentu, 1874), 33–34, 66.

all the equipment; they have posed themselves; [but] they have not been able to penetrate the secret of the photographer-medium." The reporter also quoted a letter from Bertall, a photographer similarly unable to explain how Buguet had produced his images. Though the photographer and the friends he brought to Buguet's studio "did not believe in Spiritism," they were nevertheless compelled to admit "that at least the thing was done with a great deal of *esprit*." The writer concluded his article by telling the story of Mme C., who claimed to have received the picture of a deceased relative she could identify clearly.[41] Another flattering article on the photographer appeared in the August 20 *Gazette des étrangers*. There, the reporter, Jules de Randon, described a typical experiment in Buguet's studio. As usual, Buguet allowed Randon to follow every stage of the process, which produced the image of a female spirit "whom I am quite afraid I recognize." Randon detected no signs of fraud and claimed to be convinced of the authenticity of spirit photography, which he called "a phenomenon that demands the attention of men of science."[42] Buguet himself built on this attention by commissioning a third story from the *Figaro*, which cost him 300 francs. It ran on October 15, 1874, and praised him in essentially the same terms as the *Gazette des étrangers*.

Not all the publicity Buguet inspired was positive, however. The satirical newspaper *Le Tintamarre* attacked the credulity of the journalists who had written articles favorable to the photographer. They were simply "good patsies" whom Buguet had "stuck, like two want-ads for laundresses above a public urinal." The *Tintamarre*'s writer asserted that he had received a considerably cooler welcome from Buguet than his colleagues had, largely because he refused to "swallow the pill so easily."[43] Negative publicity, however, did not appear to hurt Buguet; his business continued to thrive in the early months of 1875.

Investigation and Arrests

This increasing visibility was not entirely to Buguet's advantage because it attracted the attention of the authorities along with that of potential customers. In their efforts to publicize spirit photography, Leymarie and Buguet failed to comprehend the change in the political and social climate that had begun after the tumult of the Commune and intensified with the rise of the Government of Moral Order. Under the new regime, Spiritism,

[41] *Le Petit moniteur universel*, May 8, 1874, 1, 2.
[42] *La Gazette des étrangers*, Aug. 20, 1874, 1.
[43] *Le Tintamarre*, Aug. 30, 1874, 3.

which the authorities had once treated as an innocuous pastime unworthy of government surveillance, came to be viewed with less indulgence.

During the Second Empire, the authorities considered Spiritism to be nonpolitical and therefore innocuous. When Evariste Edoux, a Spiritist from Lyon, applied to start a journal in 1863, for example, the police officer who investigated him presented his religious sympathies as a sure sign of harmlessness. "He is not reported to have ever been involved in politics," the officer wrote, "and at this time he is up to his neck in Spiritism."[44] This kind of commitment, while perceived as eccentric, did not merit official censure: Edoux received his authorization. While similar archival evidence does not exist for Paris,[45] the case of the Davenport brothers in 1865 indicates that the same attitude held sway there. These American performers generated considerable controversy when they presented themselves as mediums instead of simple conjurors, but their claims, clearly intended to boost ticket sales, did not attract police attention or formal accusations of fraud. Instead, the emperor and empress invited the Davenports to demonstrate their powers in a command performance at Saint-Cloud. In early 1870, Mme Kardec received her *brevet de libraire* quite easily—in his pro forma report, the police investigator even assessed her husband's conduct as morally "favorable."[46] During the conflict over Kardec's succession and the creation of the Caisse générale in 1870, the dissident Spiritists approached the Paris Prefecture of Police for advice, asking the *contrôleur général* Marseille to suggest a new leader for the movement—which he did, naming Camille Flammarion as a possible successor to Kardec.[47]

With the rise of the Government of Moral Order in 1873, this tolerance diminished markedly. The new climate of political anxiety, resurgent popular faith, and growing Catholic influence placed Spiritists in an awkward position. Their movement was no longer seen as a harmless diversion, but as a potential social threat to be monitored. The investigation and prosecution of Buguet and Leymarie indicate how Spiritism's situation had changed under the new regime. In the view of the court, fraudulent spirit photography was the specific crime that had led the police to intervene, but Spiritism as a whole was culpable as well.

The first sign of trouble for Leymarie and the Spiritists appeared in mid-November 1874, when the Ministry of the Interior banned Audouard's *Les Mondes des esprits*. The Ministry gave no account of its reasons for censoring the book, but the text's enthusiastic support of Buguet probably motivated

[44] Archives municipales de Lyon, carton I2 61, pièce 634.

[45] The fires of the Commune destroyed almost all the police archives accumulated during the Second Empire.

[46] Archives nationales, F/18/1819, dr. Rivail.

[47] Archives de la Préfecture de Police de Paris, dr. BA 1243, report dated June 14, 4.

the decision: The police investigation of the photographer had begun about three weeks before the Ministry issued its order.[48] The initial report on the subject, drafted by a police officer who had happened across a copy of the *Revue spirite* full of testimonial letters, presented Buguet and Leymarie not as isolated eccentrics but as leaders of a large, well-organized new "cult": "in France, Spiritism includes about 3000 groups, each with its own president," the report maintained.[49] In the tense political atmosphere of the mid-1870s, the sheer size and coordination of the movement alarmed this officer—especially given the increasingly anti-Catholic, left-leaning nature of the ideas it espoused. The Ministry of the Interior's action against Audouard, in turn, indicates that others in the government shared his concern.

Still, in the early stages of the investigation, Guillaume Lombard, the coordinating *officier de paix*, had difficulty convincing the prosecutor's office—which the French refer to as the *parquet*—to take Buguet's case seriously. Several months after the initial report was filed, Lombard received a letter on the subject from a colleague at the Palais de Justice. Lombard's correspondent observed that "the *parquet* has already taken up this affair, which has not been pursued"; he was, however, happy to send Lombard the case dossier for his "own edification." The writer finished with an ironic bit of advice: "In closing, I strongly recommend that you contact *good spirits who will help you succeed, and provide you with good fluids to send to the courageous experimenter M. Buguet*."[50]

Building on the initial dossier, Lombard began working independently to assemble evidence against Buguet. In January 1875, he sent an officer named Geuffroy to Buguet's studio. Geuffroy posed as a customer, produced a detailed report of his sitting, and observed that the image he had received depicted no loved one he could recognize—despite its faint resemblance to his deceased father.[51] Lombard also called on the expertise of Alphonse Chevillard, a professor at the Ecole des Beaux-Arts who had written a successful anti-Spiritist pamphlet.[52] According to Chevillard, Buguet used an incompletely cleaned glass plate as the basis for his photographs—the initial, partly effaced image emerged as a spectral double exposure in the final portrait.[53] The partial cleaning would have made the plate appear fresh to observers, who would not have been able to notice the subtle traces

[48] See Archives de la Préfecture de Police de Paris, dr. BA 941, newspaper clipping dated Nov. 12, 1874.

[49] Archives de la Préfecture de Police de Paris, dr. BA 880, report fragment dated Oct. 5, 1874.

[50] Ibid., letter dated Dec. 29, 1874. Emphasis in the original.

[51] Ibid., report dated Jan. 10, 1875.

[52] Alphonse Chevillard, *Etudes expérimentales sur certains phénomènes nerveux et solution rationnelle du problème spirite, troisième édition revue, corrigée et précédée par un aperçu sur le magnétisme animal* (Paris: Dentu, 1875).

[53] Archives de la Préfecture de Police de Paris, dr. BA 880, report dated Feb. 19, 1875.

of the previous exposure. Later, using this procedure, the Prefecture's *service de la photographie* produced "spirit photographs" quite similar to Buguet's.[54]

Judging by the delay between reports, the investigation remained a low-priority project until the gathering at Kardec's tomb on March 31, 1875, which seems to have finally mobilized the *parquet* in earnest. The large crowd at the event provided disconcerting proof that Spiritism was a thriving movement, not just the peculiar concern of a small group of eccentrics.[55] Two weeks after this gathering, Lombard sent a long report to the *parquet* detailing his investigations into spirit photography. Authentic spirit photographs, images produced "by purely intellectual means," he wrote, were a physical impossibility. Nevertheless, the manner in which the photographs depicted the purported spirits made them convincing for the credulous: "in Sr. Buguet's prints, bodies and objects are 'hazy,' if you will, and reproduced in conditions that are in no way natural; as a result he is able to make others believe in the supernatural."[56] This uncanny effect, however, could be created quite easily with the use of double exposures. Lombard included several examples of the false spirit photographs produced by the Prefecture's *service de la photographie* to underscore his point.

The same day Lombard filed this report, the republican journalist Francisque Sarcey published an article in the *XIXe siècle* describing a case in which an American medium, Alfred Firman, had been exposed as a fraud by the Parisian doctor Hilarion Huguet and his wife. The Huguets, who were avid students of psychical research, suspected that Firman was less than sincere in his claims to be in direct contact with the spirits. They decided to catch him *en flagrant délit* by inviting him to hold a series of séances in their apartment. In their parlor, the Doctor built a secret cabinet in which Mme Huguet was to hide while Firman performed. At the first séance, the Huguets and their guests were content simply to watch as Firman caused "the usual phenomena" to occur—musical instruments played themselves, chairs turned upside-down, and finally the "little Indian," a black-faced, shrouded apparition about the size of a man on his knees, appeared and began to speak "childish prattle in a squeaky voice."[57] During the second séance, Mme Huguet, who had pretended to be ill, hid in the secret compartment and watched Firman execute these

[54] Ibid., report dated Apr. 15, 1875. For a good discussion of the photographic techniques used during the period, see D. v. Monckhoven, *Traité générale de photographie* (Paris: Masson, 1873), 201–248.

[55] Archives de la Préfecture de Police de Paris, dr. BA 1243, report dated Mar. 31, 1875.

[56] Archives de la Préfecture de Police de Paris, dr. BA 880, report dated Apr.15, 1875.

[57] Quoted in Hilarion Huguet, *Spiritomanes et Spiritophobes, étude sur le spiritisme* (Paris: Dentu, 1875), 19.

tricks. As the evening approached its climax, Mme Huguet jumped from her hiding place and unmasked the "little Indian," who was in fact Firman wrapped in a sheet. This exposure was clear enough that even the convinced Spiritists present willingly signed a document declaring the medium to be a charlatan—or at least to have resorted to trickery under a set of exceptional circumstances.[58] Sarcey's article inspired Lombard to broaden his investigation to include Firman, even though the medium neither advertised nor accepted formal payment for his séances.

On April 22, 1875, after having received an order from the investigating magistrate in charge of the case, Lombard and de Ballu, his principal inspector, went to Buguet's studio on the boulevard Montmartre. They presented themselves as ordinary customers. After a brief discussion with Mlle Ménessier, Buguet's young and sociable cashier—her job was to subtly probe for clues about the spirit a given customer wanted to evoke—the two police officers waited for the photographer. When Buguet entered the front room, the officers told him they hoped to obtain the portrait of Ballu's deceased father. Buguet escorted them to the posing studio, left them for fifteen minutes, and returned with a glass plate, which he then slid into the camera. Just as Buguet "gave himself over to his invocations," Lombard announced that he was a police officer and demanded to see the plate.[59]

After a brief show of resistance, Buguet handed the plate over. Shortly thereafter, Clément, a *commissaire de police*, and Lessondes, who ran the Prefecture's *service de la photographie*, arrived. They developed a print from the confiscated plate. It already "bore two quite visible images of people, whose somewhat indistinct rendering could justifiably be termed ghostly."[60] In the face of this evidence, Buguet confessed and proceeded to reveal his methods in detail. He led the officers to a second, hidden studio, in which he made the preliminary exposures. He used wooden dolls wrapped in gauze to simulate the bodies of the spirits; their faces were made from photographs pasted onto cardboard discs and affixed to the dolls by a clip. The police confiscated a box containing 240 different heads, along with the dolls and a variety of other evidence.

Two other customers were in the studio when the officers sprang their trap—Louis Darget, a cavalry lieutenant stationed in Paris, and the Comte de Bullet, a wealthy man who had been one of Buguet's most generous patrons. Bullet, terrified and embarrassed, provided the officers with Firman's address and admitted to being "a great lover of these types of séances." The officers then went to Firman's apartment at 52 rue

[58] See Leymarie, ed., *Le Procès des spirites*, 50–51.
[59] Archives de la Préfecture de Police de Paris, dr. BA 880, report dated Apr. 22, 1875, 3.
[60] Ibid., 3.

de Rome, where they found the studio of a fully equipped medium. Along with a large collection of spiritualist literature, Firman had "tambourines, trumpets, a music box, a hand-bell, etc., etc." Lombard and his fellow officers followed their *perquisition* at Firman's apartment with a trip to the offices of the *Revue spirite*, where they confiscated Leymarie's entire inventory of American and Buguet-produced spirit photographs.[61]

The police arrested Buguet, Leymarie, and Firman on May 7. All three were charged with *escroquerie*, an offense defined in the Penal Code as the earning of money "by employing maneuvers to create belief in the existence of an imaginary power, or to provoke the hope or fear of a chimerical success."[62] For the police and *parquet*, the behavior of the three accused was decidedly culpable under the law. Indeed, Spiritism itself struck them as an enterprise exclusively devoted to the creation of belief in an "imaginary power."

After their initial searches, the police began to assemble further evidence in preparation for the trial. The reports they filed indicate that more was at issue in this case than the specific charges of fraud. Spiritism, as the depredations of this unscrupulous photographer proved, the police argued, was a hazardous ideological product that needed to be contained. The *contrôleur général* Marseille, who had been so helpful to the dissident Spiritists five years before, wrote a particularly damning assessment of the movement—his views had clearly changed to reflect the new climate of opinion. In addition to a brief history of Spiritism and a description of his role as moderator in 1870, Marseille's report included a harsh assessment of the dangers of this new belief:

> From this time [1869], we nevertheless considered Spiritist doctrines to be singularly dangerous; speaking of the sheets that spread their ideas, a clever man once said: "it will soon be as necessary to prohibit these sheets from distributing their murderous legends, as it is [to prohibit] certain commercial poisons. Spiritism and absinthe cruelly ravage democracy."

Spiritism, for Marseille, was a religious epidemic that threatened irreparable harm to the French soul. Most important, Spiritism, like absinthe, actually drove people mad. Throughout the 1860s, Marseille claimed, "Bicêtre [a well-known asylum on the outskirts of Paris] grew rich with subjects who read Spiritist newspapers seriously."[63] The prosecution of Buguet, Leymarie and Firman, then, was a matter of public hygiene, an

[61] Ibid., 13, 17.
[62] Quoted in *Le XIXe siècle*, Jun. 23, 1875, 2.
[63] Archives de la Préfecture de Police de Paris, dr. BA 1243, report dated Jun. 14, 1875, 2.

opportunity for the repression of this disconcerting new belief, which, like absinthe, had been wrongly tolerated during the Second Empire.[64]

In addition to causing madness, according to the police reports, Spiritism fostered a dangerous and subversive political radicalism in its adherents. This view was clear in Lombard's report of his visit to an emergency meeting of Spiritists that took place shortly after the *perquisitions* of April 22. The group, temporarily presided over by a M. Boist, was most concerned about Leymarie, its absent leader, Lombard noted, and invoked the spirits for advice about how to help him. The Spiritists in the room clearly felt embattled. According to Lombard, Boist declared "that war was being made against the Spiritists, that the situation was dangerous for them; but that those who believe firmly in Spiritism are part of an extraordinary religion sheltered against all persecutions." In addition, Boist encouraged the Spiritists at the meeting to research the religious convictions of Delahaye, the *juge d'instruction* (investigating magistrate) in charge of assembling and organizing all evidence pertaining to the case. His beliefs, Boist stated, "could have a great influence on the Tribunal and the judgment it is called to hand down."[65]

These hints of a conspiracy to interfere with the investigation caught Lombard's attention, as did the active role taken by M. Henricy, a wood merchant from the working-class suburb of Ivry and a man of outspoken left-wing political views. Henricy, Lombard noted, had been an "orator at electoral meetings in 1873, and in the clubs in 1869, [and was] condemned to two years of prison in 1850, by the *Cour d'Assises* of the Seine, for being president of a secret society." He was now "a fanatical Spiritist" who "pushed his beliefs to the point of going to the sick in order to offer them his healing services as a medium." The presence of such dangerous characters, espousing radical political ideas to the credulous, for Lombard, indicated just how dangerous Spiritism was. At the end of his report, he asserted that "I have serious motives for thinking that this strange world is directed by a few cunning men who make a living from it, and who exploit public credulity by offering expensive medical or other consultations."[66] Since he was the leader of these "cunning men," the police viewed Leymarie as an equal partner in Buguet's fraud, even though the editor's conduct in 1874 and early 1875 indicated that he sincerely believed in the authenticity of the photographer's gifts.

[64] Marseille was not alone in his insistence that Spiritist practices caused madness. Since 1863, when the Lyon psychologist Philibert Burlet produced a study on the subject, this had been a common trope among anti-Spiritist polemicists. See Philibert Burlet, *Du Spiritisme considéré comme cause d'aliénation mentale* (Lyon: Richard, 1863). For a recent study of French psychology's various approaches to Spiritism, see Pascal Le Maléfan, *Folie et spiritisme, histoire du discours psychopathologique sur la pratique du spiritisme, ses abords et ses avatars (1850–1950)* (Paris: L'Harmattan, 1999).

[65] Archives de la Préfecture de Police de Paris, dr. BA 1243, report dated Apr. 28, 1875.

[66] Ibid.

Spiritist Reason Confronts Legal Reason

Buguet, Leymarie, and Firman were tried on June 16 and 17, 1875, before the seventh chamber of the Tribunal Correctionnel de la Seine. Both days, the courtroom was packed with spectators. Buguet and Firman fielded minimal defenses, because they had already confessed to their crimes, and repeated their admissions to the court. Their lawyers pleaded for leniency, arguing that the willingness of their victims should be considered a mitigating circumstance—Spiritists, they maintained, *wanted* to be tricked and hence deserved what they got.

Leymarie, on the contrary, assembled a much more active defense, enlisting Charles-Alexandre Lachaud, one of the most famous *avocats* of the period, to present his case.[67] Leymarie contested the police assumption that he had been Buguet's knowing accomplice. The editor insisted he had believed in the authenticity of Buguet's powers as completely as any of the photographer's satisfied clients; he was therefore as much a victim of fraud as they. To prove this assertion, Lachaud called over twenty-five witnesses to testify on his client's behalf; still others submitted letters to the *juge d'instruction.* All were convinced Spiritists who not only vouched for Leymarie's sincerity but also testified to the authenticity of the spirit photographs Buguet had made for them. While the photographer may have committed fraud occasionally, they maintained, his mediumistic powers were nevertheless genuine.

Despite this abundant testimony and Lachaud's exhaustive, sentimental closing statement, the outcome of the trial was as unfavorable for Leymarie as it was for the others accused. Buguet and Leymarie were both sentenced to one year in prison and fined five hundred francs, in addition to court costs; Firman was sentenced to six months in prison and fined three hundred francs. All three appealed the court's decision, but the Cour d'Appel, after hearing essentially the same arguments, upheld the judgment of the Tribunal Correctionnel.

The witnesses who testified on Leymarie's behalf saw this trial as an opportunity to both vouch for the integrity of an eminent co-religionist and defend the phenomena of Spiritism. As a result, the courtroom became a battleground between two incommensurable points of view—that of the Spiritists and that of the Tribunal. Each side saw itself as fundamentally rational and the other as hopelessly deluded. The Spiritists were deeply discouraged by the presiding magistrate's refusal to accept what they saw as clear evidence that at least a few of Buguet's photographs had been authentic. The *Président,* for his part, was frustrated by the obstinacy of the

[67] Maurice Garçon, *La Justice contemporaine, 1870–1932* (Paris: Grasset, 1933), 570–571; Pierre Larousse, *Grand dictionnaire universel du XIXe siècle* (Geneva: Slatkine, 1982), 10:30.

Spiritist witnesses, who refused to draw "rational" conclusions from the evidence of fraud the prosecutor presented.

For observers at the time, the most striking moment of the trial was the Comte de Bullet's testimony. Bullet's first sitting with Buguet had yielded a spirit image of the Comte's sister, who lived in Baltimore. When the *Président* questioned him about the authenticity of the picture, Bullet refused to admit that it might be false:

Q. Nevertheless you have the box of portraits here; do you not think that two women's faces could resemble one another?
A. Oh! Everyone recognized my sister, it was certainly she.
Q. Well, *monsieur*, you have been duped.
A. No.
Q. You now know Buguet's procedure? Here, look, do you not think it possible that two pictures of women could resemble one another . . . In any event, Buguet's procedure has been demonstrated.
A. I certainly saw a mannequin . . . that was shown to me; but that proves nothing: he is a medium.
Q. Yes, this doll; and beside it, you see the collection from which he took your sister's portrait.
A. *M. le juge d'instruction* showed me these heads; a mannequin; but what does that prove? He could have used them once, twice; but in my case, I summoned my sister's spirit, which appeared. For my part, I am convinced.

Here, scientific rationality seemed to have been abandoned for religious faith. Bullet believed so completely in the reality of spirit phenomena that he was willing to ignore overwhelming evidence of fraud and even to endure the humiliation of being called a "dupe" in a public trial. The *Président*'s frustration with his testimony was obvious—here was an educated member of the social elite who nevertheless seemed impervious to reasoned argument. Bullet, however, did not see his belief in these terms. Indeed, he viewed himself as a scientifically sophisticated researcher of spirit phenomena and held that his convictions were the product of systematic empirical study.[68]

The power of this new basis for belief became increasingly clear as the testimony of Leymarie's witnesses continued. Over and over again, the *Président* would present the box of cut-out heads and the doll; over and over, convinced Spiritists would refuse to accept this evidence as proof that their particular photographs had been faked. All of these witnesses—many of whom were technically trained, either as army officers or engineers—viewed their belief in scientific terms and attempted to present it to the

[68] Leymarie, ed., *Le Procès des spirites*, 27, 29.

court in a similar manner. Despite the qualifications of these witnesses, their idiosyncratic use of scientific methods and terminology had a decidedly feeble effect on the judges.

Spiritist science followed a distinctive set of rules: The Spiritists tailored their experiments to what they believed to be the elusive nature of spirit phenomena. These manifestations, they argued, were not simple, repeatable processes like chemical reactions. Instead, like Mesmeric lucidity, spirit phenomena depended on a wide variety of conditions, including the observer's emotional state, the presence of a medium, and the will of the spirits themselves. As a result, Spiritists tended to stress the importance of eyewitness accounts, and above all the *quantity* of evidence. A phenomenon may have been impossible to produce reliably in an experimental context, but if it could be shown to have occurred frequently nevertheless, Spiritists believed, its existence could be considered "proved." The Spiritist effort to take what they saw as the fundamental unpredictability of spirit phenomena into account, while retaining an essentially "scientific" approach, made their ideas seem bizarre to skeptics. Spiritist science, however, appeared perfectly rational when viewed in the context of Spiritist assumptions about the nature of the beyond.

Leymarie's testimony provided a revealing example of the dissonance between "court science" and "Spiritist science." The *Président*, in the process of asking Leymarie why he had published so many testimonials from Buguet's customers, alleged that he had printed these letters to manipulate his audience. Leymarie took a different tack, asserting instead that the testimonials, which were unsolicited, constituted a form of scientific proof in their own right, particularly since he had included them in the *Revue* only after experimentally verifying the authenticity of Buguet's photographs. The sheer number of letters proved the truth of these experimental findings: "only a crowd of testimonials makes it possible to recognize that a fact is true, that there is a strict and severe *criterium*," Leymarie told the judge. Any phenomenon supported by such an impressive number of eyewitness accounts, Spiritists argued, had to be real, even if it could not always be replicated in the laboratory. The unpredictability of spirit phenomena also served to explain those instances in which Buguet failed to produce the likeness the sitter requested. In the *Revue*, Leymarie noted, "we ceaselessly repeated . . . that M. Buguet could not always obtain a complete result."[69] Mediumism was a temperamental faculty, and no human being could presume to have the spirits at his beck and call. Hence, for Spiritists, the very inconsistency of Buguet's results was proof of their authenticity. If Buguet were a fraud, would he not have made sure every picture turned out properly?

[69] Ibid., 13.

Other Spiritist witnesses attempted to explain Buguet's powers by drawing on concepts from the physical sciences. When the *Président* questioned the reality of spirit photography in scientific terms, for example, arguing that invisible entities could not be photographed, the Spiritist Colonel Carré offered a rebuttal:

> *A.* Because you have mentioned science, allow me to mention that when you pass light through a prism, it yields the solar spectrum; at each end of it, you have invisible rays; some are perceptible because of the heat they produce, others are chemical rays; they exist, even if you do not see them . . . The rays of the sun break down in a manner that covers the spectrum, and on one end of it, you have calorific rays, which are something you cannot see, that can only be perceived with a thermometer or extremely sensitive instruments . . .
>
> *Q.* This does not disprove what I have said. We cannot, in any case, have scientific discussions here. Be seated.[70]

The judge's interruption prevented Carré from elaborating his point, but the use he made of new scientific concepts—in this case, infrared and ultraviolet light—was telling. For Carré, as for many other Spiritists, science was a revealer of the invisible. If ultraviolet and infrared light had been proved to exist, then why not spirits? Perhaps the soul was made of a substance that man had simply not yet developed the instruments to detect. Perhaps the medium, working in conjunction with a camera, *was* the necessary instrument. The world was full of invisible mysteries; only recently had science progressed far enough to begin unraveling them. Spiritism merely continued this process of innovation.

The Tribunal had little patience with such arguments. For the *Président*, Spiritism was the opposite of science, a form of "black or white magic" that harked back to a more primitive and credulous age.[71] When viewed from this perspective, the Spiritists' willingness to adopt superstitious beliefs so uncritically was a symptom of mental disease. Babinet, Chevreul, and Littré had already made this view common currency in the 1850s, and the prosecutor, Dubois, took it for granted in his statement of the facts of the case. According to him, it was all too easy to explain the spirit likenesses that so many of Buguet's clients saw in their photographs. When confronted with the blurry images Buguet presented, most of his satisfied customers succumbed to "a kind of hallucination caused by the fanaticism of their Spiritist faith." The testimony of Spiritists, no matter how socially distinguished or well educated they seemed to be, was worthless, the prosecutor argued.

[70] Ibid., 48.
[71] Ibid., 37.

Almost all the eminent people who recognized the spirits Buguet photographed for them were victims of their own "troubled imaginations."[72]

On the second day of the trial, the *Président* demonstrated that he shared this conviction. M. Bourgès, a captain in the army, traveled all the way from Marseille to testify about the authenticity of two spirit photographs he had received from Buguet. The *Président* responded in his usual manner, and then made a more general pronouncement:

> *Q.* You see this doll. Please open the box, so the witness can see the Spirits.
> *A.* One honorable person in Marseille has recognized his wife, one his daughter.
> *Q.* But that is by chance. (*Movement in the audience—Exclamations*) . . .
> *A.* I have seen people who have perfectly recognized . . .
> *Q.* There are illusions, hallucinations. You bring nothing new but your person.[73]

For the judge, the long parade of witnesses who claimed to have identified friends and relatives in their various spirit photographs simply proved the extent of human credulity. All of them had "hallucinated" the resemblances they perceived; they saw what they wanted to see, not what was there. To be a sincere Spiritist was to be insane. As a result, the Spiritists' testimony, as far as the *Président* was concerned, had no value; "hallucinations" had no place in the law.

In his closing statement, Dubois neatly summarized the opinion implicit in the *Président*'s treatment of the Spiritist witnesses:

> We all have the same impressions of this bizarre and ridiculous doctrine, and the tribunal has likely been painfully surprised to see an unconditional credulity resist all natural demonstrations, produced before witnesses in the office of the *juge d'instruction* and in this trial. It is astonishing to see this ardor for the supernatural among people who reject all religion: the substitution of unhealthy superstition for faith is as strange as it is sad.[74]

Spiritists may have believed that their doctrine joined faith and reason, but Dubois thought otherwise. He discredited all the witnesses who asserted that Buguet's photographs were authentic by opposing their testimony to the competing truth of the evidence in the trial. The witnesses may have *thought* their images were authentic, but by asserting that authenticity in the face of palpable evidence of fraud, they seemed to be in the grip of a

[72] *La Gazette des tribunaux*, Jun. 17, 1875, 1.
[73] Leymarie, ed., *Le Procès des spirites*, 57. Italics in the original.
[74] Ibid., 57.

powerful delusion. To choose to believe in the reality of a particular image, despite the evidence, became a profession of unreasoning faith—not the logical conclusion a court case demanded.

Whatever testimony willing "dupes" of this kind had to offer was beside the point. Belief in Spiritism, the prosecutor argued, while "bizarre and ridiculous," was not in itself illegal; but it also could not be invoked as a rebuttal against charges of *escroquerie*, particularly in a case as clear-cut as this.[75] In addition, the prosecutor asserted, the "effusive commentaries" that Leymarie published on Buguet's behalf were a social danger that deserved legal censure in their own right because they enflamed the devotion of the *Revue spirite*'s already credulous readership. Many of the witnesses at the trial were "unfortunates" who "have been driven to such a state of exaltation by their reading [of Leymarie's journal] that they remain convinced of the role supernatural intervention played in these photographs—despite the revelation of Buguet's fraudulent procedures, and his own confessions."[76] Anyone who willingly provoked such a dramatic recrudescence of unreason deserved to be punished, the prosecutor argued. In this way, then, he turned the witnesses who testified on Leymarie's behalf into damning evidence: Their very presence in the courtroom proved that the *Revue spirite* had created a disturbingly large coterie of fanatics. In its judgment, the Tribunal accepted Dubois's version of the case, asserting that all three of the accused had shamelessly exploited "the credulity of the idle and the poor," a crime that the testimony of the Spiritist witnesses had only made more evident.[77]

The *Procès des Spirites* in the Press

The confrontation between the Spiritist point of view and the court's resonated with many journalists. In the weeks after the trial, reports of it appeared in most major Parisian newspapers. The press was no more indulgent of the Spiritists than the *Tribunal Correctionnel* had been. Journalists who wrote more developed pieces on the trial—as opposed to those who simply strung together quotes from the transcript—tended to present the steadfast credulity of the Spiritist witnesses as the most remarkable aspect of the case. Efforts to explain and interpret these witnesses' reluctance to accept the evidence against Buguet gave rise to a polemic about the nature and importance of faith, not as a purely Catholic sentiment but as a particular way of knowing based on intuitively accepted a priori assumptions. This

[75] Ibid., 57.
[76] *La Gazette des tribunaux*, Jun. 17, 1875, 2.
[77] Leymarie, ed., *Le Procès des spirites*, 99.

controversy reveals how dramatically the political landscape of France had changed since the fall of the Second Empire, and how difficult it would be for Leymarie and his fellow Spiritists to claim the place in it they dreamed of occupying.

The polemic that the *procès des spirites* inspired, which primarily involved Catholics and republicans, was rooted both in a specific set of political circumstances and in a more general philosophical disagreement.[78] Two recent right-wing triumphs, the passage of the *loi Buffet* in July 1875, which allowed Catholic institutions of higher education to award degrees, and the well-attended inauguration of construction on the Basilica of the Sacré Cœur, led to an intensification of republican critiques of the Church.[79] The philosophical aspect of the polemic stemmed from the two camps' opposing views of human nature. For the Catholic journalists, human beings were essentially irrational creatures of their emotions. The desire to believe in supernatural forces was a human urge with the force of instinct, which they argued could never be definitively mastered. To prevent this irrationality from gaining the upper hand, it needed to be controlled by priests—who were qualified experts in the management of the human faith instinct. The republicans, on the other hand, believed that man was an essentially rational animal. Whatever weakness for the supernatural people might have had was simply the vestige of an earlier worldview, an atavistic trait to be subdued with a good dose of laic education. Rationality, for republicans, was also the sine qua non of liberty—the only citizens who could make truly autonomous decisions were those with minds free from the categorical assumptions on which religious faith depended. For the republicans who commented on the *procès*, this free rationality was the domain of men; they thought that women, by their nature, were incapable of this kind of detachment.

In a front-page article on the trial in *L'Univers*, Louis Veuillot, the most prominent Catholic journalist of the day, argued that most nonreligious newspapers had failed to understand the true meaning of the Spiritist witnesses' obstinate credulity. Left-wing journalists used accounts of the Spiritists' testimony as an opportunity to "settle the question of the supernatural with a single word." In fact, the subject could not be so glibly dismissed, Veuillot maintained:

[78] The Bonapartist newspaper *Le Gaulois* also included extensive commentary on the case. It tended to follow the Catholic lead by stressing the essential irrationality of human beings but also argued against the systematic repression of Spiritism. The best way to solve the problems this new belief posed, according to the *Gaulois*, was to ensure that the most fanatical Spiritists were placed in insane asylums. See *Le Gaulois*, Jun. 19, 1875, 3; Jun. 21, 1875, 1; Aug. 6, 1875, 3; Aug. 7, 1875, 3. Royalist newspapers ignored the trial altogether.

[79] See René Rémond, *L'Anticléricalisme en France de 1815 à nos jours* (Paris: Fayard, 1999), 178–181; Raymond Jonas, *France and the Cult of the Sacred Heart, an Epic Tale for Modern Times* (Berkeley: University of California Press, 2000).

> This belief in the supernatural, a marvelous instinct of human nature, can be distorted but not destroyed; when attacked by enemies of the true faith, it survives and is reborn in the service of error, where it inevitably bolsters either the sermons of a dervish, or the vulgar tricks of a swindler.

The credulity of the Spiritist witnesses was in no way a sign of mental deficiency or spiritual weakness but rather a misdirected expression of humanity's instinctive susceptibility to the allure of the supernatural. This predisposition made possible the "marvelous" faith of Catholics, just as it did the eccentric certainties of the Spiritists. Faith was hard-wired into the human soul. For Veuillot, this fundamental fact of human nature gave the Church a clearly defined social function. By propagating "the true faith," it managed humanity's instinctive irrationality, a quality that no amount of education or technological advancement could eradicate. In a truly well-ordered society, therefore, Veuillot asserted, priests would be called on to put their expertise to work by serving as a kind of police force of the supernatural, exposing cases of fraud and endorsing authentic manifestations of the divine. Without this structuring influence, the ineradicable urge to believe would always play into the hands of fanatics and charlatans.[80]

Spiritism, for Catholic journalists, was a telling example of what became of the faith instinct in a society that had foolishly attempted to suppress it. Depelchin, a journalist who published in *Le Monde*, asserted that, despite the seeming incongruity, Spiritism's rise in aggressively rationalistic France made perfect sense:

> That an American, thrown out of the iron circle in which his countrymen's reason prefers to confine itself by reading Edgar Poe or Allan Kardec, or a German, in his heavy ecstasies of beer and tobacco, would imagine seeing spirits, receiving blows from their tambourines, and being given their *cartes de visite* . . . can be amusing; but that Frenchmen, hucksters for skepticism, who do not believe in the living God and for whom miracles are mere parlor tricks; that Frenchmen, so proud of their reason, would be taken in by so coarse a ploy, is not amusing at all, it is deeply distressing. . . . Here is something to make even the most sensible people reel; or at least to destroy our confidence in this arrogant human reason, so jealous of its rights, which claims to account for everything.[81]

The *procès des spirites* proved that the notion of "raison humaine" on which republicans had based their hopes for France's future was merely

[80] *L'Univers,* Jun. 20, 1875, 1.
[81] *Le Monde,* Jun. 21–22, 1875, 3.

an illusion, Depelchin argued. Even the most rigorous adherence to secular, scientific ideas did nothing to diminish the human compulsion to believe. Hence, if France did not embrace Catholicism—which provided a register of truth secular thought could never attain—it would necessarily see its cherished but feeble rationality unseated by a resurgence of grotesquely deformed and misdirected belief.

According to this view, Spiritism, with its desire to unite faith and reason on reason's terms, was exactly the religious scourge that secular, republican France deserved. Indeed, Catholic journalists argued, Spiritism was much closer to the "reasoning mania" of the secular republican left than it was to the deep-seated religious conviction of devout Catholics.[82] Writing in *L'Univers*, for example, Arthur Loth argued that the tendency of republican journalists to identify Spiritists with Catholics was quite mistaken:

> We have another question to ask the *Charivari* [a satirical newspaper with an anticlerical bent]. Since it claims that believers in miracles are so closely related to the Spiritists, let it tell us why the latter prefer to associate with freethinkers; why, for example, they are more likely to subscribe . . . to the [republican] *Rappel* than to [the Catholic] *L'Univers*.[83]

Loth, as we have already seen, was correct in his assessment of Spiritist sympathies. The Spiritists saw themselves as rational and progressive; they shared the republicans' opinions on most issues and often took strongly anticlerical positions. The disturbing irrationalism that republican journalists took to be an excess of religious faith, functionally indistinguishable from that of Catholics, Loth argued, was in fact something quite different—a harbinger of the arid and dangerous religion that would inevitably arise in France if republicans succeeded in reducing the Catholic Church's power over the human faith instinct. As Veuillot argued, the destruction of "faith in the divine supernatural" would simply turn "the people" toward "the diabolical supernatural."[84]

The "people," however, were not the only ones that France's militantly secular atmosphere had made susceptible to the charms of Spiritism. Depelchin argued that the secular French education system had also rendered the technically trained elite vulnerable to this new religion. He therefore expressed no surprise that so many of the obstinately credulous witnesses who testified on Buguet's behalf were distinguished and well-educated. Indeed, he wrote,

[82] *L'Univers*, Jun. 20, 1875, 1.
[83] Ibid., Jul. 6, 1875, 1.
[84] Ibid., Jun. 20, 1875, 1.

> A critical and calculating mind does not prevent these aberrations: *Gens credulas*, one says of philosophers, and can say of mathematicians. For one as for the other, the mind, enslaved by method, is sometimes only an over-compressed spring, which can suddenly snap back and take its revenge on reason with insane ideas like those we have just seen [in Spiritism].[85]

The narrowness that a scientific education demanded, Depelchin argued, hobbled the mind. By trying to "enslave" the aspect of their character that made them most human—the nonrational aspirations of the soul toward higher realms—scientists increased their vulnerability to the seductions of the occult. The scientifically trained witnesses who testified on Leymarie's behalf, for Depelchin, were cautionary examples of the paradoxical unreason that excessive rationalism and laic education would create. If students did not absorb the precepts of faith in a structured environment, they would be dangerously prone to adopt peculiar doctrines like Spiritism; by leaving the faith instinct untrained, laic education created an intellectual atmosphere that inevitably fostered spiritual chaos.

Republican journalists saw the question of Spiritism in very different terms. For them, the faith demonstrated by the Spiritist witnesses in the Buguet trial was hardly a typical human trait; on the contrary, it was a grotesque aberration, proof of intellectual or even physical inferiority. On the front page of *Le Petit journal*, for example, a reporter who signed himself Thomas Grimm ended his article on the trial by declaring: "Decidedly, the spirit world is the world of stupidity!"[86] In his description of the crowd at the appeal, Achille Dubuc made a similar point, albeit indirectly:

> The audience is quite peculiar; it is unquestionably composed of convinced adepts, passion visible in the set of their features. The periform crania, recessed foreheads, and acute facial angles of the majority of the spectators prove the value of Lavater's science.[87]

The Spiritists were not simply naïve or emotionally weak, they were fundamentally inferior human beings; Dubuc presented this inferiority as being visible to any trained observer. Intense faith, here, was not a matter of instinct or even intellectual predilection—it was a symptom of outright mental deficiency.

[85] *Le Monde*, Jun. 21–22, 1875, 3.
[86] *Le Petit journal*, Jun. 19, 1875, 1.
[87] *Le National*, Aug. 7, 1875, 3. Johann Kaspar Lavater (1741–1801) was a Swiss thinker widely admired in nineteenth-century France for his theories about the study of physiognomy—the connections between the character of a person and the physical appearance of his or her body.

Despite the fact that twenty out of the twenty-five professed Spiritist witnesses were men, many republican journalists sought to explain their belief in terms of gender. In general, the journalists did this by describing the audience in the courtroom, which they claimed was predominantly female. Victor Cochinat, for example, discussed the prevalence of women in the audience and the close attention they paid to the proceedings. This, he wrote, was perfectly logical: "When this enchanting sex has had nothing to do with mediums, it is already stubborn enough to try the patience of a marble saint. Imagine the tenacity in error an acquaintance with the high priests of Spiritism would provoke!"[88] By nature, women were far more disposed to ardent belief than men because they were both more emotional and more stubborn, Cochinat argued. Belief in Spiritism, therefore, was a symptom of feminine—or at least effeminate—weakness of mind.

The "stupidity" and foolishness of Spiritists, for many republican journalists, were also forms of atavism. Belief in spirits and spirit contacts seemed to these writers to be a return to the most rudimentary kind of paganism, a disconcerting exception to the general progress that seemed to characterize the period. Grimm, for example, wondered at the persistence of "supernatural beliefs of bygone centuries" in this technologically advanced age.[89] Cochinat, musing at greater length, wrote:

> I can easily explain why primitive peoples, as deficient in astronomical science as they were in coats bordered with fur, prostrated themselves before Fernand Cortez when he used a solar eclipse for his own benefit, as a means of instilling fear. . . . But I could never have imagined that in the year eighteen hundred and seventy-five, decent fathers who support their families by working; lucid landlords, respected in their regions; even an artillery colonel in the French army, would believe a photographer . . . when he told them the dead could leave the other world, pose in his studio, and assist him in his little phantom-grocery concern!

These believers, many of whom were "very wily businessmen, quite reluctant to trust others," had been seduced by superstition despite all the advantages of a modern education. It was as if they had chosen to renounce all the enlightenment and freedom of the modern age to return to an abject, servile, and primitive credulity. The irrational behavior of these witnesses could be explained only in psychopathological terms: "A gentle madness troubles them," Cochinat declared, arguing that any attempt to explain the faith of Spiritists further was futile.[90]

[88] *La Petite presse,* Jun. 19, 1875, 1.
[89] *Le Petit journal,* Jun. 19, 1875, 1.
[90] *La Petite presse,* Jun. 19, 1875, 1.

The Spiritist witnesses' efforts to frame their ideas in scientific terms clearly did nothing to convince these writers, who tended to present such Positivistic aspirations as further proof of irrationality, if they mentioned them at all. Sarcey, for example, ruthlessly mocked Colonel Carré's efforts to provide a scientific explanation for spirit photography. "This use of the trappings of chemical technology to advance an idea so prodigiously absurd," Sarcey wrote, "made the entire audience roll its eyes."[91] For Sarcey, the Spiritists' scientific talk was empty mumbo-jumbo, a laughably ineffective attempt to lend credibility to a fundamentally irrational idea. While Spiritists, both stupid and desperate to believe, might have been convinced by this charade, others were most emphatically not. Even a layman could see the fundamental ridiculousness of Carré's premise, Sarcey asserted; all the scientific speculation in the world was powerless to unseat the simple common-sense of the skeptical majority, here represented by an audience very different from Dubuc's crowd of ardent, pointy-headed believers.

Scientific pretensions aside, the faith of the Spiritist witnesses at the Buguet trial was no different from that of orthodox Catholics, republicans argued. In the *Bien public*, for example, a reporter named Valère saw the Spiritists' dogged willingness to believe, despite what seemed to be overwhelming evidence to the contrary, as a symptom of a "truly worrying situation in a portion of French society." After all, he observed, the spectacle of credulity Buguet's trial afforded had taken place

> the day after Paris witnessed the solemn placement of the first stone of a temple devoted to a purely mystical cult, a revival of the mysteries of Adonis, the modern founder of which is a poor girl subject to hallucinations, whom the Faculty of Medicine would unhesitatingly place in the care of a doctor, were she examined today.

The cult of Mary, the visions of Lourdes, and the building of the Sacré Cœur, for Valère, were no less irrational than Spiritism. All these developments, he argued, pointed to a disconcerting rise in the public's willingness to seek consolation in atavistic mystical pursuits.[92]

Auguste Vacquerie, writing in *Le Rappel*, argued that republican press coverage of the Buguet trial could have serious negative repercussions for the Catholic Church, largely because of the way in which it made belief in miracles—in any form it might take—look ridiculous. This observation, he observed, had not escaped Catholic journalists:

[91] *Le XIXe siècle*, Jun. 21, 1875, 2.
[92] *Le Bien public*, Jun. 19, 1875, 2.

> M. Veuillot understands that the mockery this trial has made of faith in spirit photographs could easily extend to the faith in other apparitions that he affirms every day. To escape this sorry pass, he attempts to use Sganarelle's aphorism: "there is firewood and there is firewood."[93]

Uncritical faith in tangible manifestations of the supernatural, Vacquerie argued, was always irrational and absurd; the institution sponsoring such faith made little difference. Hence, Catholics, when forced to compare themselves to Spiritists, could do nothing but draw spurious distinctions between identical objects, as Molière's famous character had in *Le Médecin malgré lui. Le Charivari*, in a similar vein, proposed an advertising slogan for the Church to use: "The best chocolate . . . pardon, the best supernatural is the Catholic supernatural."[94]

Vacquerie observed that Spiritists did not make these spurious distinctions. In fact, they believed in the reality of "the priests' apparitions" and used them as "their primary argument" for the existence of spirit phenomena. To prove this assertion, Vacquerie quoted a letter he had received from Colonel Carré, who proclaimed himself an "apostle of Spiritism." Carré, Vacquerie wrote, "tells us he has drawn his belief from the Bible," which, he noted, "is full of spirit apparitions." Indeed, "the apostle Carré" argued that his faith in these apparitions was fundamentally the same as a devout Catholic's would have been. In a passage Vacquerie quoted directly, Carré took his comparison with Catholicism several audacious steps further: "What can Buguet's recantation do to believers? Did Saint Peter not renounce the Savior three times, under the same circumstances?" While Vacquerie made sure to note that he did not share Carré's high opinion of Buguet, he also observed that his sentiment was quite edifying because it proved "*L'Univers* believes no more strongly in its apparitions . . . than the Spiritists believe in theirs." The Church's efforts to distinguish itself from Spiritism by arguing that its apparitions were somehow superior and more legitimate was, for Vacquerie, purest hypocrisy.[95]

Other, more daringly anticlerical papers developed the analogy between Spiritism and Catholicism even further, arguing that if Buguet had been found guilty of fraud then the Church could be as well. On June 23, 1875, for example, *Le Charivari* published a skeptical article on the stigmatic Louise Lateau, who claimed to sleep for weeks on end without either eating or excreting. She would recover consciousness only if ordered to do so by her bishop. "Pious journals present these intestinal fantasies as

[93] *Le Rappel*, Jun. 22, 1875, 2.
[94] *Le Charivari*, Jun. 24, 1875, 2.
[95] *Le Rappel*, Jun. 22, 1875, 2.

miracles of divine power," the journalist E. Villiers wrote. He then recounted a story told by a priest, who claimed that even though he had been delegated secretly by the bishop, he had been unable to awaken the sleeping ecstatic. When he invoked the bishop's name directly, on the other hand, Lateau regained consciousness immediately. The fraud in this case, Villiers asserted, was clearly evident, provoking him to wonder: "Should the *police correctionnel*, who are currently attacking spirit photographers here, go take a little look at this case of providential constipation?"[96] Villiers' tone was ironic, but the point was serious. If the state's job was to protect its citizens from frauds that spurred excessive faith, the physical manifestations of Catholicism deserved the same treatment as those of Spiritism.

Sarcey drew another parallel, asserting that Buguet's case was eerily similar to one that had recently been tried in Grenoble, where a young woman was found guilty of staging fraudulent apparitions of the Virgin Mary at La Sallette:

> There was artifice in that case, as there is trickery here. In Grenoble there were obstinate believers—just like the many you have seen in this hearing—whom the *Président* could not convince of their error. In both cases, things are absolutely identical.

The clerical witnesses who insisted on the reality of the apparitions at La Sallette were fanatics blinded by their irrational beliefs, just like the Spiritists, Sarcey argued. He admitted only one difference between the two cases of spiritual fraud: "one case concerns a legal religion funded by the budget, the other an independent faith reduced to its own resources." Fanaticism was always the same, whatever doctrine inspired it, Sarcey argued. Faith could always lead people to behave irrationally, even when it took an officially sanctioned form.[97]

Perhaps because of the government's stake in the French Catholic Church, Sarcey observed, the court had done nothing to stop those who sought to profit form the gullibility of the faithful at La Sallette. Sellers of bibelots and holy water—to say nothing of ecclesiastical solicitors of donations—were allowed to continue as they had before the trial. Buguet's similar attempt to take commercial advantage of the faithful, in contrast, provoked a sharp official response. This inequality, Sarcey argued in a mock plea to the court on Buguet's behalf, indicated a patently unjust double standard. After all, the court in Grenoble had decided to leave the water-sellers and their ilk alone because

[96] *Le Charivari,* Jun. 23, 1875, 1.
[97] *Le XIXe siècle,* Jun. 23, 1875, 1.

> Simple folk must have the freedom to believe and hope as they please. These are decent, suffering people; throngs of them go on pilgrimages, believe themselves cured, and return home delighted to have made the journey. . . .
>
> Our cause is the same, *messieurs*. These are Spiritists who mourn a dead friend, and who would like to gaze upon his likeness. . . . M. Buguet . . . gives it to them by means no more strange or culpable than the ones this lady used to represent the Holy Virgin. They return home overjoyed. Is twenty francs too much to ask?

The real problem, Sarcey argued, was not the spiritual frauds like Buguet, but the stupid yet well-meaning people who proved so willing to believe him. "It is not by attacking charlatans that one eliminates them from society;" Sarcey wrote, "it is by enlightening their dupes, and, as you see, that is no easy task."[98]

Difficult as it might have been, the project of enlightening potential "dupes," for these journalists, was essential to the well-being of French society. By allowing religious faith to flourish unchecked, the government had created a situation that placed the founding principles of republicanism—individual freedom above all—in danger. Action, therefore, had to be taken. In *Le Bien public*, Valère insisted that the epidemic of blind faith he saw overtaking France, whether it was Spiritism or the cult of Mary, posed a serious social threat: "More than ever, this 'freedom of error,' so just in theory, so rightfully dear to philosophers, seems to be a danger, because it can finish by attacking the freedom of reason."[99] Deep religious conviction, for Valère, was dangerous because it robbed believers of their freedom—that most important aspect of democratic citizenship. Once a person had succumbed to faith, he became nothing more than an automaton, impervious to all logical argument and hence unable to participate in the political life of a republic.

The primary means of protecting society from the danger of faith, many republican journalists argued, was education. Vacquerie, for example, asserted that there was only one way to combat "all exploitation of lies, of superstitions," whether Spiritist or Catholic:

> What we need to fight all this is a great deal of light. What we need is truly free education. Alas! Education! We are in the process of handing it over to the clergy, whose faith in the apparitions of Lourdes and La Sallette makes it a sorry vanguard in the battle against the apparitions of Spiritism.[100]

[98] Ibid., 2.
[99] *Le Bien public*, Jun. 21, 1875, 1.
[100] *Le Rappel*, Jun. 22, 1875, 2.

Universal laic education, Vacquerie argued, was by far the best method available for purging French society of its vestigial irrationality. The Church, by teaching impressionable children to have faith, was in fact carrying out a socially destructive act, producing nothing but future "dupes." Forbidding the Church to teach the young would create "truly free education" by producing a generation capable of using its reason autonomously, no longer in thrall to the a priori principles Catholics insisted on espousing. Freedom, for Vacquerie and many of the other journalists who shared his political views, was unquestionably a secular good—religious faith in any form served only to limit it.

Spiritists Respond

Spiritists found this polemic distressing. Every camp, from right to left, appeared to misconstrue their ideas, and the Spiritists saw their practices become a kind of comic shorthand for illusion. In the weeks following the trial, for example, *Le Charivari* printed cartoons that used the trope of spirit photography to poke fun at everything from skinny dandies to Bonapartist propaganda.[101] A variety of Spiritist defenses appeared in short order. Dr. Hilarion Huguet, whose wife had testified against Firman in the trial, and L. Legas, an active Spiritist, both produced pamphlets. Marina Leymarie, wife of Pierre-Gaëtan, published a book that included a partial transcript of the trial, along with supplementary evidence, an explanatory afterword, and a large collection of letters from prominent Spiritists written in her husband's defense.[102] The *Revue spirite*, which continued to appear during its editor's absence—Leymarie had fled to Belgium to avoid his prison sentence, which was later commuted—also included many articles responding to specific aspects of the press critiques.

For these commentators, the characterization of the Spiritist witnesses in the press as "fanatics of their belief" rankled most. After all, Legas argued in his pamphlet, the witnesses' certainty about the likenesses in their photographs had nothing to do with religious sentiment:

> The question here, it seems to us, is not the relative value of a doctrine; in which case prejudice, fanaticism, early education, credulity, could explain the obstinacy of these witnesses. No! None of that is under discussion, and

[101] See *Le Charivari*, Jun. 28, 1875, 3 and Jul. 8, 1875, 3.

[102] Marina Leymarie omitted the prosecutor's statement of the facts along with some of the less flattering parts of his *réquisitoire*. For these documents, see *La Gazette des tribunaux*, Jun. 17, 1875, 579–581 and Jun. 18, 1875, 583.

> the Spiritist doctrine itself is not at issue. The problem is simpler, it . . . concerns a material fact to which only the senses can testify. Do the photographs the witnesses obtained exactly reproduce the likenesses of the people evoked? They say: Yes.

The steadfastness with which so many witnesses insisted on having perceived likenesses in Buguet's photographs was not a matter of faith, education, or innate predisposition. Their certainty was a straightforward response to unambiguous empirical evidence. Here, Legas argued, conviction was a question of reason pure and simple. The witnesses responded to Buguet's photographs as objective depictions of people they had known; nothing could have been less fanatical than their brave willingness to affirm the conclusion logic demanded.[103] Buguet lent further weight to this argument by sending a letter to the *Revue spirite* from Belgium, where he had fled after his conviction. In it, he admitted to falsifying images late in his career but insisted that the earlier "two thirds" of his output had been authentic, something the skepticism of the *parquet* had made him loath to mention at the trial.[104]

Under these circumstances, the critiques of the republican journalist Auguste Vacquerie, who presented laic education as a "cure" for Spiritism, carried a particular sting. Twelve years before, Vacquerie, the editor of the *Rappel,* had expressed a very different view of spirit phenomena. In a memoir of his time in exile with Victor Hugo (discussed in chapter 1), he had described his fascination with communications from the beyond. There, he strongly affirmed his belief in the otherworldly nature of the phenomena that occurred in séances:

> Why should it not be possible for [spirits] to communicate with man by some means, and why should this means not be a table? Immaterial beings cannot cause matter to move? Who says these are immaterial beings? They could have a body as well, subtler than ours and imperceptible to our eyes, much as light is to our touch. . . . I therefore have no reasoned objection to the reality of the phenomena of the tables.[105]

The writer of this pronouncement appeared to have little in common with the stern republican critic of 1875, who called Buguet's customers "dupes" and derided Spiritism as "superstition."[106] An anonymous contributor to the *Revue spirite* bitterly observed that Vacquerie appeared to have

[103] L. Legas, "La Photographie spirite et l'analyse spectrale comparées" (Paris: chez l'auteur, 1875), 16–17.

[104] Quoted in Leymarie, ed., *Le Procès des spirites,* iv.

[105] Auguste Vacquerie, *Les Miettes de l'histoire* (Paris: Pagnerre, 1863), 412.

[106] For "dupes" see *Le Rappel,* Jun. 22, 1875, 1.

become susceptible to "unexpected memory lapses" and deplored his cynical willingness to "join . . . the pack of barkers."[107]

As the writer in the *Revue* implied, Vacquerie's dramatic change of heart probably indicated more than simple forgetfulness. At midcentury, the majority of French leftists—Vacquerie and Hugo included—embraced the idea that religious belief was essential to the smooth functioning of society. By 1875, this attitude had changed dramatically: metaphysical speculations of the kind Vacquerie had espoused a decade before no longer appeared to have a place in the republican mainstream. This shift was a product of new political circumstances. In the 1870s, the establishment of an enduring democratic government was by no means a foregone conclusion; republicans of all persuasions were engaged in an intense struggle against the united forces of monarchism and the Church. The nature of the conflict led each side to emphasize its difference from the other and to develop its ideology in directions that heightened this difference.[108] For republicans, anticlericalism, along with a Positivistic confidence in the power of reason and scientific inquiry, became important aspects of this project.

Under the pressure of ideological conflict, therefore, republican objections to the power of the Catholic Church became not just political but also epistemological: The Church hampered social justice both with its steadfast opposition to democratic reform and with its continued insistence that faith was a valid form of knowledge. By promoting an atavistic reliance on feeling and intuition, the Church blocked human progress, which clearly depended on the triumph of reason and science. Religious belief and modernity, in this view, were mutually exclusive—and republicans saw themselves as quintessentially modern. To indulge in the visionary, explicitly theistic rhetoric of earlier republicanism, in this climate of opinion, therefore, would have seemed incongruous and retrograde.

Despite the vigorous republican claims to the contrary, Spiritists continued to see themselves as standing at the forefront of human progress. Eventually, they believed, the republicans would see the error of their ways and realize that Spiritism was a powerful complement to their rationalistic, democratic views. Georges Cochet, for example, insisted that Spiritism would emerge triumphant, largely because "physical phenomena confirm it as a fact, and therefore place it in the experimental domain of the positive sciences." Scientists, and with them the rest of mankind, Cochet insisted, would not be able to ignore the reality of spirit phenomena forever—the inevitable progress of human knowledge would eventually cause the Spiritist vision of the beyond to become as widely accepted as

[107] *La Revue spirite* 18 (1875): 410.
[108] Auspitz, 19, Rémond, 30–36.

the Copernican model of the solar system. Once this acceptance had occurred, Cochet believed, humanity would be transformed; at long last, a new, democratic spirituality would emerge to replace Catholicism, which the nineteenth century had so unambiguously consigned to "the museum of antiques as a precious memory of the Middle Ages."[109]

To all but the most devoted Spiritists, Cochet's optimism probably seemed quixotic in the wake of the disruption and scandal the *procès des spirites* had caused. While the publication of the *Revue* continued under the supervision of a replacement editor, the organization behind it had sustained a serious blow: In March 1876, only one hundred people appeared at Père Lachaise cemetary to commemorate Kardec's death.[110] This setback, as chapter 5 shows, was only temporary. The 1880s and 1890s would see a dramatic resurgence of heterodoxy, but in different, far more diverse forms.

[109] *La Revue spirite* 18 (1875): 345, 346.
[110] Archives de la Préfecture de Police de Paris, dr. BA 1243, report dated Mar. 31, 1876.

CHAPTER FIVE

Confronting the Multivalent Self, 1880–1914

Among readers of guidebooks, the Universal Exposition of 1900 is usually remembered for giving Paris the Pont Alexandre III and the Grand Palais. It is less well-known for having inspired a flurry of international conventions. The Fourth International Congress of Psychology was among the most important of these gatherings, both for its size and for its position in the larger history of the field. Psychology had only recently emerged as an autonomous scientific discipline—its first international convention had been held in Paris in 1889, a scant eleven years before. The 1900 Congress, a return to Paris after meetings in London and Munich, further cemented psychology's hard-earned status as an empirical pursuit, based not on philosophical speculation but on data gathered from controlled experiments.

When psychologists discussed the evolution of their discipline during this period, they generally spoke in terms of its changing relation to metaphysics. In his address to the Second International Congress, for example, Charles Richet asserted that psychology had come into its own in the last quarter of the nineteenth century by striving to "disengage itself from scholastic formulas, theology, metaphysics, and deductive reasoning," instead concentrating exclusively on empirical observation.[1] For commentators focused on the German-speaking world, this shift was primarily the work of Wilhelm Wundt and his intellectual descendents, who developed a "physiological psychology" based on experiments conducted on

[1] Reprinted in *Annales des sciences psychiques* 2 (1892): 343.

normal subjects.[2] The French account of psychology's abandonment of metaphysics, in contrast, gave a prominent place to the controlled study of pathological cases.[3] This current began with the clinical psychiatrists Jean-Martin Charcot and Hippolyte Bernheim, who transformed hypnosis—a set of techniques appropriated from Mesmerism—into both a treatment for hysteria and a means of analyzing the disease experimentally. For thinkers such as Richet, Théodule Ribot, and Pierre Janet, this clinical use of hypnosis served as the foundation of a new empirical approach to the psyche. French experimental psychology, therefore, emerged from a collaboration with clinical psychiatry, but also differed crucially from it. Psychiatrists, whose earliest professional organization in France dated from 1852, were predominantly affiliated with hospitals and focused on the diagnosis and treatment of mental illness, while psychologists, largely university-based, sought to discover the general principles that governed the workings of the mind.[4]

In France, the emergence of experimental psychology roughly coincided with the steadily increasing ideological polarization of Catholic right and republican left that had begun in the late 1870s. French medical psychiatry had a long-standing reputation for hostility to Catholicism, and indeed to religious faith in all forms. Charcot, for example, was famous for his ability to "diagnose" the ecstatic saints in Renaissance paintings: By looking at facial expressions and gestures, he could identify the precise symptoms of hysteria that earlier viewers had mistaken for mystical transport. For the French psychologists who gathered in Paris in 1900, the primary mission of their new discipline derived from this intellectual precedent. Their task, as they saw it, was to elaborate an empirical and secular understanding of consciousness.

At the same time, however, the range of opinions presented at the 1900 Congress revealed that the precise terms of psychology's abandonment of metaphysics had yet to be fully negotiated.[5] The fluidity of the discipline's boundaries came to the fore in the sessions of the Congress devoted to

[2] See e.g. Edwin G. Boring, *A History of Experimental Psychology*, 2nd ed. (New York: Appleton-Century-Crofts, 1950).

[3] For descriptions of this difference, see Jacqueline Carroy and Régine Plas, "The Origins of French Experimental Psychology: Experiment and Experimentalism," *History of the Human Sciences* 9, no. 1 (1996): 73–84; Laurent Mucchielli, "Aux Origines de la psychologie universitaire en France (1870–1900): enjeux intellectuels, contexte politique, réseaux et stratégies d'alliance autour de la *Revue philosophique* de Théodule Ribot," *Annals of Science* 55 (1998): 263–289.

[4] Matthew Brady Brower, "The Fantasms of Science: Psychical Research in the French Third Republic, 1880–1935" (Ph.D. diss., Rutgers University, 2005); Sofie Lachapelle, "A World outside Science: French Attitudes toward Mediumistic Phenomena, 1853–1931" (Ph.D. diss., University of Notre Dame, 2002).

[5] Ann Taves is preparing a manuscript on these disputes, which she graciously allowed me to consult in draft form.

"hypnosis, suggestion, and related questions." In these meetings, academic psychologists read papers alongside clinicians, psychical researchers, Spiritists, and Occultists. While some—notably German experimentalists—expressed their unease at the ways in which particular speakers attempted to transform metaphysics into science, others were intrigued by the prospect of experimentally proving the existence of the soul.[6] Ribot himself, in his presidential address, referred to the strange phenomena psychical researchers investigated as "the most advanced, adventurous parts of experimental psychology, and not the least seductive."[7]

Ribot's enthusiasm made considerable sense in the context of the broader development of the field: Many of the landmark works of early psychological theory drew the bulk of their data from studies of hypnosis and mediumism, which had come to serve as the empirical basis for an array of innovative models of consciousness. Clinical psychiatrists, as one speaker at the conference put it, had demonstrated how the observation and manipulation of trance states could serve as a kind of "mental vivisection," a way of digging far more deeply into the tissue of subjectivity than had previously been possible.[8] The theories of mind psychologists derived from this new source of data showed certain revealing commonalities, which the clinical psychiatrist Durand de Gros summarized in his address to the 1900 Congress. Studies of mediums and hysterics under hypnosis, he told his audience, had proved that the mind was multivalent, a system of "multiple centers of consciousness . . . arrayed hierarchically under the supremacy of the chief consciousness that each of us calls his ego."[9]

The task of psychology was to explain the functioning of these various centers, all of which existed beyond the ambit of normal waking awareness. Already, figures such as Ribot, Janet, Richet, Frederic W. H. Myers, and Théodore Flournoy had used hypnosis to reveal that the "subconscious" played an "enormous" role in the "generation of ideas, knowledge, memories, emotions, desires, determinations, and acts" that previous observers had assumed to be unproblematic products of "the speaking ego." This discovery had dramatic implications, Durand told his audience:

[6] For an analysis of these debates and their crucial role in redefining the disciplinary boundaries of experimental psychology, see Françoise Parot, "Le Banissement des esprits, naissance d'une frontière institutionnelle entre spiritisme et psychologie," *Revue de synthèse* 115, no. 3–4 (July–Dec. 1996): 417–443. For the critiques of psychical research offered by various attendees at the 1900 Congress of Psychology, including the objection of Hermann Ebbinghaus, a key figure in the development of German experimentalism, see *IVe Congrès*, 656–663.

[7] In Pierre Janet, Théodule Ribot, et al., *IVe Congrès international de psychologie, tenu à Paris, du 20 au 26 Août 1900, compte rendu des séances et texte des mémoires* (Paris: Alcan, 1901), 46.

[8] Ibid., 670.

[9] Ibid., 649.

> Our ego depends for its living, so to speak, on constant loans from subaltern egos, and only acts under their impulsion. It reigns, but does not govern. It is a constitutional Head of State that delegates the business of government almost entirely to its ministers.[10]

The self of which people were normally aware in their day-to-day lives was only a surface phenomenon; not the sole constitutive element of mental life, as previous generations of introspective psychologists had assumed but rather a figurehead that imposed the illusion of order on a welter of hidden urges and impulses. By 1900, this novel vision of subjectivity as fragmented and multivalent was well on its way to becoming common sense. For many commentators, scientific and literary, academic and popular, a vision of interior life as the product of complex, mysterious interactions between conscious thought and subconscious impulse had come to seem essential to an understanding of the "modern self."

When the organizers of the 1900 Congress chose to include psychical researchers, Spiritists, and Occultists in their gathering, then, they were responding directly to this development. On the surface, these diverse presenters seemed united in a common project: the study of consciousness, trance states, and the phenomena they made possible. In fact, however, they espoused an array of often mutually exclusive points of view and represented several different stages of institutional development. Psychologists generally accepted psychical researchers as legitimate colleagues. At the same time, however, the distinctive conception of the "subliminal self" that psychical researchers like Myers elaborated—which stressed the transcendent, rather than the "inferior," aspects of the subconscious—generated considerable controversy, and indeed defined a fundamental split within the discipline of psychology as a whole. Spiritism, for its part, seemed to be enjoying a remarkable resurgence after the lean years of the late 1870s, but this resurgence also revealed a new weakness in the movement. As independent mediums and journals flourished, the organization Kardec founded forty years before lost much of its influence. Its brand of Spiritism had come to seem increasingly out of step with the intellectual preoccupations of the time. Occultists shared the Spiritist concern with resolving the religious quandaries of the age, but broached the matter differently, drawing on precisely the new intellectual currents that proved so problematic for Spiritism. Blending contemporary psychology with ancient esoteric ideas, Occultists transformed the notion of the subconscious from a purely scientific concept into a metaphysical one. By 1900, Occultism was flourishing in France—as an array of secret societies, widely circulated periodicals, and thriving bookshops amply testified.

[10] Ibid., 650–651.

The disciplinary consolidation of scientific psychology, the rise of psychical research, the fragmentation of Spiritism, and the triumph of Occultism are linked episodes in a broader story. All four developments mark important stages in the emergence and reception of a new model of subjectivity. The idea of a "subconscious" mind only partly under the control of a "speaking ego" had the potential to undercut orthodox religious teachings as deeply as Positivism had forty years before, challenging fundamental assumptions about the soul, individual identity, conscience and free will. In the years before the triumph of Freudian psychoanalysis, psychologists addressed the questions this new vision of the self raised by engaging in an ambivalent dialogue with psychical research. Some sought to use this new notion of subjectivity as a means of purging studies of the mind of all metaphysical implications, while others hoped it would bring spiritual questions into the sphere of scientific discourse. Heterodox believers, as we will see, met this challenge by using an array of eclectic sources to invent a new and enduringly influential solution to the religious crisis of factuality: a spirituality of the multivalent self.

Psychology and the Challenge of Psychical Research

At a séance held in Geneva on February 2, 1896, the medium Hélène Smith, sitting at a table in darkness, entered a trance and saw a mysterious floating carriage appear, trailing sparks. A woman stepped out of the conveyance and began speaking. At first, Smith responded to her invisible interlocutor with consternation, unable to understand anything she said. Then, suddenly, the medium's demeanor changed, and she began to speak rapidly "in an unintelligible tongue, like Chinese." Auguste Lemaître, one of the small group in attendance, grabbed paper and pencil, transcribing the words as accurately as he could: "*Michma michtmon mimini thouainenm mimatchineg masichinof mézavi patelki abrésinad navette naven navette mitchichénid naken chinoutoufiche.*" Soon, however, the words came even faster, and Lemaître was reduced to scribbling whatever isolated fragments he could catch: "*téké . . . katéchivist . . . meguetch.*" This language, Smith later told her audience, was Martian, and the mysterious floating car had come to take her to the Red Planet.[11] The extraterrestrial visitations continued in subsequent séances, becoming steadily more complex. Certain principal characters appeared, and the language they spoke gradually acquired a coherent vocabulary, a distinctive script, and a

[11] Theodore Flournoy, *From India to the Planet Mars: A Case of Multiple Personality with Imaginary Languages,* ed. Mireille Cifali, trans. Daniel Vermilye (Princeton: Princeton University Press, 1994), 95.

grammar.[12] Mars was not the only place Smith's trances took her. Other séances saw her transported to the Ancien Régime as Marie Antoinette, or to ancient India, where she claimed to have once lived as a Hindu princess; in the latter case, she spoke another invented language, which she identified as Sanskrit, and which attracted the professional interest of the linguist Ferdinand de Saussure.[13]

These prodigies of visionary imagination would have left no trace beyond a few isolated journal articles if it had not been for one frequent attendee of Smith's séances: Théodore Flournoy, professor at the University of Geneva and founder of one of Europe's first laboratories of experimental psychology.[14] In late 1899, he published *Des Indes à la planète Mars*, a case study of Smith's mediumship based on years of detailed observation. At the 1900 Congress, Flournoy shared his findings and spoke more generally about the conclusions he derived from them. His research, he told his audience, had shown him two remarkable things: "the richness and extent" of the "subconscious play of mental faculties" among mediums on the one hand, and "the prodigious indulgence" of the "otherwise very cultivated people" who accepted séance phenomena as supernormal on the other. According to Flournoy, his studies had proved that the communications and visions mediums like Smith received were rooted in subliminal recollections of past experiences, which he called "cryptomnesia." While it was not possible to identify these recollections explicitly in every case, Flournoy continued, Spiritists were wrong to present this simple absence of evidence as proof of otherworldly intervention because the notion that spirits existed had received considerably less empirical confirmation than his alternative hypothesis.[15]

The "prodigious indulgence" of Spiritists, as Flournoy saw it, posed a serious social threat—one that it was psychology's duty to address. While Smith was remarkable in many respects, Flournoy did not wish to encourage others to follow her example. The dissociation of consciousness and "subconsciousness" that Smith purposely induced was a pathological state in Flournoy's view, but this was not his primary concern. Instead, he worried most about Smith's audience and its eagerness to transform the medium's playful, childish fantasies into visions with profound metaphysical implications. Psychologists, Flournoy argued, needed to prevent this ignorant worship of false idols, in which ill-understood phenomena were peremptorily granted a significance that had more to do with the percipient's desires than with

[12] For an analysis, see Victor Henry, *Le Langage Martien* (Paris: Maisonneuve, 1901).

[13] Sonu Shamdasani, intr., *From India*, xxxv.

[14] See Serge Nicolas and Agnes Charvillat, "Théodore Flournoy (1854–1920) and Experimental Psychology: Historical Note," *American Journal of Psychology* 111, no. 2 (Summer 1998): 279–294.

[15] *IVe Congrès*, 106, 108.

objective fact. Superstitions generated in this manner, he declared, "weigh . . . like a nightmare" on the imaginations "of our contemporaries."[16] As he saw it, the best way to save Europeans from the harmful consequences of their own unmet spiritual needs would be to foster psychical research, encouraging its development as an institutionally recognized subfield of experimental psychology.

Not all psychologists, Flournoy noted, were comfortable with this prospect. In his book, he explained their unease by drawing an analogy to current events. The debate over the existence of "supernormal" powers like telepathy, he wrote, had become "the Dreyfus case of science."[17] The wrongful conviction of the Jewish army officer Alfred Dreyfus had caused profound ideological divisions between secular republican advocates of individual liberty and Catholic conservatives eager to preserve the honor of the French military; the question of psychical research and its scientific validity resonated with equal strength. For many French psychologists, the future of the discipline seemed to turn on the outcome of this debate. Would psychology abandon metaphysics and embrace a materialistic conception of the mind, or would it instead seek to use the empirical methods of scientific inquiry to elaborate a new conception of the human soul, thereby redeeming religion for the modern age? The choice was between the militant secularism of clinical psychiatrists, who saw all religious faith as a form of pathology, and a more idealistic approach, one that sought to use science not as a replacement for transcendence but as a means of comprehending it more fully.

This debate played a crucial role in the development of French psychology in the late nineteenth and early twentieth centuries. To grasp its significance, we need to begin by looking at the emergence of psychical research as a scientific discipline and its reception in French psychological circles, which was surprisingly cordial. At first, these studies of as-yet-undiscovered powers of mind were accepted as a legitimate part of the psychological project, though the question of their ultimate significance remained controversial. Disputes about the position of psychical research in relation to the other branches of psychology, as we will see, stemmed from theoretical arguments about the nature of the newly posited subconscious or subliminal mind: Was it an inferior realm of pathological inclinations, or was it the seat of the soul? As in the Dreyfus case, the broader implications of this debate captured the imaginations of intellectuals and journalists, who saw the question of the "subliminal self" and its ultimate nature in terms of a broader critique of Positivism and a growing skepticism of reason's power to encompass every aspect of human experience.

[16] Ibid., 105.
[17] Flournoy, *From India*, 223.

The Development of Psychical Research in France

During the 1870s, as we saw in chapter 4, a growing number of British academics began to study the phenomena mediums produced. The articles published by the eminent chemist William Crookes between 1871 and 1874, in particular, attracted considerable attention, especially at Trinity College, Cambridge, where a group of scholars began to hold experimental séances of their own. The group at Trinity included Henry Sidgwick, an important moral philosopher, as well as two lesser-known young Fellows in Classics, Frederic W. H. Myers and Edmund Gurney, and Sidgwick's wife, née Eleanor Balfour. Gradually, word of this group's activities spread, and in January 1882, the spiritualist W. F. Barrett invited its members to a conference.[18] At the gathering, someone suggested the creation of a new society devoted to the study of mediumistic phenomena; the Society for Psychical Research (SPR) was founded a month later, with Sidgwick as its president. The first issue of its primary journal, the *Proceedings of the SPR*, appeared in July 1882.

Spiritualists played an important role in the Society during its earliest years, but by 1886 the Trinity group had become the dominant faction. During this period, Sidgwick, Myers, and Gurney published extensively in British scholarly and popular journals, elaborating an approach to séance phenomena that quickly came to be associated with the new field of psychical research as a whole: Building on the example Crookes had provided, it joined open-mindedness with a relatively sophisticated understanding of experimental method and a self-conscious reluctance to engage in explicit metaphysical speculation. As the Trinity group gained ground in the SPR, the size and prestige of the organization increased. Between 1883 and 1900, its membership grew from 150 to 946; the rolls eventually included such eminent figures as the naturalist Alfred Russell Wallace; physicist Oliver Lodge; future Prime Minister Arthur Balfour; philosopher Henri Bergson; psychologists Flournoy, Carl Jung, Sigmund Freud, William James, and Pierre Janet; and writers John Ruskin, John Addington Symonds, and Alfred, Lord Tennyson. Members of the SPR also figured prominently among presenters and organizers at the first four International Congresses of Psychology.[19]

The British scholars who created the field of psychical research called the phenomena they studied "supernormal"—to avoid the ideologically fraught term "supernatural"—and divided them into two groups, physical and mental. Physical phenomena were manifestations that seemed to involve the

[18] For more on Sidgwick, see Bart Schultz, *Henry Sidgwick: Eye of the Universe, an Intellectual Biography* (Cambridge: Cambridge University Press, 2004).

[19] Alan Gauld, *The Founders of Psychical Research* (London: Routledge and Kegan Paul, 1968), 137–147.

mind's direct action on the material world, including telekinesis, levitation, and the production of apparitions or of mysterious substances called ectoplasms. Mental phenomena, in contrast, were largely a matter of one mind influencing another or trance-induced alterations in consciousness. All of these diverse forms of mental activity, for psychical researchers, seemed to call existing materialistic conceptions of the psyche and its functioning into question. This categorization of phenomena—along with the technical nomenclature members of the SPR introduced to accompany it—proved extremely influential.

Throughout Europe and the United States, scholars who wished to approach these uncanny phenomena from a scientific perspective followed the British example. In France, this process of emulation began in earnest with the 1891 founding of a new journal, the *Annales des sciences psychiques.* The *Annales,* under the direction of a doctor named Xavier Dariex, was published by the prestigious firm of Félix Alcan. Alcan's list was very different from those of most French publishers who produced books on the "supernormal" during this period. Instead of mixing titles on psychical research with Occultist books or compilations of spirit communications, Alcan published studies of the paranormal alongside works of philosophy and social science at the vanguard of fin-de-siècle thought. Alcan was the primary French publisher of such luminaries as Bergson, the criminologists Cesare Lombroso and Gabriel Tarde, Charcot, and the social critic Max Nordau.[20] The texts on psychical research that Alcan chose were generally the most intellectually rigorous in the field: Works by French writers such as Richet and psychologist Gustave Geley appeared alongside translations of texts by Myers, Gurney, and others.

Dariex, as he explained in the journal's first issue, created the *Annales* in order to keep French science on the intellectual cutting edge. "The British, the Americans, and even the Germans" had begun exploring psychical research, while the French remained comparatively unwilling to address it. This reluctance, as Dariex saw it, reflected a dangerous tendency among French students of the mind to sacrifice progress to prejudice. Some eminent French psychologists, psychiatrists, and philosophers had joined the SPR, but there was no analogous group in which they could participate at home. Consequently, Dariex wrote, "we would like to do what has already been done in England, namely to create a neutral center for this kind of study, a center that will have no aims other than the search for absolute truth—naked truth, not truth clothed according to the tastes of one doctrine or another."[21] There certainly were Spiritist journals in France, but if they remained the only French-language source for information about

[20] See the list of titles included with the annual table of contents for the *Annales* 5 (1895).
[21] *Annales* 1 (1891): 19.

developments in psychical research, the field would always be compromised in the eyes of academic scientists. The journal Dariex created, therefore, made its bid for intellectual legitimacy by modeling itself on the publications of the SPR.

In further emulation of British precedent, Dariex enlisted the support of a variety of medical doctors and academics. Preeminent among these was Charles Richet, a professor at the Faculté de médecine in Paris. When the first issue of the *Annales* appeared in 1891, Richet was already a major figure in the field of physiology—his discovery of anaphylaxis would earn him the Nobel Prize in 1913. Richet had also made a name for himself in psychology: In 1875, he published a paper on "provoked somnambulism" that was among the first efforts to use hypnosis as a technique for gathering psychological data.[22] In subsequent years, he published more studies of hypnotic phenomena and was widely acknowledged alongside Charcot and Bernheim as a pioneer in the field. Richet's work, however, had a different focus than that of his better-known colleagues. His primary interest was not in treating mental disease but in using hypnosis as a tool of psychological investigation; by the 1880s, he had turned his attention to the study of mediums.

Richet readily endorsed the *Annales* and expressed his hopes for the future of French psychical research in a long introduction to the first issue. The new journal, he wrote, would be furthering the crucial project that the SPR had begun: bringing "certain mysterious, elusive, unknown phenomena into the framework of the positive sciences." There was nothing unique about this process in the case of psychic phenomena, Richet argued; all sciences went through an "occult" stage at their beginnings. Hypnosis had begun as Mesmerism, chemistry as alchemy. Scientific progress would eventually work the same transformation on the study of mediumistic phenomena. The way to spark this change, in Richet's view, was to introduce a new approach to the subject. Spiritist and Occultists journals assumed that almost all reports of marvelous phenomena were authentic and then theorized freely about their significance. Dariex promised to reverse this process: His new journal would shy away from the rich banquet of uncanny stories that Spiritist and Occultist publications offered their readers, limiting itself instead to the reporting of facts concerning telepathy. In addition, the *Annales* would avoid publishing any theoretical speculation until the authenticity of the phenomena themselves had been fully proved. At this stage, Richet argued, "a little fact, well-studied in all its details" had "infinitely more value than the most learned dissertations in metaphysics or hyperphysics."[23]

[22] Henri F. Ellenberger, *The Discovery of the Unconscious: The History and Evolution of Dynamic Psychiatry* (New York: Basic Books, 1970), 314.
[23] *Annales* 1 (1891): 2, 7.

In practice, however, this ascetic focus proved impossible for Dariex to maintain as a long-term editorial strategy. Beginning in the latter part of 1891, the journal included reports on a steadily widening range of phenomena, both mental and physical. The catalyst for this shift was a series of séances held in Naples in June 1891, which were devoted to the study of phenomena produced by the Italian medium Eusapia Paladino (fig. 14). Paladino, who had been a domestic servant and amateur spiritualist before her discovery by the Italian psychical researcher Ercole Chiaia, was capable of producing spectacular phenomena under what appeared to be the strictest experimental controls. In the Naples séances, according to reports published in the *Annales*, she caused a table to levitate, produced mysterious raps, and sent a bell flying through the air without physical contact. These manifestations were not distinctive on their own—spiritualist journals were full of such accounts, and had been for years. What made Paladino's phenomena remarkable was their audience. Chiaia had arranged the June 1891 sittings for the famous criminologist Cesare Lombroso, who had been an outspoken critic of psychical research. After the first séance, however, Lombroso pronounced himself "quite ashamed" of his earlier resistance and reversed his opinion.[24]

For Dariex, the conversion of a man of science as eminent as Lombroso amply justified broadening the journal's approach. In a response to a reader's criticism published in 1897, Dariex emphasized the significance of the Paladino séances—particularly those held in 1891 and 1892. "Since a long and patient study has led us—with our friends—to admit the authenticity of the phenomena stemming from Eusapia's mediumship," he informed his readers, "we think we may now permit ourselves to cease keeping the *Annales psychiques* so rigorously closed, and to welcome new facts—even when they are not presented in scientific form and have value only as diverting anecdotes." Dariex himself had never seen Paladino produce true supernormal phenomena, but the eminence of the "men of science" who *had* done so—or at least believed they had—was in his view more than enough proof for any rational observer.[25]

Throughout the 1890s, this confidence seemed justified. Many considered psychical research to be a legitimate branch of psychology. As the German psychical researcher Albert von Schrenck-Notzing pointed out in the *Annales*, figures as eminent as Gilles de la Tourette and Hippolyte Bernheim had published statements admitting the possibility of mental action beyond the limits of the body.[26] Such phenomena seemed far-fetched

[24] Ibid., 132.
[25] *Annales* 7 (1897): 370, 66–67, 70.
[26] Ibid., 71–72.

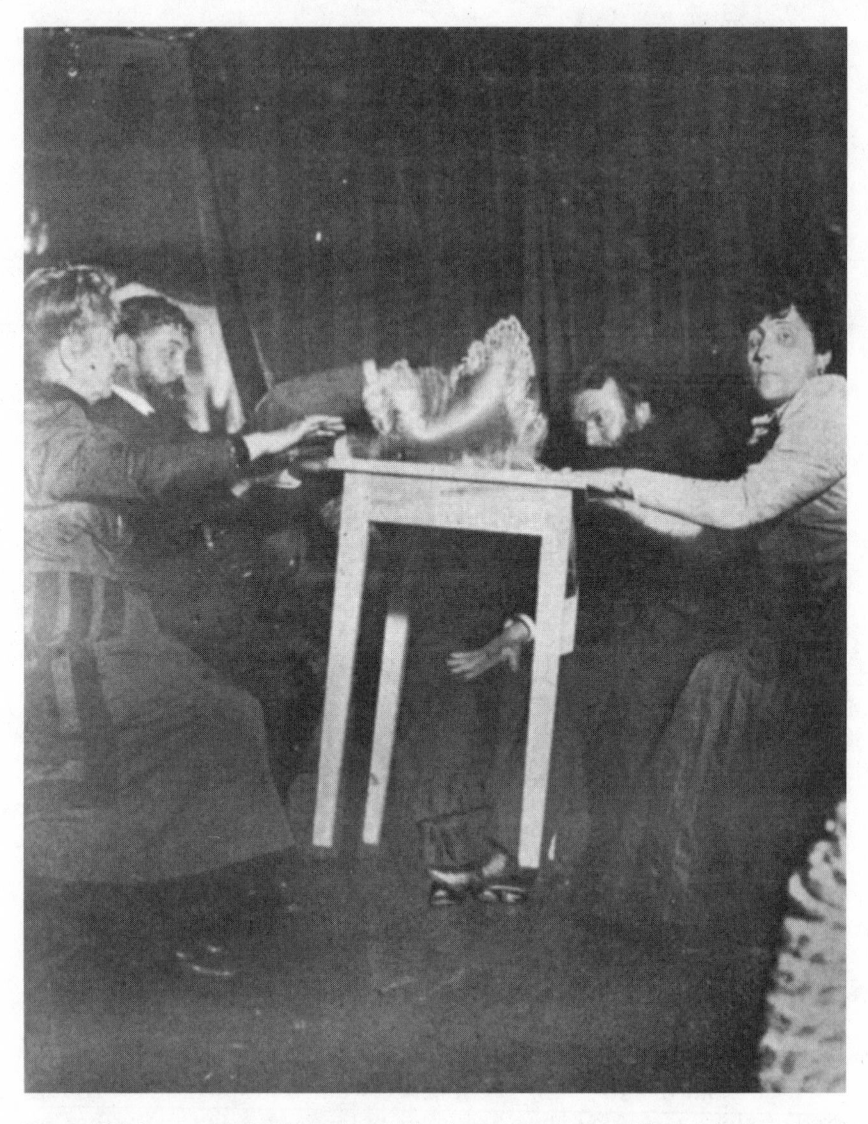

Fig. 14. The famous medium Eusapia Paladino in a séance in Paris. The original caption of the image, published in Camille Flammarion's popular *Les Forces naturelles inconnues* (1907), reads "Complete levitation of a table: photograph taken in the living room of M. Flammarion, November 12, 1898. (The first leg, to the left, is 18 centimeters above the floor, the second is 13, the right rear leg is 8, the left rear, 14). An assistant uses a cushion to hide the medium's eyes from the bright magnesium flash. This medium (Eusapia) is held in a position that renders it impossible to perform any suspicious movements." (Collection of the author.)

to less imaginative observers because as yet, the "intermediary causal psychic factor" remained unknown. Once this mystery was solved, however, telepathy and telekinesis would come to seem no stranger than automatic writing. When an essay in one of the period's most prestigious philosophical journals accused psychical researchers of practicing "modern mysticism"—abandoning scientific rigor for an all-too-human obsession with unanswerable questions—Richet defended his work in similar terms.[27] His current studies of telepathy, he argued, were no different from the studies of hypnosis he had carried out sixteen years before. At the time he undertook those experiments, Richet asserted, phenomena like automatic writing had seemed utterly implausible.[28] By 1891, however, the notion that consciousness could splinter in trance states had been accepted as a central principle of experimental psychology. Hypnosis, a technique once condemned as a form of pseudoscientific "mysticism," had revealed vast and previously unsuspected realms of mental life, making them accessible to rigorous scientific study. Psychical research simply took this process a step further.

This conception of the psychological project was by no means idiosyncratic. In his inaugural address as president of the first French institute devoted to experimental psychology, the Institut psychique internationale—later the Institut général psychologique—Janet accorded a similarly important role to psychical research. One of psychology's primary missions as a discipline, he argued, was to provide novel perspectives on "philosophical and religious problems." Studies of "telepathy, telekinesis, clairvoyance, and mediumism" played a crucial role in this enterprise, because they "seem to relate to the deepest powers of thought," and as such promised to shed new light on "human nature."[29] While Janet later testified to the increasing difficulty of reconciling psychical research with the physiological psychology pursued by other members of the institute, the organization's journal nevertheless continued publishing studies of supernormal phenomena into the 1930s.[30]

[27] See M. Rosenbach, "Etude critique sur le mysticisme moderne" in *Revue philosophique* 34 (1892).

[28] Reprinted in *Annales* 2 (1892): 295.

[29] *Bulletin de l'Institut psychique international* 1 (1900–1901): 5, 4.

[30] For Janet's complaints, see Ibid., 135; for a more detailed history of the *Institut général*, including an account of its development during the twentieth century, see Lachapelle, "A World outside Science." Janet dealt with the growing pressure from students of psychical research by establishing a separate Société de psychologie under the auspices of the Institut in 1901. According to Serge Nicolas, this organization excluded "all those with an interest in the paranormal." Nicolas, *Histoire de la psychologie française, naissance d'une nouvelle science* (Paris: In Press Editions, 2002), 171.

Psychical Research and the Pre-Freudian Multivalent Self

In large part, the French psychological community's continuing acceptance of psychical research during the 1890s and early 1900s stemmed from the crucial role studies of mediums and trance states played in the development of the discipline as a whole. As Henri Ellenberger has shown, such studies underpinned one of the most dramatically innovative aspects of the fin-de-siècle psychological project—the elaboration of new theories of consciousness that shattered the earlier notion of a unitary self.[31] The strange feats of mediums such as Hélène Smith and Eusapia Paladino seemed to reveal a vast zone of mental life that existed beyond the sphere of ordinary waking awareness, which the experimental techniques of empirical psychology were perfectly suited to explore. Introspection, that old standby of philosophers and metaphysicians, was powerless to illuminate this newly discovered form of mental activity: Only experimental inquiry, and the gimlet eye of the trained specialist, could truly plumb its depths. Subconscious phenomena, as psychologists conceived them in this pre-Freudian era, were best studied with the controlled use of experimental techniques like hypnosis, and it was mediums, along with hysterics and Mesmeric somnambulists, who made the whole undertaking possible.

Even as psychologists reached a broad consensus about the importance of subconscious mental processes, however, they differed sharply when it came time to ascribe a value and scope to this newly theorized category of intellectual expression. Some considered all such activity to be pathological, while others saw some forms of it as beneficial parts of normal mental life. More dramatically still, some psychologists insisted that the mind was bounded by the limits of the individual body; while others, like Richet, were willing to admit the possibility that the spectrum of mental phenomena might include such things as telepathy and telekinesis. In France, this debate coalesced around the period's two most influential conceptions of the subconscious: Pierre Janet's "psychic disaggregation" model, advanced in his 1889 monograph *L'Automatisme psychologique*, and Frederic W. H. Myers' theory of the "subliminal self," first introduced to French-speaking audiences in an 1897 translation of an article that had appeared in the *Proceedings of the SPR* five years before. Janet, following the precedent set by the philosopher Hippolyte Taine and by psychiatrists like Moreau de Tours and Charcot, treated subconscious mental activity as a symptom of mental disease, resolutely disconnecting it from any

[31] Ellenberger, *Discovery*, 254–330, 749–784. Jan Goldstein insightfully analyzes this conception of the unitary self in relation to the thought and influence of Victor Cousin. See Jan Goldstein, "The Advent of Psychological Modernism in France," in *Modernist Impulses in the Human Sciences*, ed. Dorothy Bass (Baltimore: The Johns Hopkins University Press, 1994).

"supernormal" manifestations.[32] For him, the subconscious was by definition a sphere of "inferior" mental activity. Myers, in contrast, argued that the "subliminal" encompassed the most exalted aspects of mental life: not only the creative inspirations of genius but also the uncanny powers certain mediums seemed to possess. The numerous other theories of the multivalent self that emerged among French psychologists during the fin de siècle generally built on one of these approaches or attempted some degree of synthesis.

The conflict between these two strikingly different visions of the psyche tended to mirror the contrasting hopes psychologists brought to psychical research—the factor that made it so similar to the Dreyfus case in Flournoy's estimation. For one group, the subconscious was merely a welter of degenerate and unruly psychic forces, and it was the task of psychical research to expose this fact. Janet, for example, believed that scientific studies of mediumism would serve as a form of social hygiene, easing the "nervous curiosity" that led so many anxious, rudderless spiritual seekers to misinterpret "certain phenomena that are in reality purely psychological."[33] For the other group, the subconscious was the terrain where scientists would finally discover the soul, and the task of psychical research was to serve as the portal to this transcendent realm. The psychologist Gustave Geley, for example, argued that a detailed experimental study of man's "subconscious being," as manifested through supernormal phenomena, would eventually demonstrate that the immortality of the soul was a material fact.[34]

The distinctions between the models of the subconscious espoused by Janet and Myers emerged most clearly in their contrasting views of mediumism. For Janet, the ability to act as a medium was a form of mental pathology. A healthy psyche, in his view, was above all unified and tightly organized. The mind automatically gathered sense impressions and stored them in memory; consciousness, and with it the awareness of a unified, coherent identity, emerged from the process of synthesizing these impressions into a meaningful whole. This process of picking, choosing, and interpreting memory and sense data required considerable effort. Some individuals suffered from a hereditary condition Janet called "psychological poverty" and therefore lacked the strength to perform this work.[35] Among these weak minds, the personality had a

[32] Pascal LeMaléfan, *Folie et spiritisme, histoire du discours psychopathologique sur la pratique du spiritisme, ses abords et ses avatars, 1850–1950* (Paris: L'Harmattan, 1999), 66–80.

[33] *Bulletin de l'Institut général psychologique* 1 (1900–1901): 4.

[34] Gustave Geley, *L'Etre subconscient* (Paris: Alcan, 1898): 156–158.

[35] Pierre Janet, *L'Automatisme psychologique, essai de psychologie expérimentale sur les formes inférieures de l'activité humaine* (Paris: CNRS, 1973), 417–431. This text was originally published by Alcan in 1889.

tendency to fly into fragments or "disaggregate." Without a strong synthetic capacity to keep order, memories and sense impressions would begin to coalesce around multiple nodes, all beyond the reach of the subject's conscious awareness, giving rise to an array of disturbing symptoms—hysterical blindness, for example, or a pathological vulnerability to hypnotic suggestion. Mediums, Janet argued, were "degenerates" who intentionally exacerbated the weakness of their synthetic capacities, mistaking signs of the disintegration of their own personalities for otherworldly intervention.[36]

Strongly expressed, apparently autonomous subconscious mental activity, in Janet's view, was inferior by definition. The highest intellectual achievements were necessarily the work of the conscious mind—acts of will, not passively received gifts from elsewhere. Synthesizing a coherent identity out of the vast welter of sense impressions was itself a creative process:

> consciousness, which is a reality at the highest level, is . . . a constructive activity. This activity, as we can see if we contemplate its nature ourselves, is above all a synthetic activity, joining a more or less numerous array of given phenomena into a new phenomenon different from its constituent parts. This is a true act of creation, since, whatever point of view we assume, . . . the act that joins these heterogeneous elements into a new form does not emerge from the elements themselves.

A genius, Janet believed, was able to take this synthetic ability a step further through a prodigious feat of mental strength—he could not only create and maintain a unified self but also invent new forms of synthesis, thereby suggesting entirely new ways of understanding the world. Subconscious phenomena were of lesser value because the will did not shape them. To believe that the uncanny phenomena mediums produced were signs of humans' higher nature was to willfully ignore such awkward realities as the banality of the vast majority of spirit communications, which Janet presented as a direct product of their "inferior" subconscious origins. Automatic writings by mediums struck Janet as having much in common with similar texts produced by children, hysterics, and Mesmeric somnambulists. In all cases, the writers were individuals with "personalities that are weak and incapable of governing their words." A healthy and productive mind, in Janet's view, was above all a mind in control of itself. Mastery of the ongoing labor of consciousness was a necessary prerequisite for true "liberty and progress."[37]

[36] Pierre Janet, "Le Spiritisme contemporain," *Revue philosophique* 33 (Jan.-June 1892): 425.
[37] Janet, *Automatisme psychologique,* 452, 392, 455.

Myers began his posthumously published magnum opus *Human Personality and Its Survival of Bodily Death* by pointing out the shortcomings of this view of the subconscious. He conceded that in many ways the self *was* "colonial"—a collection of primitive, recalcitrant fragments held together by a domineering, always-outnumbered capacity for synthesis—but considered theories like Janet's to be reductive. In particular, Myers took issue with the way this new conception, which transformed the unitary self into a fleeting illusion produced by sheer force of will, seemed to close off any possibility that the mind might have a transcendent spiritual dimension. That dimension could exist only if individual identity was ultimately founded on some kind of stable, immutable intellectual core. Supporters of theories like Janet's, he observed, "have frankly given up any notion of an underlying unity—of a life independent of the organism—in a word, of a human soul."[38]

Myers sought to restore this "unity" by developing a broader definition of subconscious mental processes, which he referred to as "subliminal." He explained this expanded conception by drawing an analogy to light. Physicists, in their studies of the sun's rays, had discovered that infrared and ultraviolet light existed beyond the ends of the visible spectrum. Psychologists, similarly, were in the process of discovering that the aspects of mental life accessible to ordinary waking awareness accounted only for a fraction of "an undiscovered illimitable ray." At the lower, infrared end of the psychological spectrum, there were the "organic processes" that sustained life, like digestion and the circulation of the blood, as well as certain forms of pathological behavior that Janet and others had noticed in hysterics. It was "the faculties that lie beyond the violet end of our psychological spectrum," however, that interested Myers most and that directly challenged Janet's theory. "The actinic energy which lies beyond the violet end of the solar spectrum," he wrote,

> is less obviously influential in our material world than is the dark heat which lies beyond the red end. Even so, one may say, the influence of the ultra-intellectual or supernormal faculties upon our welfare as terrene organisms is less marked in common life than the influence of the organic or subnormal faculties. Yet it is *that* prolongation of our spectrum upon which our gaze will need to be most strenuously fixed. It is *there* that we shall find our inquiry opening upon a cosmic prospect, and inciting us upon an endless way.

[38] Frederic W. H. Myers, *Human Personality and Its Survival of Bodily Death* (London: Longman, 1903), 1:11. Italics in the original. For the French translation see Myers, *La Personnalité humaine, sa survivance, ses manifestations supranormales,* trans. S. Jankelevitch (Paris: Alcan, 1906). French translations of Myers' 1892 articles on the Subliminal Self first appeared in *Annales* 7 (1897).

The "colonial" model of the mind that Janet, Ribot, and other like-minded psychologists elaborated, Myers argued, concerned itself exclusively with the infrared end of the mental spectrum. A truly effective model needed to take the ultraviolet into account as well.[39]

This broader conception, Myers argued, made it patently irrational to think of nonconscious intellectual activity as by definition pathological or degenerate. Some forms of it may have been, but others were signs of something higher. The highest manifestations of subliminal intellection, in Myers' account, were the creative inspirations of genius and—more dramatically still—telepathic communication between minds. The fact that telepathy could occur—which Myers believed psychical researchers had definitively shown—proved that man "is not a planetary or a transitory being; he persists as very man among cosmic and eternal things."[40] Telepathy, in other words, allowed Myers to present the "violet" end of the subliminal spectrum as the physical location of the immortal soul.

This approach to the subconscious aspects of mental life led Myers to present mediums, and spiritualism in general, in a decidedly positive light. While he expressed reservations about the otherworldly origins of many spirit communications, he also argued in favor of the practice of mediumism. The renunciation of conscious control that Janet found so dangerous struck Myers as a potential form of salutary liberation. The subliminal, for Myers, was not a principal of disorder but of transcendence. In the presentation he made to the 1900 Paris psychology congress, Myers stated his case emphatically:

> I claim that this substitution of personality or spirit-control, or possession, or pneumaturgy, is a normal forward step in the evolution of our race. I claim that a spirit exists in man, and that it is healthy and desirable that this spirit should thus be capable of partial and temporary dissociation from the organism; itself then enjoying an increased freedom of vision, and also thereby allowing some departed spirit to make use of the partially vacated organism for the sake of communication with other spirits still incarnate on earth. I claim that much knowledge has already thus been acquired, while much more is to follow.[41]

Instead of presenting the medium's trance as a form of degeneracy, Myers argued the opposite: Mediums were at the vanguard of human progress, evolving faculties that would eventually enable mankind to achieve as yet undreamt-of levels of spiritual awareness. Normal consciousness, which

[39] Myers, *Human Personality*, 17–18.
[40] Ibid., 26.
[41] *IVe Congrès*, 114.

for Janet marked the *summum* of human creativity and achievement, was for Myers just a way-station in an ongoing journey.

For a reader used to the current disciplinary boundaries and assumptions of experimental psychology, Myers' remarks seem surprisingly far-fetched. This was not the case for the psychologists in attendance at the 1900 Congress, however, who took Myers' ideas quite seriously. As chair of the gathering, Ribot, whose own work posited a "colonial" self with no room for the soul, even nominated Myers to be an honorary president.[42] The audience supported his candidacy warmly. In part, this cordial reception was probably an acknowledgment of the British thinker's failing health: He died shortly after his trip to Paris. The lack of criticism may also have been a result of the language barrier, since Myers presented in English, and most of his intellectual opponents spoke French or German. At the same time, however, the general tolerance of Myers' theory reveals the remarkable breadth of admissible approaches to the subconscious at this stage in psychology's development as a discipline. Provided it was addressed with sufficient rhetorical sophistication, the notion that telepathy and other "supernormal" faculties might exist beyond the sphere of waking awareness, and might serve as proof of the reality of the immortal soul, remained a tenable if increasingly controversial position.

Janet and Myers were not the only psychologists to address the question of the subconscious during this period, of course, but before the ascendancy of Freud, their two approaches were the poles around which new ideas in the field coalesced. In France, the years around 1900 saw a proliferation of novel theories of mind based at least in part on empirical data gathered from studies of mediums. Some of these theories shared Janet's assumptions, stressing the fragmentary nature of the psyche and placing abstract thought in an ordinary waking state at the summit of intellectual activity. For Ribot, Joseph Grasset, and Alfred Binet, for example, the phenomena mediums produced had nothing to do with the transcendent, but were instead either signs of pathology or simple fraud.[43] Others, however, took Myers as their point of departure. Thinkers like Richet, Geley, and Paul Gibier—all of whom had prominent academic or clinical positions—argued that mediums possessed a new kind of mental faculty that marked a step forward in human evolution.[44] Flournoy, for his part, developed an alternative view, drawing from both of these approaches: He agreed that

[42] For Ribot's view of the "colonial" psyche, see Théodule Ribot, *Les Maladies de la personnalité* (Paris: Alcan, 1921 [1884]), 152–172; for the nomination as honorary president and the approval of the audience, see *IVe Congrès*, 48.

[43] See Ribot, *Maladies*; Joseph Grasset, *Le Spiritisme devant la science* (Paris: Masson, 1904); for Binet, see LeMaléfan, *Folie et spiritisme*, 88–90.

[44] Charles Richet, *Traité de métapsychique* (Paris: Alcan, 1922); Geley, *Etre subconscient*; Paul Gibier, *Analyse des choses, essai sur la science future* (Paris: Dentu, 1890).

the subliminal could be a source of creative inspiration and admitted the possibility of telepathy, but he presented mediumism as a pathological state. The approaches each of these figures adopted, in turn, shaped their attitudes toward psychical research. Ribot, Grasset, and Binet remained cautious when discussing the question of "supernormal" phenomena, while Richet, Geley, and Gibier viewed such manifestations as the key to any true understanding of the psyche.

The fin-de-siècle proliferation of new theories of mind was not just a matter of intradisciplinary debate among psychologists. These novel ideas attracted attention in a wide variety of different quarters, ranging from the rarefied world of academic philosophy to the pages of the *grande presse.* As H. Stuart Hughes has shown, the question of consciousness, its nature and its limits, was a central preoccupation of the period and stemmed from a growing concern with the epistemological shortcomings of Positivism.[45] When Henri Bergson, for example, addressed Janet's Institut général in 1901, he ended a complex meditation on dreams and the contingency of sense impressions by declaring the importance of continued study of the "secret depths of the unconscious," including telepathy—a project that in his view would be "the principal task of psychology in the coming century."[46]

In an article first published in the *Figaro illustré* in 1891, Camille Flammarion—now France's most famous scientific popularizer—expressed a similar point of view in far more accessible terms. It was characteristic of "our fin-de-siècle," he wrote, that "the mind tires of the affirmations of philosophy that qualifies itself as positive." This skepticism, in his view, was itself a product of scientific progress. The epistemology of Positivism depended on the assumption that the information provided by the senses was transparent and reliable. New discoveries in psychology, physiology, and physics, however, had rendered sense perception more problematic:

> When analyzing the testimony of our senses, we find that they deceive us absolutely. . . . We touch solid bodies: there is no such thing. We hear harmonious sounds: the air is only transporting waves that are themselves silent. We admire the effects of light and the colors that bring the spectacle of nature alive before our eyes: in fact, there is neither light, nor color, but only obscure ethereal movements which, striking our optic nerve, give us a sensation of luminosity. We burn our foot in the fire: unbeknownst to us, the sensation

[45] H. Stuart Hughes, *Consciousness and Society: The Reorientation of European Social Thought, 1890–1930* (New York: Knopf, 1958), 15–16.

[46] Henri Bergson, "Le Rêve," in *Mélanges* (Paris: Presses universitaires de France, 1972), 462. Originally published in the *Revue de l'Institut général psychologique* 1 (1900–1901): 97–122.

> of burning resides solely in our brain. We speak of heat and cold: in the universe, there is neither heat nor cold, only movement. Thus our senses deceive us. Sensation and reality are not the same.

The senses were not a path to direct knowledge of the world, as Positivists had so confidently assumed. Not only did they mediate experience in a manner that obscured its true nature, they were also woefully limited. A vast range of frequencies, for example, existed between the highest sound the ear could hear, and the reddest light visible to the eye. In Flammarion's view, this simple fact was a powerful argument for the reality of mediumistic phenomena, which he saw as inhabiting the vast swath of nature that lay beyond the capacity of the senses to perceive. From this perspective, he argued, "the action of one mind upon another, at a distance," became no more extraordinary "than the action of a magnet on iron, the pull of the moon on the earth, the transport of the human voice by electricity, the revelation that the chemical constitution of a star can be analyzed from its light, or any of the other marvels of contemporary science." It would only be a matter of time, in turn, before this new science, already freed from its Positivistic obsession with the world as perceived by the senses, would begin to elaborate a much deeper and more nuanced understanding of consciousness and human potential.[47] By the early years of the twentieth century, this conception of the multivalent self—along with its connection to the questions raised by psychical research, the hopes it inspired, and the anxieties it caused—had become common currency among educated French men and women.

The Crisis of Spiritism

When he undertook a historical survey of Spiritist literature in 1892, Pierre Janet noticed that writers in the field had developed a surprising blind spot. Until the late 1870s, Spiritist books and periodicals had been filled with detailed accounts of the séances in which automatic writings were produced, but by the early 1890s, these once-crucial bits of evidence had become rare. Instead, journalists within the movement devoted their attention to other matters, like internecine doctrinal polemics, essays on metaphysical topics, and secondhand descriptions of spectacular physical phenomena, usually taken from English-language periodicals. More strikingly still, this turn away from firsthand empirical observation went along with a total neglect of recent psychological discoveries. Silence, in this case, was eloquent. Spiritist journalists, Janet suggested, had "entirely

[47] Reprinted in *Annales* 2 (1892): 79, 80, 85.

unconsciously" begun to avoid describing exactly what happened in automatic-writing séances because such phenomena had now become scientifically explicable in a way that discredited the assumptions on which Spiritism depended. These writers, as Janet saw them, were "like those deeply convinced, very sincere monks who no longer have the courage to expose their idols to the jeers of the profane." The production of written spirit communications still remained central to the movement, but could no longer be touted to outsiders as an authentic wonder.[48]

Certainly, the tenor of the Spiritist press had changed markedly in the years since 1875, when Kardec's successor as editor of the *Revue spirite*, Pierre-Gaëtan Leymarie, was convicted for his role in selling fraudulent spirit photographs (see chapter 4). During its first fifteen years, Spiritism had been notable for its homogeneity. While some dissidents, like Zéphyre-Joseph Piérart, had attempted to elaborate alternative approaches, such figures had occupied a marginal place in the heterodox landscape: Allan Kardec's theories, and the organization he founded to support them, remained dominant. In the early 1880s, this unity broke down. Some leaders began to look beyond Kardec for new approaches, while others chose to focus on contrasting aspects of the founder's original ideas. Disputes about money and institutional priorities exacerbated these differences, as did the assertiveness of a few charismatic independent mediums. This fragmentation made Spiritist journalism considerably more contentious than it had been in the past, since each group defended its views in polemics directed at the others, but it also marked a dramatic increase in the movement's general vitality. Judging by the sheer quantity of periodicals, societies, and independent mediums, the decentralized, argumentative, exuberantly diverse Spiritism of the 1880s and 1890s attracted an unprecedented number of adherents.

The central disagreement among Spiritists during this period stemmed from conflicting views of the movement's nature: Was it a "moral" enterprise, a source of consolation for those in mourning, or was it a "scientific" project, the primary goal of which was to elaborate an empirical approach to the beyond? The dominant spokesmen for each of these views emerged as central figures in fin-de-siècle Spiritism as a whole: Léon Denis became the leading exponent of the moral conception, Gabriel Delanne of the scientific. Moral Spiritists generally accepted Kardec's teachings as articles of faith, while scientific ones viewed them as hypotheses still in need of definitive proof. For most Spiritist commentators, this division was gendered: Supporters and opponents of both approaches equated the moral with femininity and the scientific with masculinity.

[48] Pierre Janet, "Revue générale, le spiritisme contemporain," *Revue philosophique* 33 (Jan.–June 1892): 426–427.

Throughout the 1880s and well into the 1890s, as Janet observed, neither of these factions took much notice of the developments that were transforming the field of psychology. Instead, both accepted the principle that automatic writing was necessarily of otherworldly origin, even as new theories of mind rendered this assumption increasingly tenuous. At the end of the 1890s, however, developments in the field of psychology became impossible to ignore. The key turning point came in 1899, when Camille Flammarion, still the movement's best-known fellow traveler, publicly asserted that spirit communications were in fact "unconscious reflections," either of the medium's own ideas or of the ideas of others present at the séance.[49] The press coverage this admission generated forced Spiritists into a belated engagement with the multivalent self, which eventually transformed the movement. Kardec had built his vision of consoling science—and scientific consolation—on the fundamental assumption of Positivism: that there was a coherent, indeed permanent self and that such a self's perceptions of the world were transparently accurate. In the first years of the twentieth century, new intellectual developments made these assumptions seem problematic. By the early 1930s, with the triumph of Freudian psychoanalysis and the decline of psychical research, Spiritism's former connections with science had disappeared. Belief in spirits once again became largely a matter of faith.

Reason, Emotion, and the Rise of Two Spiritisms

The collapse of centralized Spiritism dated from the early 1880s, when the moral authority Leymarie had acquired as a martyr for the Spiritist cause began to erode. His shortcomings as a Spiritist leader derived from the same traits that had been his undoing in his dealings with Buguet: an incorrigible enthusiasm for dramatic and novel ideas, coupled with a lack of sensitivity to appearances, particularly when money was involved. Leymarie's difficulties started when the Romantic Socialist Charles Fauvety became president of the independent organization formed in the wake of Kardec's passing, the Société scientifique d'études psychologiques. Fauvety's unabashedly left-wing political views, which he espoused volubly in the pages of the *Revue spirite*, appealed strongly to Leymarie. Other Spiritists, however, did not share this enthusiasm, particularly when Fauvety argued in favor of what one horror-struck critic, the medium Berthe Fropo, described as "*Free Marriage*, without even the mayor and his sash, or the laws that protect Society from adultery and immorality." Ideas like these, Fropo declared, made it seem as if "these people's mission is to return us

[49] See Camille Flammarion, "Les Problèmes psychiques et l'inconnu," *Annales politiques et littéraires*, May 7, 1899, 291–293.

to barbarism." Many other Spiritists, including Kardec's widow, appear to have shared Fropo's reservations.[50]

Leymarie joined this fondness for utopian politics to a lively interest in new spiritual ideas, which he expressed with increasing boldness after the death of Kardec's widow in early 1883. In the spring of that year, for example, Leymarie delivered an address that questioned whether it was necessary to "always sing the same praises of the Spiritist *maître*."[51] Kardec's texts had provided the movement's point of departure, he told his audience, but they were not the only available sources of knowledge about the beyond. In his quest for further enlightenment, Leymarie explored a variety of alternatives, including Anglo-American Theosophy and the ideas of the Bordeaux lawyer Jean-Baptiste Roustaing, whose writings controversially suggested that Christ had not been a human being but rather an unusually powerful full-form spirit materialization.[52] Intellectual restlessness of this kind angered many committed Spiritists. As one disgruntled former contributor to the *Revue spirite* observed, Leymarie's decision to join the Theosophical Society was "something like a Protestant pastor becoming a devotee of the Catholic, Apostolic, and Roman Church."[53]

Leymarie's financial management of the Société de la caisse générale et centrale du spiritisme—called the Société scientifique du spiritisme after 1883—deepened these tensions. His tastes grew increasingly grandiose, and he appeared to be willing to alter his ideas in order to fund them. In 1881, he moved the society headquarters, which included his family's lodgings, from a modest building on the rue de Lille to a more impressive one on the rue Neuve-des-Petits-Champs, with a rent of 4,600 francs per month.[54] More disturbingly still, Leymarie began receiving donations from a wealthy Bordeaux businessman, Jean Guérin, who was an ardent

[50] Berthe Fropo, "Beaucoup de lumière" (Paris: Imprimerie Polyglotte, 1884), 36, 22. Italics in the original. Fauvety espoused controversial social ideas more vigorously in a separate journal, *La Religion laïque*, free copies of which were distributed to *Revue spirite* subscribers in 1881. See Fropo, "Beaucoup de lumière," 23. Leymarie and his allies considered Fropo's attack threatening enough to issue a detailed rebuttal. See Société scientifique du spiritisme, "Fictions et insinuations, réponse à la brochure 'Beaucoup de Lumière' " (Paris: Librairie des études psychologiques, 1884).

[51] Quoted in ibid., 30–31.

[52] See Jean-Baptiste Roustaing, *Spiritisme chrétien, ou révélation de la révélation, les quatre Evangiles suivis des commandements, expliqués en esprit et en vérité par les évangelistes assistés des apôtres—Moïse*, 3 vols. (Paris: Librairie centrale, 1866).

[53] *Le Spiritisme* 1, no. 20 (Dec. 15, 1883): 2. While Spiritists, Theosophists, and Occultists made common cause at international spiritualist congresses during this period, their journals tell a different story. Spiritists and Occultists occasionally contributed to one another's publications, but as the angry response to Leymarie's interest in Theosophy indicates, these exchanges did not bridge fundamental philosophical rifts. To varying degrees, Spiritist journals attacked Theosophical and Occultist ideas as retrograde and irrational. For the congresses, see e.g. *Compte rendu du Congrès spirite et spiritualiste international, tenu à Paris du 9 au 16 septembre 1889* (Paris: Librairie Spirite, 1890); and *Compte rendu du congrès spirite et spiritualiste international, tenu à Paris du 16 au 27 septembre 1900* (St. Amand: Imprimerie Daniel-

disciple of Roustaing. Several Spiritists noted that once he had acquired this new source of income, Leymarie's enthusiasm for Roustaing's ideas increased markedly—even though Kardec himself had condemned them outright in 1868.[55]

By the end of 1882, Leymarie's flamboyant political radicalism, enthusiasm for novel spiritual systems, and un–self-conscious pursuit of material opulence had created a considerable amount of discontent in the Société de la caisse générale. A group of particularly irate members, led by Fropo, the young engineer Gabriel Delanne, and an aspiring lecturer from Tours named Léon Denis, decided to found a new society, the Union spirite Française, which would also produce an inexpensive journal. Despite its clear philosophical differences with Leymarie, this organization began as an affiliate of the Société de la caisse générale. The Union spirite broke away from its parent society, however, when Leymarie opposed the creation of a journal that might compete with the *Revue spirite.*[56] In March 1883, the first issue of the new journal appeared. It was an eight-page biweekly called *Le Spiritisme,* edited by Delanne, which sold for a very reasonable 10 centimes. After three months, it had 660 subscribers.[57]

The new publication gradually became a major force in French heterodox life, serving as the focal point for a rapidly expanding number of national spirit societies and local study circles. Between 1889 and 1893, five new national spirit societies appeared: the Comité de propagande spirite (1889), the Société du spiritisme scientifique (1890)—not to be confused with Leymarie's Société scientifique du spiritisme (1883)—the Société fraternelle spirite (1891), and the Fédération spirite universelle

---

Chambon, 1902). For Spiritist critiques of Theosophy and Occultism, see e.g. *Le Spiritisme* 2, no. 11 (Aug. 1–15, 1884): 1–7; *La Lumière* 3 (Oct. 1884–Feb. 1886): 153–155, 163–164, 169–171; *Le Spiritisme* 6 (1888): 33–34; *Le Spiritisme* 8 (1890): 129–133, 161–163, 177–180; *Le Spiritisme* 9 (1891): 2–6, 17–22, 25–26, 33–37, 49–52, 81–84, 97–104, 113–116, 124–135, 161–167, 177–185; *Le Spiritisme* 10 (1892): 81–85; *La Lumière* 4 (Jan. 1891–Dec. 1892): 25–29, 37–40, 49–52; *Le Progrès Spirite* 4 (1898): 70–71.

[54] Fropo, "Beaucoup de lumière," 22.

[55] Quoted in ibid., 33.

[56] Ibid., 20. Leymarie, Denis and Delanne had reconciled by 1889, when they joined forces to support the Congrès spirite international, but five years previously their relations had been quite strained. As the official organ of the Union spirite, to which both Denis and Delanne belonged, *Le Spiritisme* publicized Fropo's allegations and critiqued the ideas of Roustaing. See e.g. *Le Spiritisme* 1, no. 17 (Nov. 1–15, 1883): 1–3, 6–7; *Le Spiritisme* 2, no. 7 (June 1–15, 1884): 1–2. The journal also printed several sharp *ad hominem* attacks against Leymarie, whose actions one contributor described as "disgusting." For the quote, see *Le Spiritisme* 2, no. 9 (July 1–15, 1884): 4. For the other attacks, see *Le Spiritisme* 2, no. 9 (July 1–15, 1884): 1–5; no. 11 (Aug. 1–15, 1884): 8–11; no. 19 (Dec. 1–15, 1884): 7–8; no. 20 (Dec. 16–31, 1884): 8–10. Articles attacking Leymarie continued to appear occasionally well after this initial spate. See e.g. *Le Spiritisme* 4 (1887): 295–296.

[57] *Le Spiritisme* 1, no. 12 (Aug. 15, 1883): 2.

(1893). Though the membership of these groups often overlapped—the indefatigable A. Laurent de Faget, for example, presided over two of them—none was formally affiliated with any other, and none had any provincial branches. At the same time, an increasing number of private, independent study circles began to appear in Paris.[58] A similar proliferation both of formal societies and of more intimate groups took place elsewhere in France, particularly in the South: in 1891, for example, Delanne listed twelve groups in Lyon, five in Marseille, three in Carcassonne, and two in Bordeaux.[59]

While all of these groups claimed to be following Kardec's example, they each conceived of that example in a different way. A burst of philosophical innovation, therefore, accompanied the growth of these diverse organizations. One of the most influential new approaches came from a group in Marseille that called itself the "Immortalists," founded in the mid-1880s by the journalist Marius George. In his periodical, *La Vie posthume*, George urged Spiritists to abandon mysticism in favor of reason, which in his view demanded the rejection not only of prayer and metaphysical speculation but also of faith in a Christian God. All were elements of a "useless and mystical creed," which Kardec had temporarily imposed to make his ideas palatable to the movement's "first converts," who had still labored "under the yoke" of "old dogmas."[60] For Spiritism to become the rational doctrine its apologists wished it to be, George argued, it would need to be stripped to its essence—the simple fact that spirits could communicate with the living.

George and his contributors discussed their "war on mysticism" in gendered terms.[61] Female believers, they argued, were largely responsible for the superstition and irrationality that continued to mar the purity of Spiritist belief. Immortalist A. Martelin expressed this view in an acid description of a "pietist séance" held by a rival society in Marseille: At the climax of the meeting, he wrote, a young female medium was possessed by the spirit of John the Baptist, who then ceded her body to Christ himself. After delivering a sermon in verse, the medium incarnating Christ lay back on a table and, writhing in pain, reenacted the final moments of the Passion. Commenting on this bit of religious theater, Martelin devoted particular attention to the audience's reactions:

> All the women, without exception, threw themselves on their knees while pouring forth copious tears. As for the men, they remained seated, with

[58] See the list of "groupes spirites parisiens" in ibid., 2, no. 17 (Nov. 1, 1884): 12; and ibid., 2, no. 20 (Dec. 15, 1884): 11–12, which includes a list of provincial groups as well.

[59] Ibid. 9 (1891): 110–111.

[60] *La Vie posthume* 2, no. 7 (Jan. 1887): 145; 1, no. 1 (July 1885): 2.

[61] Ibid. 1, no. 1 (July 1885): 4.

contemplative and respectful expressions, but it was easy to see their embarrassment; they made a point of not looking at one another.[62]

Spiritism, as Martelin described it, seemed to be pulled in different directions by its female and male adherents. The women at this gathering were unapologetically emotional and unapologetically Christian. Spiritism, from their perspective, seemed to provide a more intense version of orthodox religious experience. The men, on the other hand, found such extravagant displays of feeling disturbing: Behavior of this kind was what they had come to Spiritism to escape. The rationalizing project of Immortalism, therefore, was emphatically masculine. For George and Martelin, it fell to male Spiritists to check the mystical impulses of their female counterparts.

These tendentious pronouncements led others to take Spiritist ideas in a very different direction, attacking *La Vie posthume* in lengthy polemics and embracing the forms of speculative and emotional discourse Immortalists eschewed. Instead of viewing women and their enthusiasm as a threat, these innovators reveled in what they perceived to be the distinctive moral power of femininity, which they presented as an integral part of a larger, increasingly eclectic religious project. In the pages of the medium Lucie Grange's long-lived journal *La Lumière*, for example, articles touting the moral value of feminine sentiment appeared alongside pleas for a universal communion of love through prayer, essays on esoteric Christianity by the notorious ex-priest Joseph-Antoine Boullan, and spirit communications from the legendary mage Hermes Trismagistus.[63] The journalist Eugénie Potonié Pierre presented Spiritism, Fourierism, and feminism as linked projects in her periodical, *L'Humanité intégrale*. She also accorded a prominent place to communications by the medium Rufina Noeggerath, who did not submit her automatic writings to any expert's approval, and un–self-consciously signed them with famous names ranging from Sakyamuni to Robespierre.[64] The predominantly male contributors to *Le Spiritisme* and its successor *Le Progrès spirite* also embraced a version of this emotional, speculative approach. Though these two journals hewed closer to Kardec's texts than Grange, Potonié Pierre, and Noeggerath, they shared interests in feminist causes and the uplifting power of regular communication with the beyond.

[62] Ibid., 2, no. 11 (May 1887): 249.

[63] *La Lumière* 6 (Jan. 1891–Dec. 1892): 29–30; 4 (Nov. 1886–Feb. 1888): 211–218; 6 (Jan. 1891–Dec. 1892): 3; Lucie Grange, *La Mission du nouveau-spiritualisme, lettres de l'esprit de Salem-Hermès, communications prophétiques* (Paris: chez l'auteur, 1896). In *La Lumière*, Boullan signed his articles "Dr. Johannès," alluding to the famous character modeled after him in Joris-Karl Huysmans, *Là-Bas* (Paris: Plon, n.d.). See Bibliothèque nationale, fonds Lambert, dr. 30, letter from Boullan to Huysmans dated June 14, 1891.

[64] *L'Humanité intégrale*, January 1897; Rufina Noeggerath, *La Survie, sa réalité, sa manifestation, sa philosophie, échos de l'au-delà* (Paris: Flammarion, 1897).

By the mid-1890s, this proliferation of novel ideas had changed the shape of the movement. What was once a relatively centralized network had become a diverse array of independent periodicals and organizations with sometimes radically different points of view. In 1897, an eager Spiritist could subscribe to five national journals, and could supplement them with a host of local publications—to say nothing of the ever-expanding book and pamphlet literature. While few authors or groups took exactly the same approach, and some charted markedly idiosyncratic courses, the bulk of the innovation followed a pair of well-defined paths.[65] As the lecturer Emma Lequesne predicted in 1884, rationalist critiques did not silence mediums or diminish the emotional intensity of séances. Instead, they led to the emergence of two Spiritisms: a sentimental "consoling Spiritism" on the one hand and an austere "scientific Spiritism" on the other.[66]

Denis, Delanne, and the Problem of the Multivalent Self

During this period of ferment, Léon Denis and Gabriel Delanne emerged as the movement's key figures, personifying the growing divide between moral and scientific approaches (figs. 15 and 16). In keeping with the new organizational ethos, both men led societies, though neither devoted the bulk of his time to such matters. Denis wrote books and toured the provinces delivering lectures. He became known for his ability to move large audiences by expounding on the consoling power of Spiritism in a florid, self-consciously literary style. While Delanne was not above the occasional rhetorical flight in his own numerous books and articles, his approach was more methodical. Rather than seeking to move his audiences, he sought to reason with them, applying the insights of psychical research to Spiritist practice. Both men presented themselves as synthesizers and explicators of Kardec's teachings, but their conceptions of his thought differed dramatically. The full extent of this difference, which the two men downplayed throughout their careers, became clear in the late 1890s, when Delanne finally began to address the problems new psychological theories posed.

For Denis, the chief benefits of Spiritism were its popular appeal and its capacity to provide consolation in a manner consistent with anticlerical republican ideals. The discoveries of psychical researchers provided an aura of up-to-date empirical legitimacy, but the specific act of communicating with deceased loved ones struck him as being of far

[65] For an example of idiosyncracy that generated a considerable amount of controversy in its time, see Arthur d'Anglemont, *Omnithéisme, Dieu dans la science et dans l'amour,* 6 vols. (Paris: Comptoir d'Edition, 1891–1896).

[66] *Le Spiritisme* 2, no. 6 (May 1884): 3–5.

Fig. 15. Gabriel Delanne. (Collection of the author.)

more immediate value. In the first of his many books, he declared that Spiritism,

> of all systems, is the only one that provides objective proof of the survival of being and a means of corresponding with those we improperly term the dead. With its help, we can converse with those we loved on earth and thought forever lost; we can receive their teachings, their advice. . . . Spiritism . . . shows everyone a more worthy and elevated goal. It brings a new sentiment of prayer, a need to love, to work for others, to enrich our intelligence and heart.

In passages like this one, the reference to "objective proof" served primarily as a rhetorical device, not a methodological injunction. Denis used the trope of empiricism to indicate the modernity and universality of Spiritism and to differentiate it from all the "dreams of an unhealthy mysticism" and "myths born of superstitious beliefs" that had preceded it.[67] This desire to overcome superstition, however, did not involve an abandonment of the tangible, emotionally intense experience of the sacred that Denis continued to place at the movement's core. Kardec's new

[67] Léon Denis, *Après la mort, exposé de la doctrine des esprits* (Paris: Editions Jean Meyer, n.d.), 343–344.

Fig. 16. Léon Denis. (Collection of the author.)

doctrine, he argued, had not served to "eliminate religious sentiment and the notion of God from the human heart," but instead had "laicized them, elevated them, purified them" by embracing the achievements of modern civilization.[68]

Denis also readily acknowledged "the immense role that woman has played" in Spiritism. Where Kardec had striven to emphasize the movement's masculine character, Denis feminized it. As he told a large audience in Paris in 1898, women benefited from a "refined and delicate organization," which gave them "the privilege of vibrating more intensely to the breath of the ideal." It was therefore natural, in his view, that they would be more susceptible to "spiritual influences," both as mediums and as practitioners of "the doctrines of charity, solidarity, and love that are the new spiritualism's acts of faith."[69] The masculine detachment that psychical researchers used in their study of the beyond provided an important foundation, but for Denis it was the moral authority of women that gave Spiritism its true power.

[68] *Le Spiritisme* 7, no. 6 (June 1889): 83.

[69] Quoted in *Revue scientifique et morale du spiritisme,* 3, no. 5 (Nov. 1898): 259. Statements of this kind, explicitly linking mediumism and femininity, were rare in the French Spiritist literature before the fin de siècle—another aspect that differentiates the French case from the Anglo-American one.

While Delanne agreed with Denis about Spiritism's ability to redeem religion for the modern age, he had a very different conception of the process by which that redemption would be achieved. For Delanne, Spiritism's primary value was intellectual, not experiential. As he saw it,

> Our doctrine is distinguished from all others by its special character of certainty. In our explanations, nothing is hypothetical. At every moment, the theory is based on phenomena, which provide its strength and authority.

This uncompromising approach forced him to acknowledge the methodological shortcomings of Kardec's texts. When Kardec elaborated his philosophy, Delanne observed, he had used the methods of his time, not always drawing a sufficiently sharp distinction between reasoned philosophical speculation and empirical induction. The notion of expiatory reincarnation, for example, was logically well-founded but did not have the support of compelling experimental data: This central tenet of Kardec's doctrine, therefore, remained a mere hypothesis. Delanne believed that his task was to correct such shortcomings by bringing a new empirical rigor to Spiritism. While this would involve a short-term renunciation of the experiential elements Denis valued so highly, in the long term, it would lead to a stronger doctrine, one in which every principle could be based on "the experimental study of nature."[70]

Like Leymarie in the halcyon years before the *procès des spirites,* Delanne sought to accomplish this goal by drawing on psychical research. Where the older man had been content to cite and popularize the work of *éminences* like Crookes, however, Delanne pushed the endeavor further, drawing on the scientific training he had received as an engineer to elaborate a distinctively Spiritist approach to the experimental study of supernormal phenomena. In 1896, he established a new journal for this purpose, the *Revue scientifique et morale du spiritisme.* To give his journal an appropriately rigorous tone, Delanne self-consciously avoided the automatic writings that featured so prominently elsewhere in the Spiritist press; instead, he published articles that bore a closer resemblance to those in the *Annales des sciences psychiques.* He also established an organization devoted to the objective study of supernormal phenomena, the Société française d'étude des phénomènes psychiques. By the end of the 1890s, he had become an important presence in the community of French psychical researchers.

Spirit communications, as Delanne himself acknowledged, were a major obstacle to the thoroughgoing rationalization he envisioned. These texts, however emotionally uplifting they may have been, usually could not serve as unambiguous proof of the reality of spirit intervention.

[70] Ibid., 3, no. 7 (Jan. 1898): 386, 388.

Communications from famous writers or historical figures, for example, often failed to exhibit "the style that initially made their names," while messages from loved ones tended to lack specific detail that could irrefutably prove the identity of the spirit in question.[71] In both cases, as thinkers such as Janet, Myers, and Flournoy had argued, spirit intervention seemed to be an explanatory hypothesis that required a decidedly unscientific multiplication of causes.[72] Looking within the automatic writer's own mind for an explanation, in contrast, had the advantage of logical directness.

To earn the intellectual legitimacy he craved, therefore, Delanne recognized that he would need to engage with the new psychological theories underpinning these objections. In the early years of his journal, he was content to publish articles that attacked the concept of subconscious intellectual activity in general terms. By the summer of 1899, however, the notion of the subconscious was gaining credibility in the eyes of the general public and a more direct response was required. Discussions of subjects like hysteria, hypnotic suggestion, "maladies of the will," and the irrational behavior of crowds had become common in journalism; they contributed to a broader sense that the older, Positivist understanding of the self—and the nature of perception more generally—had failed to encompass much that was crucial in human experience.

Camille Flammarion demonstrated how far the intellectual tide had turned against Spiritists in May 1899 when he published an article on spirit communications in the widely read biweekly *Annales politiques et littéraires.* An objective analysis of documented spirit messages, Flammarion argued, could lead to only one conclusion: "in the majority of cases . . . the communications of [séance] tables reflect the thoughts of one or several of those in attendance." Victor Hugo's table had communicated in Romantic verse, for example, while the table at the *Démocratie pacifique* expressed Fourierist ideas. This observation led Flammarion to make a confession of his own. In the early 1860s, he had produced several automatic writings that he had believed were dictated by the spirit of Galileo and had submitted them to Kardec, who then incorporated them into his book *Le Genèse selon le spiritisme.* Now, though, Flammarion doubted the authenticity of the signature. These communications, he argued, were merely "the reflection of what I knew, of what we thought at that time, about the planets, the stars, cosmogony, etc." His alleged Galileo had told him, for example, that Saturn had eight moons, when by 1899 astronomers had discovered a ninth. If the communication had in fact come from the posthumously enlightened spirit of Galileo, Flammarion reasoned, it would have provided better

[71] Ibid., 2, no. 5 (Nov. 1897): 310.

[72] See, e.g., Flournoy's address to the 1900 Psychology Congress, *IVe Congrès,* 108.

astronomical data. Flammarion unsettled his Spiritist admirers even further by generalizing this observation. "All of my experiments to verify *the identity of a spirit*," he confessed, "have failed."[73]

For Delanne and other Spiritists, Flammarion's confession was explosive. The daily press covered the famous scientific popularizer's "defection" from the Spiritist cause extensively; the torrent of journalistic commentary attested to just how much intellectual currency new psychological theories of the multivalent self had acquired.[74] Flammarion's public confession, therefore, not only challenged the legitimacy of one of Spiritism's canonical texts but also forced believers to confront the extent to which new psychological theories undercut the premise on which all of Kardec's books depended for their authority: the notion that the "speaking ego" was a stable, indivisible entity. Flournoy's study of Hélène Smith, which appeared several months later to widespread acclaim, deepened the growing sense of crisis. The unitary, reasoning subject that Kardec had taken for granted now seemed an outdated concept.

If Spiritism were to survive in the face of this unprecedented challenge, Delanne believed, it would need to find a way to acknowledge the new theories of the multivalent self while simultaneously making a case for the possibility of spirit intervention. Delanne attempted to accomplish this ambitious project in his book *Recherches sur la médiumnité*, which initially appeared as a serial in the *Revue scientifique et morale du spiritisme*. He began this voluminous study by positing a radically new approach to spirit communications. Recent work in psychology, he wrote, had forced him to recognize

> *that the automatic character of the writing is insufficient to serve as the sole criterion of mediumism.* If a person writes without being conscious of what his hand traces on the paper, it does not necessarily follow that his hand is under the influence of spirits; the hand can certainly write of its own accord, as an array of indisputable examples have shown.[75]

Most communications, in other words, were not the work of spirits, but instead were of either unconscious or telepathic origin. A few exceptional messages, however, did come from the beyond: Those written in languages unknown to the medium or those that far exceeded the medium's normal intellectual capacity, for example, could still logically be ascribed to spirit intervention. By admitting the reality of subconscious mental activity in this

[73] *Annales politiques et littéraires*, May 7, 1899, 292, 293. Italics in original.

[74] For "defection," see *L'Eclair*, July 9, 1899, 1. For a sense of the extent of the coverage, see the scrapbook in the collection of the FCF, which includes numerous clippings from seventeen large-circulation newspapers.

[75] Gabriel Delanne, *Recherches sur la médiumnité* (Paris: Editions de la BPS, 1923 [1900]), vii. Italics in original.

way, Delanne transformed authentic spirit communication from something quite common into a marked rarity—only a precious few French mediums spontaneously wrote in Greek or automatically produced long works of historical scholarship without the benefit of any terrestrial research. Reconciliation with psychology, therefore, came at a tremendous price: To preserve the possibility of otherworldly intervention, Delanne had to deprive ordinary believers of the experience of personal consolation that made Spiritism attractive in the first place.

While Spiritists remained quick to acknowledge Delanne as a *frère en croyance* (brother in belief) throughout his long life, few chose to adopt his rigorous approach. Instead, many faced the challenges psychology posed by following the example of Denis, who broached the subject in a more ambivalent manner. Like Delanne, Denis admitted that the medium's own mind could exert a subconscious influence on spirit communications. At the same time, however, Denis insisted that authentic communications were common and easily recognizable, especially when they came from departed loved ones. Mediums frequently revealed intimate details known only to the deceased and his or her family, Denis asserted, but these instances were often too personal to be publicized. Few Spiritists wanted to expose "the most sacred sentiments, the most intimate secrets of their hearts" to public scrutiny.[76] What Delanne had seen as a paucity of evidence, then, struck Denis as an indirect demonstration of the reality of spirit intervention—the emotional impact of these messages was simply too strong to be inauthentic.

As alternative explanations for trance phenomena forced believers to ask uncomfortable questions about the evidentiary value of spirit communications, the practices of mediums changed. Most strikingly, in the séances Denis led and in those reported in journals like the *Progrès spirite*, automatic writing gave way to trance speech. As new psychological theories made text into an increasingly slippery form of evidence, acting took its place. When speaking through *médiums à incorporations* (mediums who embodied spirits) at his group's séances, Denis observed, the most evolved souls tended to communicate in a lofty, impersonal manner. To "convince skeptics," however, these exalted beings called in a cast of "lesser Spirits," who played a variety of character roles: "a street vendor, a village blacksmith, a talkative old maid, and many more."[77] According to Denis, the uncanny accuracy of these portrayals proved that they could only be the products of otherworldly intervention. These characters were real spirits, in other words, because they *felt* real. During and after the First World War,

[76] Léon Denis, *Dans l'invisible, spiritisme et médiumnité,* nouvelle édition (Paris: Librairie des sciences psychiques, 1911), 389.

[77] Ibid., 390.

as we will see, this approach came to define the movement as a whole: In the final analysis, the palpable experience of the sacred that Spiritism offered its believers proved more attractive than any promise of full acceptance by the mainstream scientific community.

The Occultist Synthesis

In his 1892 historical survey of Spiritist literature, Janet did not limit himself to authors who claimed to perpetuate Allan Kardec's legacy. Over the course of the two decades since he had begun studying the subject, Janet noted, "the old classical Spiritism" had received increasing competition from a group of "new schools more pretentious and more obscure." The most important of these new currents was a movement called "Occultism," which drew upon the traditions of alchemy and Renaissance Hermeticism as they had been adapted and developed by a small group of French thinkers in the eighteenth and nineteenth centuries. Occultists, Janet observed, had founded a number of organizations, wrote prolifically, and had all "promoted themselves to the exalted rank of mage." For Janet, the philosophy of this new movement, which seemed to owe more to "high scholasticism" than to modern thought, was an ironic consequence of scientific psychology's growing intellectual prestige. The best thinkers of the age had turned away from metaphysics, but ordinary people remained "preoccupied by the mysteries of providence and the life to come." Occultist journals, whose articles were "long, muddled, very obscure, full of personifications and metaphors," addressed this need by creating a "popular metaphysics," which bore the same resemblance to academic philosophy as "popular dramas and novels" bore to more cultivated forms of literature.[78]

Shortly after Janet's article appeared, the *Revue philosophique* published an emphatic rebuttal from a young doctor named Gérard Encausse. Under the pseudonym Papus, Encausse had emerged as France's leading Occultist, and he strongly objected to Janet's characterization of the new movement as a backward-looking "popular" phenomenon. Janet, in his effort to "seek out the ridiculous aspects," had overlooked the sophistication of Occultism and its exponents. Encausse himself, for example, was a formally trained and active practitioner of scientific psychology:

> I have just finished my medical studies at the Paris *faculté*; during my service as a hospital extern, I received a bronze medal . . . In addition, I have at this point published sixteen volumes or treatises; two of them have prefaces graciously

[78] Janet, "Spiritisme," 418, 433, 440, 441.

contributed by M. Ad. Franck, a member of the *Institut*. Currently, I direct the hypnotism laboratory established by Dr. Luys at the Hôpital de la Charité.[79]

Janet, in other words, had been incorrect to assume that a desire to revive the Hermetic tradition necessarily entailed a willful ignorance of the latest developments in contemporary thought. On the contrary, for Encausse, psychological studies and the pursuit of esoteric knowledge were complementary endeavors. Occultism, as he and his colleagues would assert repeatedly in more sympathetic venues, was a powerful tool for understanding the broader implications of the same phenomena Janet and other psychologists used as the basis for their novel, if more narrowly conceived, theories of mind.

Enthusiastic engagement with the intellectual concerns of the moment, in fact, contributed strongly to Occultism's appeal during the fin de siècle. Beginning in the mid-1880s, works by writers such as Joséphin Péladan transformed magic and esotericism into key elements of the Decadent sensibility, and by 1892, Occultism had become a notable presence on the French cultural scene. That year, Encausse edited a special issue of a leading Symbolist literary magazine, *La Plume*, while Péladan—who differed on key points of doctrine but shared a similar attitude—organized the first Salon de la Rose-Croix, featuring works by artists who used images drawn from esoteric traditions "to ruin realism, reform Latin taste, and create a school of idealist art."[80] This ever-increasing cultural cachet led to the steady expansion of the various organizations Encausse and his fellow Occultists coordinated, which included three major societies, the Ordre Martiniste, the Ordre Kabbalistique de la Rose-Croix, and the Groupe indépendant d'études ésotériques. Adherents of these groups and others like them received a disproportionate amount of attention from journalists in search of glamorous and novel subject matter.

Where Spiritism remained wedded to Kardec's musty blend of Romantic Socialism and Positivism, Occultism was self-consciously au courant. An inexhaustible stream of Occultist books, pamphlets, and journals affirmed such fin-de-siècle nostrums as the decadence of modern European civilization, the epistemological shortcomings of Positivism, and the importance of unconscious mental processes. More strikingly still, Occultist texts imbued these ideas with a reassuring metaphysical significance by reinterpreting them in light of a much older esoteric tradition. For the Occultists, potentially disconcerting new ideas about the insufficiency of human reason, the

[79] Papus [pseud. of Gérard Encausse], "Correspondance", *Revue philosophique* 33 (Jan.–June 1892): 574–575.

[80] Joséphin Péladan, "Salon de la Rose+Croix, règle et monitoire," quoted in Robert Pincus-Witten, *Occult Symbolism in France: Joséphin Péladan and the Salons de la Rose-Croix* (New York: Garland, 1976), 211.

fragility of free will, and the multivalent nature of the self became the philosophical basis for a timeless-seeming approach to the sacred. Occultists shared the Spiritist focus on concrete phenomena, but as Alex Owen has observed in the British context, they viewed those phenomena in active, not passive terms: as products of individual cultivation and mental discipline, rather than signs of intervention by the deceased.[81] Though the Occultist movement waned after the First World War, the path to transcendence it introduced, with its stress on individual "seekership" and the exploration of mental regions beyond the sphere of ordinary consciousness, would become an important element of twentieth- and twenty-first-century heterodoxy.

The Emergence of Organized Occultism

Like Spiritism before it, organized fin-de-siècle Occultism began with an American incursion: the arrival of the Theosophical Society in France. The first meeting of the Society took place in New York City on September 17, 1875. Its two founders, the lawyer Henry Steele Olcott and the charismatic Russian émigrée Helena Petrovna Blavatsky, had both been involved with American spiritualism—one as a journalist and frequent participant in séances, the other as a medium. The new society they created, however, was much more ambitious than a typical spiritualist circle. Its members sought to discover nothing less than the essence of religion itself, which they believed took the form of a unified body of esoteric knowledge subtending all the world's diverse and seemingly irreconcilable faiths. Though they deplored the naïveté of spiritualists, the Theosophists did not dismiss supernormal phenomena altogether. In their view, an ability to produce uncanny manifestations at will was one of the most dramatic benefits conferred by a thorough knowledge of the primordial "Wisdom Religion," the hidden heart of all human creeds.[82]

Blavatsky was the new society's primary spiritual authority and shaped the esoteric doctrine its members sought to discover. She claimed to have spent seven years in Tibet receiving spiritual instruction from a group of enlightened sages. These "Mahatmas," Blavatsky claimed, remained in telepathic communication with her, and in 1877 inspired her first book, *Isis Unveiled.* In addition to offering a critique of contemporary scientific thought—especially Thomas Huxley and Charles Darwin—the book stressed the importance of the Orient as a wellspring of spiritual knowledge. Hinduism and Buddhism, Blavatsky wrote, constituted "the double source from which all religions sprung." Westerners in search of the

[81] Alex Owen, *The Place of Enchantment: British Occultism and the Culture of the Modern* (Chicago: University of Chicago Press, 2004), 114–147.

[82] Helena Petrovna Blavatsky, *Isis Unveiled: A Master-Key to the Mysteries of Ancient and Modern Science* (New York: Bouton, 1893), 1:515.

truth, then, would need to turn to the East. "White-skinned people," she maintained, would never be able to achieve the "intuitive perception of the possibilities of occult natural forces in subjection to the human will" that "the Orientals" had developed as a matter of course, though this disadvantage could be partially overcome with careful study and training.[83]

These ideas proved intriguing enough to make *Isis Unveiled* a publishing success, but even so, the Theosophical Society grew slowly during its first five years. The real breakthrough came in 1879, when Olcott and Blavatsky left New York for Calcutta. Once there, they founded a journal and won a growing number of adherents among the Anglo-Indian elite. This powerful constituency gave a new impetus to the previously obscure organization. By the early 1880s, the Theosophical Society had become an international phenomenon, with branch lodges in Burma, Ceylon, Britain, and the United States.[84] Tentatively, the new society also spread across the English Channel. Leymarie, a friend of Blavatsky's from her early days as a medium, made an abortive attempt to start a Parisian branch in 1879; five years later, there were two small Theosophical lodges in France, boasting a collective membership of fifty-three.[85]

Theosophy's French presence grew steadily during the second half of the 1880s, as small heterodox periodicals began to discuss the implications of Blavatsky's ideas.[86] Articles in the mass-circulation press spread the word further, giving colorful accounts of the new society, which was presented as a striking, distinctively novel response to "the needs for idealism that have appeared today in the West amidst the excesses of the century's exact sciences."[87] By 1887, the Theosophical Society had truly arrived: That year, Félix-Krishna Gaboriau, an ambitious student of Blavatsky's, published a translation of A. P. Sinnett's *Occult World*, the first book on the teachings of the Mahatmas to appear in French; he also used his own money to establish a Theosophical journal, the *Lotus*.[88] Several months later, he joined a group of young enthusiasts in founding the Isis Lodge, a new, more dynamic Parisian branch of the mother society.

The *Lotus* and the Isis Lodge enjoyed greater success than any of the Theosophical Society's previous ventures in France, but they also proved

[83] Ibid., 2:639, 635.

[84] For an account of the Theosophical Society's early years, see Peter Washington, *Madame Blavatsky's Baboon* (New York: Schocken, 1995), 26–69.

[85] For Leymarie's efforts, see Joscelyn Godwin, "The Beginnings of Theosophy in France" (London: Theosophical History Center, 1989), 7; for the membership number, see Charles Blech, *Contribution à l'histoire de la Société Théosophique en France* (Paris: Editions Adyar, 1933), 35–36.

[86] See, e.g., *La Revue des hautes etudes*, 1 (Sept. 1886–Feb.1887).

[87] *La Vie moderne*, June 4, 1887, 365.

[88] A. P. Sinnett, *Le Monde occulte, hypnotisme transcendant en orient,* intro. and trans. Félix-Krishna Gaboriau (Paris: Carré, 1887).

inordinately conflict-ridden, largely as a result of Blavatsky's pugnacious distaste for Western religion and culture. Writing in the *Lotus*, she attacked Christianity with unstinting virulence, deriding it as the blasphemous worship of "an absurd and grotesque anthropomorphic fetish."[89] Blavatsky had no kinder words for France. "The Revolution of 1789," she declared, "has only produced a single clearly visible result: a false fraternity that says to its neighbor: 'Think as I do, or I will beat you; be my brother or I will strike you down!' "[90] Unless the French renounced their naïve faith in Christianity and their arrogant sense of the importance of their own tumultuous history, they would never be able to follow the path of true wisdom as revealed by the esoteric traditions of the East.

While Gaboriau embraced this uncompromising vision whole-heartedly, many other members of the Isis Lodge were less willing to swallow their cultural pride, particularly if it meant assigning spiritual primacy to India, a nation whose alleged weakness had led it to be "oppressed in the most shameful manner."[91] Instead, they tended to prefer an alternative conception of the esoteric tradition primarily inspired by Martinism, a philosophy first elaborated by a pair of late-eighteenth-century French thinkers, Martinès de Pasqually and Louis-Claude de Saint-Martin. In the aftermath of the Revolution, these ideas inspired a small group of writers, the most important of whom were Antoine Fabre d'Olivet in the early part of the nineteenth century, Eliphas Lévi in the years after 1848, and Joseph-Alexandre Saint-Yves d'Alveydre in the 1880s. By the fin de siècle, the works of these authors had come to define a distinctively French current of esoteric thought, which synthesized elements of classic Renaissance Hermeticism, Mesmerism, Freemasonry, and ritual magic. As David Allen Harvey has observed, exponents of this "invented tradition" espoused a mythic vision of history that accorded France a privileged place in the spiritual development of humanity as a whole and looked to Christian esotericism as a means of overcoming the Revolution's legacy of social, religious, and ideological conflict.[92]

Gérard Encausse quickly emerged as the leading advocate of the Martinist tradition in the Isis Lodge. He had probably discovered this neglected intellectual current sometime in 1886, about a year after he enrolled at the Paris Faculté de médecine.[93] At the time, he was in the midst of an intellectual crisis: His old belief in the "materialist faith" had

[89] *Le Lotus* 2 (Oct. 1887–Mar. 1888): 13.

[90] *Le Lotus* 1 (Mar. 1887–Sept. 1887): 334.

[91] *La Revue des hautes etudes* (Sep. 1886–Feb. 1887): 1:52.

[92] David Allen Harvey, *Beyond Enlightenment: Occultism and Politics in Modern France* (DeKalb: Northern Illinois University Press, 2005), 7; for "invented tradition," see 62–90.

[93] Encausse claimed to have begun his medical studies in 1882; documents found by Marie-Sophie André, however, make a strong case for the later date. See Marie-Sophie André and Christophe Beaufils, *Papus, biographie, la belle époque de l'occultisme* (Paris: Berg International, 1995), 29.

begun to wane. Intensive medical studies, particularly of hypnosis, he later wrote, had shown him that materialism was a radically incomplete way of looking at the world, and that transcendental forces played crucial roles in all aspects of life. Encausse found a means of addressing this problem at the Bibliothèque nationale, where he first encountered the dusty volumes of Saint-Martin, Lévi, the mid-nineteenth-century alchemist Louis Lucas, Fabre d'Olivet, Saint-Yves, and others. Increasingly, he began to neglect his medical books for "works by alchemists" and "old tomes of magic spells."[94]

In 1887, this project of spiritual discovery forced Encausse out of the library. Many of the thinkers he studied stressed the importance of direct personal contact with a teacher. Books, they maintained, provided but a portion of a larger body of knowledge, which could be comprehended fully only by a reader who had been initiated into the tradition through a secret society. Encausse would eventually claim to have begun his initiation in 1882, as a youth at the feet of the dying Henri Delaage, who, according to legend, was among the last men with a direct connection to the Martinist secret societies of the late eighteenth century.[95] In fact, however, the young medical student is considerably more likely to have started his initiatic journey with a visit to the Isis Lodge. When the Theosophical Society approved the group's bylaws in October 1887, Olcott named Encausse "Delegate of Adyar," a minor officer's position.[96] Soon, he began delivering lectures to the rest of the lodge—not on Blavatsky, but on the French tradition he studied so passionately at the Bibliothèque nationale. Encausse quickly developed into an engaging public speaker, who expressed himself with what one observer later described as "the slightly vulgar facility that moves a middlebrow audience."[97] He also began writing books and pamphlets of his own, which he signed "Papus," a pseudonym borrowed from Lévi's translation of the *Nuctemeron*, a text ascribed to the third-century Greek mage Apollonius of Tyana.[98] As Papus, Encausse adopted a new persona: The ordinary medical student became a dashing, self-possessed master of arcane wisdom.

At the same time, Papus continued to pursue his medical studies. Around 1887, he began work as an extern at the Hôpital de la Charité, under the supervision of the psychiatrist Jules-Bernard Luys, a student of Charcot who had broken with his teacher by developing an alternative

[94] *L'Initiation* 29 (Oct.-Dec. 1895): 196, 198.

[95] For this story, see Philippe Encausse, *Sciences occultes, Papus, sa vie, son œuvre* (Paris: Editions OCIA, 1949), 57–59.

[96] André and Beaufils, *Papus*, 38.

[97] Victor Emile Michelet, *Les Compagnons de la hiérophanie, souvenirs du mouvement hermétiste à la fin du XIXe siècle* (Paris: Dorbon, 1937), 33.

[98] See Eliphas Lévi [pseud. of Alphonse-Louis Constant], *Dogme et rituel de la haute magie* (Paris: Niclaus, 1948), 2:385–401.

approach to hypnosis, one that induced trance by more gentle means than the abrupt shocks—loud bells and gongs, for example—used by the famous *Napoléon des névroses*. When Papus joined Luys's clinical staff, Luys was studying the influence of medicines at a distance. In the trials Luys conducted, hypnotized patients would exhibit the physical effects of a drug simply by coming into contact with a sealed glass vial containing a sample dose.[99] While these experiments impressed Papus, Academic circles remained skeptical. Luys continued to test the boundaries of scientific orthodoxy for the next several years, pursuing new studies that used "magnetic crowns" to transfer diseases away from his patients.[100] During this period, the older psychiatrist's working relationship with Papus grew steadily closer. At the end of 1888, at Luys's behest, Papus became editor in chief of the *Revue d'hypnologie*. In 1890, when his professor's health began to fail, the young extern officially became head of the *laboratoire d'hypnologie* at the Hôpital de la Charité.[101] For Papus, these activities were closely related to his extracurricular readings: By revealing the astonishing capacity of the mind to act beyond the limits of the body, researchers like Luys had begun to rediscover "the magic of 2000 years ago."[102]

Charisma and this unusual ability to combine occult science with medical knowledge led Papus to become an influential presence both in the Isis Lodge and in the larger milieu of Parisian heterodoxy. He quickly developed a circle of like-minded friends, the most important of whom was a young nobleman from Lorraine, Stanislas de Guaïta. Guaïta had discovered the French Hermetic tradition independently, through his friendship with the writer Joséphin Péladan, whose scandalous, best-selling 1884 novel *Le Vice suprême* had made magic and esotericism fashionable. Together, Guaïta and Péladan had even founded a new secret society, inspired by the German mystical tradition of the Rosicrucians, the Ordre Kabbalistique de la Rose-Croix. By inducting Papus into the society, Guaïta played a crucial role in building his friend's cultural prestige. The florid style Guaïta developed in his own writings also furthered this project, cementing the connection between the study of things occult and the glamorous cutting edge of the *décadence littéraire*.

As Papus and his friends grew more confident in their ideas, tensions in the Isis Lodge escalated. Blavatsky and Gaboriau responded to the challenge this ever-more influential group of dissidents posed by launching an all-out polemical attack. In mid-1888, Papus later claimed, the situation had

[99] André and Beaufils, *Papus*, 31–32.
[100] See the obituary for Luys published in *L'Initiation* 37 (Oct.-Dec. 1897): 125–126.
[101] André and Beaufils, *Papus*, 77–79.
[102] *L'Initiation* 7 (Apr.-Jun. 1890): 187.

grown so perilous that his "invisible masters" intervened, formally ordering him to "combat the anti-Christian influence" that the Theosophical Society had come to represent.[103] After several months of increasingly heated debate, the Isis Lodge collapsed, and Papus became the most influential member of a new group, the Hermès Lodge, which Olcott formally recognized in September 1888.

From his new position, Papus quickly set about laying the groundwork for a separate organization based on the Martinist tradition. Guaïta had already created one secret society, to which Papus had added a second, the *Ordre Martiniste.* He built on this foundation by establishing a new periodical, *L'Initiation,* which he staffed with members of the Hermès Lodge. In 1890, Papus officially broke with the Theosophical Society, creating a new movement—which he referred to as Occultism—and a new umbrella organization, the Groupe indépendant d'études ésotériques. According to its founding statutes, this society had four goals: (1) to publicize Occultism; (2) to cultivate potential members of secret societies; (3) to train Occultist lecturers; and (4) to study the phenomena of "Spiritism, Mesmerism, and Magic." It would accomplish these tasks by sponsoring a diverse range of activities, including lectures, concerts, courses in kabbalah and Hermeticism, and séances. To become a member, all one needed to do was subscribe to the journal.

However it might have bruised Papus's amour propre, Janet's description of Occultist literature as "popular metaphysics" had an element of truth. *L'Initiation* struck a shrewd balance between the accessible and the recondite—with an added dose of commercialism. Articles discussing techniques of divination and providing colorful descriptions of haunted houses appeared alongside Symbolist poems and dense essays on topics like Schopenhauer's pessimism and its relation to Eduard von Hartmann's philosophy of the unconscious. Beginning in 1889, to ensure that new readers would not feel disoriented, Papus added a page to the front of the journal suggesting a few "useful readings for initiation." The list began with a selection of introductory texts for "people who want a general sense of this question, but do not have the time to read a great deal," and continued through four further degrees of difficulty.[104] All of these books, Papus observed helpfully, could be ordered from the Parisian publisher Carré. To attract further readers, Papus offered frequent "free gifts" to his subscribers, including discounts on selected new books, a complimentary subscription to a Mesmerist periodical, and a "splendid portrait" of an unspecified subject,

[103] *L'Initiation* 29 (Oct.-Dec. 1895): 287.

[104] *L'Initiation* 4 (Jul.-Sep.1889): third unnumbered page inside the front cover of the September issue.

hand-colored to order.[105] Each of these gifts, Papus declared, perhaps a trifle optimistically, "is, in itself, worth the price of an issue."[106]

These promotions appear to have worked. Several years after its founding, *L'Initiation* had 8,000 subscribers, a considerable number for a special-interest periodical of its kind. [107] The Groupe indépendant thrived as well. By March 1890, the group had 367 members; three months later, the number had increased by an additional 205.[108] In May 1890, Papus moved the organization's headquarters to a building on the rue de Trevise, where he also established a lending library and an occult bookstore. That same year, the Groupe indépendant began a second journal, the inexpensive four-page weekly *Le Voile d'Isis.* By January 1892, the regular print run of the new publication had reached 10,000 copies, and the Groupe indépendant counted seventeen branches throughout France.[109]

As this organization expanded, so did the two secret societies that functioned under its aegis. The larger of the two associations, the Martinist Order, was easy to join and extensively publicized in *L'Initiation.* Aspiring members simply wrote to the Order's Supreme Council—itself part of the Groupe indépendant—and requested to be assigned an initiator. The initiator would then ask the aspiring member to write a brief spiritual autobiography. The next stages involved a dialogue between initiator and initiate, in which the initiator suggested readings and then responded to the initiate's interpretations of them. As this conversation continued, the initiate could be admitted to higher ranks in the Order at his initiator's discretion. The third grade was the Order's highest, and entitled the recipient to follow his or her name with the initials S. I. To become an initiator him or herself, the S. I. needed to copy and interpret three manuscripts, which described certain key symbols and rituals of the Order.[110] For members who lived near Martinist lodges, initiation and advancement also involved participation in secret rituals, but these material components were not absolutely necessary.[111] In the provinces, isolated seekers could join the Order and advance within it entirely by correspondence.

Though Papus invented its specific organizational structure, the Martinist Order derived its ideals and rituals from the secret societies established

[105] For the book discount, see ibid., 12 (Jul.-Sep.1891): back cover of the September issue; for the free subscription, see ibid., 6 (Jan.-Mar.1890): 91; for the portrait, see ibid., 7 (Apr.-Jun.1890): 280.

[106] Ibid., 4 (Jul.-Sep.1889): inside front cover of the September issue.

[107] Archives historiques de l'Archevêché de Paris, carton 4 E 18, dr. "Société des sciences psychiques," letter from the Abbé Timothie Ferdinand Brettes to the Archbishop of Paris, Mar. 29, 1897.

[108] See *L'Initiation* 6 (Jan.–Mar. 1890): 187; and ibid., 7 (Apr.–Jun. 1890): 280.

[109] Ibid., 12 (Jul.–Sept. 1891): 87.

[110] For copies of these manuscripts, see *Fonds Papus de la Bibliothèque municipale de Lyon* (FP), cote 5490.

[111] For a manuscript describing several of these secret rituals in detail, see FP, cote 5490.

by Pasqually and Saint-Martin in the second half of the eighteenth century. Martinism, as Papus glossed it, was based on "two great principles: preservation of the initiatic tradition of Spiritualism, characterized by the Trinity, and the defense of Christ beyond the confines of any sect."[112] The Order implemented these ideals by serving as "a school of moral knighthood." The teachings of this "school" were twofold: first epistemological and then more directly moral. Members of the Order sought to develop their "spirituality . . . by the study of the invisible world and its laws, by the practice of devotion and intellectual aid, by the creation in each mind of a faith made all the more solid by its basis in observation and science." Members expressed this faith through "altruism," performing acts of charity under the cover of strictest anonymity. In this way, the Order's concern with secrecy served a moral function, ensuring the essential disinterestedness of every act of charity it anonymously performed.[113]

The second secret society affiliated with the Groupe indépendant, Stanislas de Guaïta's Ordre Kabbalistique de la Rose Croix, took a very different approach. Where the Ordre Martiniste was well-publicized and easy to join, the Ordre Kabbalistique sought to maintain both secrecy and high barriers to entry. The Order received regular mentions in *L'Initiation* from 1889 onward, but these announcements provided considerably less information than those concerning Martinism. The Ordre Kabbalistique, in fact, did not make its entry requirements public until mid-1892. As described in *L'Initiation*, the prerequisites appeared intimidating indeed. To be admitted to the Order, an aspiring member had to earn a *baccalauréat*, *licence*, and *doctorat* in kabbalah. In addition to producing a thesis worthy of publication in *L'Initiation*, candidates for these degrees had to pass a series of progressively more demanding oral examinations, in which they demonstrated their knowledge of subjects such as hypnotism, practical magic, Mesmerism, Neoplatonic philosophy, Hebrew, Sanskrit, alchemy, Freemasonry, Buddhism, and Zoroastrianism.[114] These barriers to entry served to keep the group quite small. Despite its modest size, however, the Ordre Kabbalistique, with its exclusivity and ostentatious secrecy, played an important role in this constellation of groups—it represented the highest degree of knowledge and distinction to which a member of the Groupe indépendant could aspire.

The secrecy of the Ordre Kabbalistique makes its basic teachings more difficult to reconstruct than those of the Martinists. In 1901, however, the

[112] Papus [pseud. of Gérard Encausse], *Martinésisme, Willermosisme, Martinisme et Franc-Maçonnerie* (Paris: Chamuel, 1899), 42. The text from the 1840s that Papus used in his own initial studies remains a good introduction to the thought of Saint-Martin. See Jacques Matter, *Saint-Martin, le philosophe inconnu* (Le Tremblay: Diffusion Rosicrucienne, 1992).

[113] Papus, *Martinésisme*, 47.

[114] For a complete list of subjects, see *L'Initiation* 16 (Jul.–Sept. 1892): 176–180.

memoirist Georges Vitoux published the Order's secret constitution, which conveys a sense of how Guaïta envisioned his society's mission. To the profane outsider, the constitution declared, the Order would appear to be "a dogmatic and visible society for the diffusion of Occultism," much like the Groupe indépendant.[115] In fact, however, the Order had a rather more exciting mission:

> In reality . . . , *it is a secret society of action* devoted to individual and mutual support; to the defense of its members; to the multiplication of their vital forces by reversibility; *to ruining adepts of black magic*; and finally, to THE STRUGGLE TO REVEAL THE ESOTERIC MAGNIFICENCES THAT ABOUND IN CHRISTIAN THEOLOGY, BUT OF WHICH IT IS UNAWARE.[116]

For Guaïta, this language was not empty posturing—as he saw it, membership in his organization involved the highest spiritual stakes. In the course of an 1893 conflict with the allegedly Satan-worshipping defrocked priest Joseph-Antoine Boullan, for example, Guaïta claimed to have used his considerable powers to fend off late-night astral salvos of "a supreme violence," while one of his friends found himself "cataleptized to his bed and nearly subjected to the attacks of a succubus."[117] According to some, as newspapers widely reported at the time, Guaïta's response to these attacks in Paris led directly to the former priest's death in Lyon. Clearly, for its members, the Ordre Kabbalistique was no idle undertaking. Joining it meant embarking on a process of thoroughgoing self-transformation, one that entailed not only a dangerous struggle with the forces of evil but also transcendent experiences of unparalleled intensity and power.

As chief organizer of Occultism, Papus ensured that the material trappings of these societies reflected the glamour and drama of their immaterial aspects. Where Spiritists had tried to make their séances as much like everyday life as possible, Papus and his fellow Occultists strove to cultivate a distinction between the sacred and the profane. The meetings of both the Ordre Martiniste and the Ordre Kabbalistique were full of pageantry, with costumes, secret images, and appropriate décor. The publisher and poet Lucien Mauchel, for example, described the trappings that accompanied examinations for the doctorate of kabbalah. The event took place in a room "hung with red cloth, barely lit"; the examiners, separated from the student by a thin red curtain, wore red robes and white *pschents*—the initiatic headdress of the Martinist Order.[118] As the

[115] Quoted in Georges Vitoux, *Les Coulisses de l'au-delà* (Paris: Chamuel, 1901), 185.
[116] Ibid., 185–186. All emphasis in original.
[117] Letter from Guaïta to Péladan quoted in Christophe Beaufils, *Joséphin Péladan, essai sur une maladie du lyrisme* (Paris: Jérôme Millon, 1993), 126.
[118] Quoted in Philippe Encausse, *Sciences occultes,* 54.

Groupe indépendant grew more prosperous, Papus went to still-greater lengths to create an appropriately mysterious atmosphere. One journalist, for example, provided a revealing glimpse of the organization's second headquarters, on the rue de Savoie:

> The rooms for courses, meetings, and lectures have mysterious inscriptions running along their walls, between Hermetic symbols, astrological signs, and tables of Hebrew and Sanskrit letters . . . Imagine a miniature *Musée Guimet*, in an atmosphere of miracle and prophecy.[119]

This flamboyant décor—and the equally colorful description the journalist produced—seemed to promise spiritual seekers entry into a strange and exciting world. What was more, as the journalist observed, any reader, male or female, could join these "extraordinary men" in their mysterious pursuits—all he or she needed to do was subscribe to their journal and attend their lectures. By the mid-1890s, publicity of this kind had made Papus, Guaïta, Péladan, and several other Occultist *hommes de lettres* into celebrities (fig. 17).

The Occultist movement lasted into the second decade of the twentieth century, though its emphasis changed in the late 1890s. Two decisive events helped bring about this shift: Guaïta's untimely death in 1898, and Papus's growing interest in a healer from a village near Lyon, Philippe Nizier Vachod, known as Maître Philippe. After Guaïta's death, Papus increasingly turned away from the magical practices that had so fascinated his friend. Instead, he began to elaborate a more emotional and mystical conception of esotericism. In large part, this shift was due to the influence of Philippe, who Papus would later call his "spiritual teacher."[120] Philippe, a charismatic former medical student of peasant stock, rejected magic and derived his healing powers from simple meditation on the Gospels. As Papus's interests changed, the Ordre Martiniste came to occupy an increasingly important place in the Occultist movement; the Christian ideals it espoused meshed well with its leader's new sensibility. After an ambitious attempt at reorganization in 1897, which would have turned it into a formally organized Université libre des hautes études, the Groupe indépendant gradually dwindled.[121] By the mid-1900s, the Martinist Order had become the central organization of French Occultism, and it would remain so until Papus's death in 1916.

[119] *Le Matin*, Nov. 6, 1899, 3.

[120] See Philippe Encausse, *Sciences occultes*, 207.

[121] For an extensive discussion of the proposed curriculum and structure of this new institution, see *L'Initiation* 34 (Jan.–Mar. 1897): 170–171, 258–263; 35 (Apr.–Jun. 1897): 77–78, 157, 277; 36 (Jul.–Sep. 1897): 170–171, 280; 37 (Oct.–Dec. 1897): 117, 220.

Fig. 17. Gérard Encausse, better known as Papus, depicted in the costume of a mage by the artist Octave Guillonnet. (Collection of the author.)

Exploring the Occultist Self

As the title of its principal journal, *L'Initiation,* indicated, the Occultist movement's goal was to create "initiates." An *initiate,* according to the definition Papus elaborated, was anyone who chose to undertake a study of esoteric wisdom and occult science. The process of initiation, in turn, was simultaneously social and individualistic. A shared quest for essential truth necessarily spurred initiates to come together in a "fraternity of intelligence," Papus wrote, but membership in this fraternity—which he extended to women as well as to men—did not require adherence to a specific set of teachings. Instead, the role of the "initiatic society," as Papus conceived it, was to "encourage the student to create a personal doctrine of his own." Papus did not believe that this freedom was absolute, however: initiatic societies were obligated to provide certain "general principles" to help their members avoid "fundamental errors."[122] The goal of the

[122] *L'Initiation* 2 (Jan.–Mar. 1889): 196, 198.

organizations the Occultists founded, therefore, was to create an intellectual framework to guide what was ultimately a process of independent self-cultivation.

For the most part, the Occultists derived their "general principles" from Renaissance Hermeticism and the later French tradition it inspired. Papus, Guaïta, and other fin-de-siècle Occultists readily embraced the Hermetic concept of a lost golden age—a period in the distant past when superior understanding had enabled human beings to control nature in ways that far exceeded the capacities of contemporary science. Many of these now-forgotten achievements, Occultists often argued, stemmed from the ancients' mastery of the hermeneutic technique of analogy. According to the Hermetic tradition, the earthly world corresponded to the celestial in ways obscure to the profane but visible to those who had received proper training. Knowledge of these correspondences, their significance, and the myriad ways in which they could be manipulated conferred marvelous powers: A well-trained astrologer, for example, could predict the future on the basis of planetary movements, while a practitioner of ceremonial magic could use specific objects, symbols, and rituals to harness cosmic forces and bend them to his or her will.

Papus supported this vision of the material power of analogy—one as old as Hermeticism itself—with physiological and religious conceptions that owed a substantial debt to the ideas of Mesmerists and French thinkers in the Martinist tradition. Human beings, Papus argued, were tripartite entities, composed of a physical body, an immaterial soul, and a mediating "astral body" made of the same subtle matter that Mesmerists called the "universal fluid," and Spiritists, the *périsprit.* As a substance capable of giving material form to ideas, the "astral fluid" acted as the bridge between all-too-material humanity and the immaterial spirit world. This physiological model underpinned a Christian conception of man's destiny and potential rooted in the philosophy of Pasqually and Saint-Martin. Though fallen, Occultists believed, the human soul was made of the same substance as God. Thus, disciplined cultivation of the mind's astral powers through the practice of ritual magic, meditation, and prayer could eventually lead still higher, returning the immaterial soul to a state of unity with the divine.

While related in kind, Stanislas de Guaïta observed, the transcendent "reintegration" that these initiates strove to achieve was fundamentally different from that experienced by orthodox Christian ecstatics. Where the "saint's" union with the divine was passive, a product of grace, the Occultist's was "active"—a conscious act of will.[123] The initiate's power, here, was the product of intellectual discipline, long practice, and total self-mastery. As David Harvey has observed, Guaïta and other Occultists

[123] *La Plume* 78 (Jul. 15, 1892): 320.

feminized conventional, "passive" ecstasy and associated their form of "active" transcendence with masculinity.[124] In practice, however, gender boundaries within the movement were more fluid. Papus supported feminist causes in *L'Initiation*, and women were welcome in both the Groupe Indépendent and the Ordre Martiniste, though they appear to have been in the minority.

An Occultist initiate undertook a demanding process of intellectual cultivation, with the goal of actively transforming his—or less frequently, her—own consciousness, first by developing a capacity to act in the "astral plane," and then by coming to understand the soul's true relation to the divine. If developed to the full, this broadened awareness could confer astonishing powers. As Guaïta put it in his contribution to the special issue of *La Plume*,

> The chief task of Initiation can be summarized . . . as the art of becoming a genius by artificial means; with this key difference, in any case—that natural genius provides inspiration at certain moments, frequent or infrequent, when the Spirit sees fit to descend; while acquired genius is, at its highest level, the ability to force inspiration and communicate with the Great Unknown at each and every time one wishes.[125]

The highly trained "Adept" could use his "artificial genius" to do more than simply tap the well of creative inspiration at will, however, as Guaïta observed elsewhere. Unlike the ordinary genius of the painter, sculptor, composer, or poet, the artificial genius of the Occultist did not depend on the crude material support of conventional artistic media; instead, it made its presence felt by more direct means. The Adept's thorough knowledge of the astral plane enabled him to transform the products of his own imagination into tangible entities capable of acting at a distance in material ways—materializing as apparitions, for example, or causing harm to an enemy.[126] For Occultists, then, the project of expanding the limits of individual consciousness through knowledge and training led directly to a distinctive experience of transcendence in which the initiate's own imaginings became uncannily palpable.

The conception of consciousness that served as the foundation for this vision of transcendence resonated strongly with new psychological theories of the multivalent self. Where Spiritist writers struggled against these theories, their Occultist colleagues embraced the idea that a rich array of mental processes occurred beyond the sphere of ordinary waking awareness.

[124] Harvey, *Beyond Enlightenment*, 100–104.
[125] *La Plume* 78 (Jul. 15, 1892): 319.
[126] See Stanislas de Guaïta, *Au Seuil du mystère* (Paris: Durville, 1915 [1886]), 94–95.

Echoing the German philosopher von Hartmann, they referred to these aspects of psychic life as the "unconscious." In a synthesis of a complex and influential series of articles by Papus's friend Albert Faucheux, who published under the pseudonym Félicien-Charles Barlet, Donald MacNab asserted that normal consciousness was "placed between two unconsciouses," one lower and one higher.[127] The inferior unconscious was the realm described in the work of Janet and other "modern psychologists." The "superior" unconscious, in contrast, was the "seat of intuition, the channel by which we become aware of eternal and necessary truths." MacNab followed a group of other regular contributors to *L'Initiation*, including Papus, Guaïta, and Barlet, by associating this higher unconscious with the astral body and used analogy to move a step further: As the inferior unconscious, rooted in the physical body, supported ordinary waking consciousness from below, so the superior unconscious, rooted in the astral body, supported a still higher form of consciousness—the "transcendental subject"—above. This most exalted aspect of the self, which lay far beyond the compass of ordinary awareness, was the divine essence Occultists strove to "know, love, and serve from within." This interior knowledge was not a product of reason, but of intuition and imagination, qualities Occultists cultivated by exploring the astral realm.[128]

Taking advantage of his engagement in both fields, Papus frequently compared the Occultist understanding of consciousness with the models being developed by contemporary psychologists, largely to the detriment of the latter. As he told his readers in 1895,

> After having devoted our closest attention to the theories of Occultism—Western as well as Eastern—on the one hand, and the most recent phenomena of hypnosis on the other, we arrived at this conclusion: only Occult Science could provide a scientific explanation of the facts of psycho-physiology that would soon be discovered. From that time, we could have produced canny pastiches of the theories of Occultism, avoiding words like astral body and astral light, and assured our scientific future, albeit at the expense of our good conscience.

Here, a critique of psychology became a broader indictment of the intellectual narrowness of contemporary science, with its stress on materialism and a vision of "progress" that entailed the steady elimination of any possibility of transcendence. In their efforts to "disguise astral phenomena

[127] For the Barlet articles, see *Le Lotus* 1 (Mar.—Sept. 1887): 27–33, 78–87, 154–167, 203–216, 282–293, 338–348; and *L'Initiation* 1 (Oct.–Dec. 1888): 1–22. For MacNab, see *L'Initiation* 5 (Oct.–Dec. 1889): 234.

[128] *L'Initiation* 5 (Oct.–Dec. 1889): 234, 235, 244–245.

as phenomena of 'Telepathy,' 'Telepathic Hallucinations,' 'psycho-physiological manifestations,' " the "adepts of scientific schools" had blinded themselves to the most remarkable powers of the human mind. At the same time, however, Papus noted, certain psychologists were rediscovering the reality of the astral despite themselves: Hippolyte Baraduc, for example, used specially treated photographic plates to capture images of mysterious energy fields that seemed to vary with his subject's mood; similarly, the various experiments Luys conducted with medication at a distance and the use of magnets to transfer diseases appeared to prove the existence of vital forces that were surprisingly easy to dissociate from the physical organism.[129]

By developing this notion of the astral, Occultists were able to transform the multivalent self into a vehicle for tangible spiritual experience. "Psychological *magnétiseurs*" like Ribot and Janet, Guaïta asserted, had constructed "an entirely superficial but perfectly rigorous theory, which provides an exact accounting of phenomenal appearances." This theory, however, was radically incomplete because it ignored the "mysterious laws that govern the astral tides."[130] The ancient tradition Occultists explored, in other words, allowed them to perceive the extraphysical dimension of the unconscious mental processes that psychologists had confined to the limits of the body. Once this element of traditional knowledge had been added, new theories of mind ceased to be a threat to cherished metaphysical assumptions and instead became a path to a fuller understanding of the soul and its transcendent powers.

This appropriation and reconfiguration of the latest psychological theory emerged most clearly in Occultist critiques of Spiritism. In general, Occultists argued, the spirits of the deceased did not appear at séances; instead, whatever phenomena occurred were the unconscious imaginative products of the people in attendance. According to Occultists, these unconscious creations took the forms of an array of more or less monstrous invisible entities, including *egregores* (products of a group's collective imagination); parasitical, malignant astral beings called *larves*; and elemental spirits. Mediums, therefore, put themselves in a perilous moral and psychological position. According to Guaïta, for example, a medium was "a man (or a woman) ill with a vital incontinence," who allowed his or her astral forces to be steadily drained away by the entities he or she unconsciously created. This lack of awareness, Guaïta believed, was what differentiated the medium from an Occultist Adept: The medium, not having been trained to expand the ordinary limits of his or her consciousness, became a victim of the same imaginative force that gave the Adept such remarkable

[129] *L'Initiation* 26 (Jan.–Mar. 1895): 97, 98, 99.

[130] Stanislas de Guaïta, *Le Serpent de la Genèse, premier septaine: Temple de Satan* (Paris: Trédaniel, n.d.), 430.

power. For Occultists, then, the regions of mental life beyond the reach of ordinary consciousness were not essentially pathological realms as they were for many psychologists; instead, they were a vast spiritual continent awaiting discovery. While the monsters that dwelled there could destroy the unwary, they could also provide powerful assistance to the explorer who arrived equipped with the proper knowledge and moral resolve.

At the beginning of the twentieth century, as interest in Myers' work spread beyond psychological circles, writers familiar with the Occultist conception of the multivalent self began developing it in new directions. Writing in 1907, for example, the journalist Jules Bois, who had been a chronicler of the Occultist milieu since the late 1880s, made the multivalent self into a crucial element of what he saw as a distinctively "modern" conception of spiritual experience. Occultists, Spiritists, psychical researchers, and "spiritualists of Myers' school," Bois argued, had revealed the "obscure capacities in which our ego is secretly rich." By doing so, these thinkers had created a dramatically new conception of transcendence, even if their willingness to indulge in speculation about otherworldly entities and psychic forces had occasionally led them astray. Thanks to their efforts to plumb the depths of the psyche, "the *Beyond* is replaced by the *Within*, the miracle *outside and above us* by the miracle *in us and by us*." This new conception of the self and its possibilities, Bois wrote, led him to what he saw as a quintessentially modern creed:

> Believe—and if your faith requires a basis, believe within yourself, in the deep and inexhaustible ocean on which we are all merely ripples. But the entire vastness of the sea is contained in a bit of froth; our "individual ego" is the "divine ego" in miniature. Through one, the other is open to us. Only we will no longer act as idolaters, seeking this Ineffable outside of ourselves; we know that it does not lie within our grasp or our gaze, but in the abysses of our thought and the depths of our love—that is the home of the "greater glory of man" to which we aspire.

Here, Bois joined the Occultists' spiritualized unconscious with Myers' "subliminal self," producing a conception of belief as a personal, interior matter. The "fraternity of intelligence" that even Occultists had sought in their efforts to master the "Ineffable" now struck Bois as superfluous. The "modern miracle," as he saw it, was simply a matter of exploring and expressing "our interior being, much richer, much more fertile, much more original than our superficial personality." This view would come to define heterodoxy to an ever-increasing degree over the course of the twentieth century.[131]

[131] Jules Bois, *Le Miracle moderne* (Paris: Société d'Editions littéraires et artistiques, 1907), xi, xii–xiii, xv, 164. Italics in the original.

Epilogue: The Emergence of a New Heterodoxy

At three o'clock on a Sunday in mid-December 1919, a spirit society called L'Œuvre de la rénovation sociale met in the seventeenth *arrondissement* to commemorate the birth of Christ. The festivities began with a speech "on the Spiritist doctrine," delivered by the society president, E. F. Bolopion. After introductory prayers, some mediumistic healing, a brief "meditation to the Virgin" with piano accompaniment, and a "historical discourse on the birth and life of Jesus," the spirit contacts began. This society's mediums did not perform automatic writing; instead, they conveyed messages from the beyond "by incarnation," acting and speaking for the spirits. Saint Paul appeared first, followed by a repentant sinner. The audience celebrated his profession of faith by singing a hymn of praise.

Then a "Soldier Spirit" appeared, writhing in pain. A German had ripped open his stomach with a bayonet, and he "still suffered horribly." The members of the society rushed to his aid, informing him that he had died, and "healing him spiritually" by working to detach his soul from the terrible residue of physical pain. Soon, the members saw their efforts rewarded: The soldier leapt up and embraced his saviors, jubilantly shouting "*Vive la France!*" Once his initial joy had calmed, he explained the predicament from which the society's attentions had delivered him. Death in the trenches was often sudden, and seemed to come from nowhere. As a result, he said, many dead soldiers "still remained attached to their physical Bodies" and thus continued to feel their wounds. In his own case, his mother had made matters worse by abandoning prayer: The horrors of the war had left her convinced that God did not exist. Before leaving the group, the young man urged the audience to heed his experience as a lesson. Dead soldiers would continue to suffer until they had received enough prayers to make them aware of the true nature of their situation. Mothers, then, had a duty not to renounce their faith when confronted with the brutal absurdity of the war. Maternal tenderness would be the saving grace of every soldier killed in battle.[1]

This story, told in the handwritten, mimeographed pages of a small, erratically appearing journal called *Le Bon berger*, provides a powerful glimpse of the changes the First World War wrought in French Spiritism. New horrors demanded the creation of more intensely emotional forms of consolation. Spiritist practice overwhelmingly followed the example set by exponents of the "moral" approach in the 1890s, growing more emotional and varied: In society meetings, trance-speech became the standard means by which spirits communicated, and groups like Bolopion's used it as the basis for novel ritual forms, joining elements of a religious service with those of a theatrical production. At the same time, the process of decentralization that had begun in the 1880s continued. Small societies proliferated, and the larger ones exerted less authority.

As this organizational change occurred, the act of engaging in dialogue with the beyond lost its close association with Kardec's ideas and norms, and became a technique of consolation that individuals employed in a growing number of highly personal ways. In 1918, for instance, Cécile Monnier began to receive automatic writings from the spirit of her son Pierre, a soldier killed in the Argonne in 1915. Monnier insisted that her dialogues with the beyond were spontaneous products of divine grace and involved "nothing resembling Spiritism." She had been raised a strict Calvinist but had difficulty accepting the austere conceptions of salvation and the afterlife with which she had grown up. For her, the continuing contact with Pierre proved the reality of a more liberal Christian God. Unfazed by Monnier's rejection of Spiritist ritual and the idea of reincarnation, Pierre-Gaëtan Leymarie's son Paul published a small compilation of her communications in 1920, with considerable success.[2]

In part, this uncoupling of Spiritist practice from Spiritist doctrine stemmed from an absence of leadership. No new philosophers or charismatic advocates emerged to replace Léon Denis and Gabriel Delanne, who despite advancing age and failing health remained central figures in the movement well into the 1920s. Denis, blind and frail, presided over the

[1] The account of this meeting and all quotations come from *Le Bon berger* 1, no. 1 (December, 1919): 3–5.

[2] Quoted in Cécile Monnier, *Lettres de Pierre*, intro. by Jean Prieur (Paris: Fernand Lenore, 1980), 1:x. Pierre's complete letters fill seven volumes. The compilation of excerpts is Cécile Monnier, *Je suis vivant* (Paris: Leymarie, 1920).

last great Spiritist *Congrès,* held in the Salle Wagram in 1925, but he would live only until 1927; Delanne died in 1926, continuing to publish his journal until the end. The *Revue spirite* survived—and indeed still exists today—but the efflorescence of periodicals that followed the end of the war had largely ceased by 1930.[3]

Delanne's death also contributed to a broader change in French psychical research, which eventually stripped Spiritism of its remaining scientific prestige. In 1919, the physician Rocco Santoliquido and the psychologist Gustave Geley established a new organization for the scientific study of psychic phenomena, the Institut métapsychique international (IMI). This organization had the support of Charles Richet, the preeminent French scientist in the field, and was extraordinarily well-funded: When it opened, its opulent facilities included a library, a lecture hall, and a fully equipped laboratory for the study of psychic phenomena. Despite Richet's involvement, at its inception the IMI was closely tied to Spiritist circles: Delanne served on its board; Geley, the director, was sympathetic to the cause; and the bulk of the Institute's money came from the wealthy wine wholesaler and devoted Spiritist Jean Meyer. Over the course of the 1920s, however, as Sofie Lachapelle has shown, these connections weakened.[4] Geley died in 1924, Delanne in 1926; the new director of the Institute, Eugène Osty, rejected both Kardecism and the "spirit hypothesis" more broadly. This situation inevitably led to tensions with Meyer, who saw the Institute as a vehicle for the propagation of Spiritism. When Meyer died in 1931, therefore, he left control of the Institute's funds in the hands of his personal secretary and medium, the Spiritist M. Forrestier. The IMI subsequently contested the will, which predictably enough led Forrestier to reduce the financial support he provided.

Though the Institute continued its work in these straitened circumstances, its profile steadily diminished over the course of the 1930s. Twenty years before, apologists had been able to present Kardec's ideas as a set of legitimate, if contentious, hypotheses that would one day receive the scientific attention they deserved; by 1931, this prospect seemed implausibly

[3] Unfortunately, publishers of Spiritist periodicals during this period seem to have been less vigilant than their predecessors about submitting them for the *dépôt légal,* so often even the Bibliothèque nationale lacks complete collections. For periodicals from the early 1920s, published in Paris unless otherwise noted, see *Les Annales du spiritisme* (Rochefort-sur-Mer 1921–1934); *Le Bon berger* (1919–1926), *L'Ere mystérieuse* (Rouen, 1921), *L'Ici-bas et l'au-delà* (1923), *Le Petit fraterniste* (Arras, 1922), *La Revue scientifique et morale du spiritisme* (1896–1926), and *Le Spiritisme kardéciste* (Lyon, 1918–1920). Two periodicals that were prominent during this period continued well after it: *La Revue spirite* changed its name to *Renaître 2000* in 1977, then became the *Revue spirite* again in 1996; *La Tribune psychique* lasted from 1893 to 1983.

[4] Sofie Lachapelle, "Attempting Science: The Creation and Early Development of the Institut Métapsychique International in Paris, 1919–1931," *Journal of the History of Behavioral Sciences* 41, no. 1 (2005): 1–24.

remote. If other groups devoted to similar pursuits had continued to function, the decline of the IMI would not have affected Spiritism's claims to scientific legitimacy so strongly. Unfortunately for followers of Kardec, however, the situation in France was part of a broader development in European intellectual life. During the late 1920s and into the 1930s, psychical research drifted further and further toward the margins of the scientific world. The famous scientists who once gave the field its prestige had died, and there was no new generation of comparable eminence to continue the project.

In the French context, according to Matthew Brady Brower, the burgeoning of psychical research in the early 1920s and its subsequent decline need to be considered in relation to the reception of psychoanalysis.[5] Throughout the early twentieth century, French psychologists ignored Freud's theories, preferring a shifting array of home-grown alternatives. During the fin de siècle, the tendency was to reject psychoanalysis in favor of what Elisabeth Roudinesco calls "l'inconscient à la française," a conception that associated subconscious mental activity with hereditary degeneration, as we saw in chapter 5.[6] By the 1920s, however, French psychological interest in phenomena such as dissociation and automatism had waned, a shift caused by growing skepticism of hypnosis as a reliable tool for experimental study of the mind. Most important, beginning in 1901, the clinical psychiatrist Joseph Babinski published a series of papers arguing that the symptoms of hysteria Charcot had famously identified were in fact the results of hypnotic suggestion—unintentional byproducts, in other words, of the allegedly objective technique the older man had developed to study the disease.[7] The psychical research practiced by the IMI, Brower argues, grew out of this skepticism: Spectacular, palpable manifestations like ectoplasms and spirit materializations seemed "to give material, and above all, visible form to unconscious thoughts," rendering them susceptible to objective laboratory study in a way that trance phenomena were not.[8] By the late 1920s, however, several of Freud's works had become available in French translation, and a group of committed advocates had emerged. The psychoanalytic conception of the unconscious, coupled with the growing influence of experimental psychology conducted according to the German model, made the concerns of the IMI—and of psychical research as a whole—appear naïve and reductive.

[5] Matthew Brady Brower, "The Fantasms of Science: Psychical Research in the French Third Republic, 1880–1935" (Ph.D. diss., Rutgers University, 2005).

[6] Elisabeth Roudinesco, *La Bataille de cent ans* (Paris: Seuil, 1986), vol. 1.

[7] Mark S. Micale, "On the 'Disappearance' of Hysteria: A Study in the Clinical Deconstruction of a Diagnosis," *Isis* 84, no. 3 (Sep. 1993): 496–526.

[8] Brower, "The Fantasms of Science," 346.

Organized Occultism fared even worse in the years after the First World War. Papus enlisted in the French Army as a medical officer in 1914, and died two years later from a disease contracted at the front. Without the twin benefits of his charisma and his gift for synthesizing seemingly irreconcilable ideas, the various organizations he once led entered a period of rapid decline. Occultism did not offer its adherents the kind of immediate consolation Spiritism could provide, and in the interwar period the distinctive ethos that had once conferred such glamour on the movement now seemed to be a fusty vestige of a spent cultural trend. The Ordre Martiniste did not disappear, and in fact still exists today, but it has not reclaimed the central position in French heterodox life that it commanded at the fin de siècle.

Interwar Developments: Traditionalism and Surrealism

During the 1920s and 1930s, then, the old organizational and ideological structures of nineteenth century heterodoxy broke down. The popularity of Spiritism and Occultism steadily diminished, as did the intellectual prestige of psychical research, but all nevertheless left deep traces in French culture. Traditionalism and Surrealism, two important but very different developments of the 1920s and 1930s, give a sense of the complex and surprisingly diverse ways in which interwar thinkers responded to the legacy of nineteenth-century heterodoxy.

Traditionalism began to take shape in 1906, when twenty-year-old René Guénon decided to abandon his preparatory studies in mathematics at the Collège Rollin and devote himself to the pursuit of esoteric wisdom. He initially joined the Ordre Martiniste, but by 1909 he had broken with Papus to pursue other initiatic paths. After the First World War, Guénon took over the journal *Le Voile d'Isis,* which had once been run by Papus, and began elaborating a philosophy of his own. His writings attracted a small but steadily growing group of like-minded thinkers and readers during the 1920s and 30s. Guénon's books remain in print today, and *Le Voile d'Isis,* which changed its name to *Etudes Traditionnelles* in 1933, continued to appear into the 1990s.[9]

Traditionalism, as Guénon elaborated it, owed something to Occultism, but departed from it in crucial ways. Like the Occultists, Guénon stressed the importance of the perennial philosophy—a body of essential truths present in all orthodox religions. Unlike his heterodox predecessors, who had to a greater or lesser extent embraced the present and the idea of progress,

[9] Mark Sedgwick, *Against the Modern World: Traditionalism and the Secret Intellectual History of the Twentieth Century* (Oxford: Oxford University Press, 2004), 21–69.

Guénon also elaborated a forceful critique of modernity. The modern West, as he saw it, was characterized by "inversion." Intellectual, technological, and cultural innovations that struck the profane as progressive advances—from mass democracy to certain forms of modern art—appeared to the initiate as destructive steps backward. Modernity, in short, was profoundly dangerous because it fostered a malign cultural orthodoxy that encouraged people to view signs of catastrophe as positive developments. After the trauma of the First World War, this dramatic alternative to nineteenth-century progressive bromides resonated powerfully with many readers.[10]

Guénon applied the idea of "inversion" to heterodox innovation as well. As products of modernity, he argued, Spiritism, Occultism, and Theosophy were grave spiritual dangers, forms of "counter-initiation" that seemed to lead toward enlightenment but actually led away from it.[11] True spiritual understanding, in his view, could only come from long-established orthodox religious traditions. Since Western culture epitomized modernity, Catholicism had the least spiritual potential. Serious seekers, he argued, would need to turn to the East, seeking initiation from a master who could claim an unbroken connection to an ancient religious tradition. It would fall to these elite initiates, in turn, to save the West from the ravages of modernity. Gradually, Guénon came to see Sufism as a particularly powerful form of traditional initiation. He joined a Sufi order in 1911, and by 1930 was living in Cairo as a devout Muslim. Over the course of the 1930s, Westernized versions of Sufism became a feature of European heterodox life, largely as a result of Guénon's influence.[12]

At the same time as Traditionalism took shape, the Surrealists appropriated the legacy of nineteenth-century heterodoxy for their own ends. In 1919, André Breton and his friend Philippe Soupault began to produce literary texts using automatic writing. Their understanding of this technique drew on the work of psychologists such as Janet, Flournoy, and Myers as well as Spiritist literature, but filtered both through their interpretation of the Freudian unconscious. Automatic writing, Breton wrote in 1933, was indeed "a vehicle of revelation for one and all," but according to the conception he and the other Surrealists developed, that revelation was emphatically secular and psychological.[13] As Breton put it in 1924, "I formally refuse to admit that communication of any kind exists between the living and the dead."[14] Spiritist automatic writers, Breton argued, always produced inferior and

[10] Ibid., 24–25.

[11] See René Guénon, *L'Erreur spirite* (Paris: M. Rivière, 1923); *Le Théosophisme, histoire d'une pseudo-religion* (Paris: Nouvelle Librairie Nationale, 1921).

[12] Sedgwick, *Against the Modern World,* 21–28, 74–93.

[13] André Breton, "Le Message automatique," in *Œuvres complètes* (Paris: Gallimard 1992), 2:385.

[14] André Breton, "Entrée des médiums," in *Œuvres complètes* (Paris: Gallimard 1992), 1:276.

misleading work because they approached the act of writing with a preexisting "hope to obtain a communication from the 'beyond,' to procure the assistance of a deceased great man who speaks in the tones of a classroom recitation."[15] Surrealist automatic writers, in contrast, used Spiritist techniques while abandoning all expectations about content—except, perhaps, the expectation that the resulting text would be digressive, exuberant, and startling in its imagery.

Nevertheless, for Breton, gifted mediums like Hélène Smith were pioneers, explorers of consciousness who provided a template for an innovative understanding of art and its relation to life. Despite his aesthetic scruples, Breton had a surprising affinity for the experiential aspect of Spiritism—the distinctive feeling of mystery, and even transcendence, that could occur in séances. The difference between Surrealist creation and Spiritist mediumism was in the form that experience took. According to Breton, "Spiritism seeks to dissociate the psychological personality of the medium, while Surrealism proposes nothing less than the unification of that personality."[16] Trance phenomena gave Spiritists a novel and strikingly palpable form of religious experience by engendering a sense of communion with otherworldly beings. For Surrealists, in contrast, these phenomena were a form of self-communion, a means of experiencing an utterly new kind of unfettered and aestheticized subjectivity.

Toward the "New Age"

The 1940s and 1950s were lean years for heterodox belief and practice in France, but the 1960s and 70s witnessed a resurgence of innovation rivaling that of the fin de siècle in its exuberance and widespread popularity. A comprehensive history of this phase of French religious life has yet to be written; the brief account that follows should be taken not as a systematic overview but as an attempt to sketch a few revealing aspects of a complex and varied phenomenon.[17] As in the nineteenth century, the heterodox resurgence of the 1960s seems to have begun with American ideas transformed by distinctively French anxieties and aspirations. In this case, one American contribution took the form of UFO religions, like the Aetherius Society and Unarius, both of which were founded in the mid-1950s. The French response, as Mircea Eliade has suggested, grew out of a building discontent with a postwar "cultural milieu . . . dominated by a few ideas

[15] Breton, "Message automatique," 385.

[16] Ibid., 386.

[17] Jean Vernette is the most prolific scholar of the subject, but his work has a strongly Catholic apologetic focus. See Jean Vernette, *Jésus dans la nouvelle religiosité* (Paris: Desclée 1987). For a less partisan overview, see Nicole Edelman, *Histoire de la voyance et du paranormal du XVIIIe siècle à nos jours* (Paris: Seuil, 2006), 221–253.

and a number of clichés: the absurdity of human existence, estrangement, commitment, situation, historical moment, and so on."[18]

UFOs and a revolt against Existential pessimism might seem ill-matched at first glance, but their combination proved remarkably attractive to the French public. The authors Louis Pauwels and Jacques Bergier discovered this in 1960, when their book *Le Matin des magiciens* became a surprise best-seller. In October 1961, Pauwels, whose career in journalism included a stint at the mass-circulation glossy *Marie Claire*, built on the success of his book by starting a magazine called *Planète*, devoted, as its cover proclaimed, to "the chronicles of our civilization, invisible history, new beginnings in science, great contemporaries, the future world, [and] lost civilizations."[19] By the mid-1960s, the magazine's circulation had reached 100,000, and the writings of Pauwels and Bergier had inspired a variety of novel social activities: There were heavily attended lectures at the Odéon and at selected Club Meds; travel agents offered "*Planète* vacations" to Indian Ashrams and laboratories at the Massachusetts Institute of Technology.[20]

As Eliade observed, Pauwels sought to develop an alternative to what he presented as the arid, pessimistic intellectual climate of France in the immediate postwar years. He did this by proposing a vision of science made spiritual, according to which advancing technological sophistication and an ever-more-refined understanding of the workings of the universe would eventually lead mankind to an utterly new "awakened state." Turning the postwar anguish of the Existentialists on its head, Pauwels declared that the terrible events of 1939 to 1945 had created the conditions necessary for this exciting step forward in human evolution:

> In this war all the channels of communication between the different worlds opened wide, and let in a powerful draught. Then came the atomic bomb and projected us to the Atomic Age. A moment later, the rockets ushered in the cosmic era. Everything became possible. The barriers of incredulity, so firmly planted in the nineteenth century, had been severely shaken by the war. Now they were about to collapse altogether.

This transformed world, as Pauwels presented it, would have little place for an anthropomorphic God or a clearly defined idea of the soul, but

[18] Mircea Eliade, "Cultural Fashions and History of Religions," in *Occultism, Witchcraft, and Cultural Fashions: Essays in Comparative Religions* (Chicago: University of Chicago Press, 1976), 10.

[19] See Grégory Gutierez, "Le Discours du réalisme fantastique: La revue Planète" (Mémoire de maîtrise de Lettres modernes spécialisées, Université Sorbonne—Paris IV, 1998).

[20] Wiktor Stoczkowski, *Des Hommes, des dieux et des extraterrestres: Ethnologie d'une croyance moderne* (Paris: Flammarion, 1999), 52–53.

would instead offer a proliferation of wonders, made perceptible at last by the disintegration of "the barriers of incredulity." In addition to a new acceptance and cultivation of psychic powers, Pauwels suggested that the coming "ultra-consciousness" might also involve contact with advanced civilizations from other planets.[21]

For Pauwels and Bergier, then, highly evolved extraterrestrials were simply one scientific marvel among many. At the same time, however, *Le Matin des magiciens* was the first widely circulated book to present a body of archeological evidence in support of the "ancient astronaut hypothesis"—the notion that, far in humanity's past, extraterrestrials with superior technology had been present on Earth. The evidence Pauwels and Bergier marshaled in 1960 has now become quite familiar: the Nazca lines in Peru; the mysterious stone heads on Easter Island, seemingly too heavy to have been erected without the help of advanced technology; star maps in Central Asian caves with enigmatic lines connecting Venus to the Earth.[22] Over the course of the 1960s, this emerging conception of extraterrestrials and their role in human development grew increasingly popular. The first author to develop the argument fully was French writer Robert Charroux, who drew heavily on Pauwels and Bergier; the second was Erich von Däniken, whose 1968 *Chariots of the Gods* owed a considerable debt to its French predecessors, and spread the "ancient astronaut hypothesis" across the West.[23]

Since French writers shaped the current heterodox understanding of extraterrestrials in crucial ways, it is fitting that the world's largest "flying saucer religion," the Raëlian movement, also has its origins in France. Founded in 1974 by former journalist and race-car driver Claude Vorilhon, it counts 35,000 adherents worldwide.[24] According to Vorilhon, members of a group of extraterrestrials, called the Elohim, revealed the primary tenets of the religion to him in 1973 and 1975—on the latter occasion, they also gave him the name "Raël." The Elohim intended Earth to be a laboratory and used their superior understanding of biotechnology to create all the forms of life currently on the planet. In 1973, the Elohim decided that the time had come for humankind to learn its true origins and to reshape terrestrial society in ways that would reflect this powerful knowledge. The Elohim chose Raël to convey this message; once he has won enough adherents, they will return to Earth and initiate a golden age by sharing their superior technology

[21] Louis Pauwels and Jacques Bergier, *The Morning of the Magicians,* trans. Rollo Myers (New York: Stein and Day, 1963), ix, 213.

[22] Ibid., 110–126.

[23] Stoczkowski, *Des hommes,* 45–59.

[24] Bryan Sentes and Susan Palmer, "Presumed Immanent: The Raëlians, UFO Religions, and the Postmodern Condition," *Nova Religio* 4, no. 1 (Oct. 2000): 86.

with human beings.[25] In the early 2000s, Raël chose to spread knowledge of mankind's true nature by publicly advocating human cloning as a means of attaining immortality.[26] Like Pauwels and Bergier, Raël has relied far more heavily on science than on metaphysics in his effort to reclaim the consolations of religion for the late twentieth century: according to Raëlianism, the self and the physical body are coterminous. Immortality has no metaphysical dimension at all, and has instead become a simple matter of technology, either on the distant planet of the Elohim—where Raëlian initiates are said to be cloned after death—or on Earth itself.

Though 18 percent of the French population in 1993 expressed a belief in "extraterrestrial visits to the earth," it is unlikely many such people would choose to join a group like the Raëlians.[27] Organizations with fixed, exclusivist doctrines—which the French call *sectes*—attract a small minority of those interested in "alternative religion."[28] Though these groups have received considerable attention from French politicians, who have passed some of the most stringent anticult laws in Western Europe, heterodox activity in France usually occurs outside any organized context.[29] Where in the nineteenth century there had been a constellation of defined movements, albeit an increasingly fluid one, in the late twentieth, there is what the sociologist Françoise Champion has called a "nébuleuse mystique-ésotérique," a blurry-edged social field out of which particular movements form and disintegrate, where many concepts are available for individual believers to appropriate and modify as they see fit.[30]

This "nébuleuse" includes some ideas and practices with roots in the nineteenth century and many others that have arrived more recently. Spiritist groups still exist in Paris, Tours, Lyon, and elsewhere, and Kardec's books remain steady sellers in French occult bookstores.[31] Over the course of the

[25] Ibid., 86–105. Many of Raël's writings are available for download on the group's website, www.rael.net.

[26] See Brigitte McCann, *Raël, Journal d'une infiltrée* (Outremont, QC: Stanké, 2004).

[27] Guy Michelat, "Parasciences, sciences et religion," *Le Débat, histoire, politique, société* 75 (May-Aug. 1993): 92.

[28] Françoise Champion and Martine Cohen, "Recompositions, decompositions. Le renouveau charismatique et la nébuleuse mystique-ésotérique depuis les années soixante-dix," in ibid., 81.

[29] See James T. Richardson and Massimo Introvigne, " 'Brainwashing' Theories in European Parliamentary and Administrative Reports on 'Cults' and 'Sects,' " *Journal for the Scientific Study of Religion* 40, no. 2 (June 2001): 143–168.

[30] Françoise Champion, "La 'Nébuleuse mystique-ésoterique': une décomposition du religieux entre humanisme révisité, magique, psychologique," in *Le Défi magique*, ed. François Laplantine and Jean-Baptiste Martin (Lyon: Presses Universitaires de Lyon, 1994): 1:315–326.

twentieth century, Spiritism has also grown deep roots in Brazil and elsewhere in Latin America; there, Kardec's ideas have become part of a rich syncretic mixture that includes African and Christian elements.[32] The strong Latin American presence at the Fourth Congrès spirite mondial, held in Paris in October 2004, demonstrated the extent to which Spiritism is now a trans-Atlantic phenomenon, with lines of influence moving in both directions.[33] Occultism has also left its mark on present-day French heterodoxy. The post-war years saw a revival of interest in esoteric secret societies, which led to a re-emergence of Martinism and a proliferation of neo-Rosicrucian groups laying claim to the mythic tradition of the Knights Templar.[34] Alongside these older forms, a vast array of new alternatives has also appeared. As Champion notes, many of these center on a specific practice or set of practices, such as visionary "shamanic" trances, various types of meditation, the manipulation of crystals, channeling, Reiki, or past-life regression.[35]

The current French heterodox scene is by no means unique: A similar diversification is visible throughout the West and has been since the 1960s. Colin Campbell has usefully characterized the social space in which these various activities occur as the *cultic milieu*. People who are part of this milieu, he has suggested, hold a few basic assumptions in common. They share a sense of their own status as members of an "underground," supporters of practices and forms of knowledge that go against "prevailing religious and scientific orthodoxies"; they are also receptive to syncretism

[31] In Paris, the most visible of these groups is the Union des sociétés francophones pour l'investigation psychique et l'étude de la survivance (USFIPES), which advocates scientific Spiritism in the tradition of Gabriel Delanne. See its website, www.usfipes.org. For discussions of Spiritism in twentieth-century France, see Marion Aubrée and François Laplantine, *La Table, le livre et les esprits, naissance, évolution et actualité du mouvement social spirite entre France et Brésil* (Paris: Lattès, 1990), 273–331; Christine Bergé, *La Voix des esprits, ethnologie du spiritisme* (Paris: Métaillié, 1990). The information on sales comes from Bernard Renaud de la Faverie, owner of the now-defunct Parisian bookstore La Table d'Emeraude, personal communication, February, 2000.

[32] Aubrée and Laplantine, *La Table, le livre et les esprits*, 103–269; Diana deGroat Brown, *Umbanda: Religion and Politics in Urban Brazil* (New York: Columbia University Press, 1994); David J. Hess, *Spirits and Scientists: Ideology, Spiritism, and Brazilian Culture* (University Park: Pennsylvania State University Press, 1991). Little work has been done on the early history of the diffusion of French Spiritism in Brazil, which probably began in the 1870s. The sociologists Jeremy Stolow and Carly Machado have begun a study that promises to shed light on this important question.

[33] Of thirty presenters at the Congrès, twelve were from Brazil, including two of the three plenary speakers; four more were from other Central and South American countries. See the Congrès program at www.spiritist.org/apoio/CEM_fra_conferencistas.pdf (accessed June 12, 2007).

[34] For an account of the neo-Templar current as it developed after the First World War, see Massimo Introvigne, "Ordeal by Fire: The Tragedy of the Solar Temple," *Religion* 25 (1995): 269–273.

[35] Champion and Cohen, "Recompositions, decompositions," 87.

and acknowledge fundamental elements of truth in a wide range of different religious ideas and practices. Most important, the culture of the cultic milieu emphasizes "an ideology of seekership"—a sense that "truth (or enlightenment) is an esoteric commodity only to be attained after suitable preparation and a 'quest,' " and that whatever truth an individual believer finds will be a highly personal synthesis constructed from an array of diverse materials encountered in the course of the journey.[36]

This milieu was already in place in the 1960s, both in France and elsewhere. During the mid-1970s, its character changed further, establishing the contours of the heterodox landscape as it exists at this writing. Increasingly, individual seekers came to see their activities as part of a coherent undertaking—one without clear doctrinal boundaries, but with a strong social meaning rooted in a shared sense of the insufficiency of Western scientific and religious orthodoxies. The cultic milieu, as Wouter Hanegraaff has observed, became conscious of itself as a "movement."[37] This self-consciousness is a driving force of what many commentators still call "New Age" spirituality—though a growing number of seekers reject that term.

After World War II, then, the structures of religious heterodoxy this book has described gave way to something new. The interest in exploring innovative—or long-neglected—methods of experiencing the sacred has persisted, as has the dream of a grand synthesis of faith and reason made possible by phenomena that seem to render the metaphysical concrete. The context in which this exploration occurs, however, has changed. The old promise of an organized *fraternité d'intelligence* that spiritualist Mesmerists, Spiritists, and Occultists offered, which was based on collective adherence to a particular set of philosophical principles, has lost much of its allure. Instead, spiritual seekers have grown more independent, preferring to avoid exclusive philosophies and the organizational commitments they entail in favor of a less dogmatic individualism. Finding enlightenment, as a heterodox believer, increasingly involves "finding oneself."

Reconciling Faith and Reason

Though the slow, telling emergence of a French cultic milieu is a development with important implications for our understanding of Western culture, this book has sought to address an array of other questions as

[36] Colin Campbell, "The Cult, the Cultic Milieu, and Secularization," in *The Cultic Milieu: Oppositional Subcultures in an Age of Globalization, ed.* Jeffrey Kaplan and Heléne Lööw (Walnut Creek, CA: AltaMira Press, 2002).

[37] Wouter J. Hanegraaff, *New Age Religion and Western Culture: Esotericism in the Mirror of Secular Thought* (Albany: State University of New York Press, 1998), 517–522.

well. The extraordinary burst of heterodox religious innovation that occurred in France between 1848 and 1930, and the complex responses it evoked, does much to complicate our understanding of French history as a whole during this crucial period. Scholars have understood it as an era that saw the triumph of a militantly secular republic, and with it, an anticlerical ideology rooted in Positivism. Technological innovation, industrialization, the growing importance of secular education, and urbanization all worked to create the material background for these political and intellectual developments. One of the defining struggles of the late nineteenth and early twentieth centuries, in this view, was between the reactionary Catholic right, with its pessimistic conception of human nature and insistence on hierarchy, and the secular republican left, with its optimistic vision of progress driven by a free, dutiful, and rationalistic citizenry.[38] This description is in many respects accurate, but it also leaves a number of revealing details out of the picture.[39]

The alternative religious currents discussed here, in particular, seem to have no place in this schema. In part, this situation is the work of these heterodox thinkers and believers themselves: spiritualist Mesmerists, Spiritists, and Occultists sought to synthesize what many had come to see as mutually exclusive ways of knowing.[40] Their attempts to reconcile faith and reason placed these movements in an awkward position, a no-man's land outside the conceptual boundaries of science, orthodox religion, and superstition. This marked lack of congruity with accepted categories, in turn, has led scholars either to ignore these movements or to dismiss them as peripheral curiosities.

The previous chapters have sought to remedy this neglect by demonstrating the vitality of French heterodoxy and by showing how important its ideas and practices are to an understanding of the period's social, cultural, and intellectual history. Several of the thinkers who figured prominently in these pages—Kardec, Denis, and Papus—were among the most widely read popular philosophers of their time. The movements they founded or developed generated an enormous amount of additional text, in the form of books, pamphlets, and journals. Like the artists of the early

[38] See Maurice Agulhon, *The French Republic, 1879–1992, trans.* Antonia Nevill (Oxford: Blackwell, 1993).

[39] Philip Nord has made this point very convincingly by arguing for the persistence of a visionary current—albeit one stripped of the overt theism that had characterized midcentury Utopian Socialism—in the aggressively laic ideology of the Third Republic. See Nord, *The Republican Moment: Struggles for Democracy in Nineteenth-Century France* (Cambridge, MA: Harvard University Press, 1995), esp. 15–30.

[40] See David B. Wilson, "On the Importance of Eliminating *Science* and *Religion* from the History of Science and Religion: The Cases of Oliver Lodge, J. H. Jeans and A. S. Eddington," in *Facets of Faith and Science*, ed. Jitse M. van der Meer (Lanham, MD: University Press of America, 1996), 1:27–48.

twentieth-century avant-garde, heterodox believers were a marginal group that nevertheless pioneered certain ideas, sensibilities, and forms of social behavior now fully incorporated into the Western cultural vocabulary. The numerous *feuilletons*, clerical diatribes, and popular novels these beliefs and practices inspired demonstrate the extent to which they captured the French imagination by furnishing provocatively innovative—if not always broadly accepted—answers to questions that struck many as timely and urgent.

The spiritual seekers this book describes did not want to see the triumph of either a militantly antireligious, materialistic republicanism or a clerical authoritarianism. Instead, they wanted to construct a new religious system capable of serving as the moral foundation for a "modern" society. Their goal, then, was not to take France back to a pre-Revolutionary age but to add a new dimension to the gains the Revolution had already made. The multifarious attempts to bring about this peaceful, progressive synthesis, in turn, sharpen our sense of the complex anxieties and aspirations that underlie what Maurice Agulhon has called nineteenth-century France's "stubborn political 'war of religion.' "[41] For many during this period, heterodox movements exerted a powerful attraction by seeming to resolve an intensely felt intellectual and emotional dissonance—a crisis of factuality in religious life. This book has sought to explain the nature of this crisis, the historical importance of the solutions heterodox thinkers and believers espoused, and some reasons why the experiments performed in these laboratories of faith have proved so compelling.

[41] Agulhon, *French Republic*, 1.

Bibliography

Archival and Manuscript Sources

Archives historiques de l'Archevêché de Paris
Archives municipales de Lyon
Archives nationales de France
Archives de la Préfecture de Police de Paris
Bibliothèque nationale de France, fonds Lambert
Bibliothèque municipale de Lyon, fonds Papus (FP)
Fonds Camille Flammarion de l'observatoire de Juvisy-sur-Orge (FCF)

Heterodox Journals

Catholic and general-interest periodicals and newspapers are not included here. Unless otherwise mentioned, the place of publication is Paris.

Les Annales des sciences psychiques
Les Annales du spiritisme (Rochefort-sur-Mer)
L'Avenir
Le Bon berger
Le Concile de la libre pensée
L'Echo d'outre-tombe (Marseille)
L'Echo du merveilleux
L'Ere mystérieuse (Rouen)
L'Etoile
L'Ici-bas et l'au-delà
L'Initiation
Le Journal du magnétisme
Le Lotus
La Lumière
La Lumière pour tous, journal de l'enseignement des esprits

Le Petit fraterniste (Arras)
Le Progrès spirite
Le Progrès spiritualiste
La Religion laïque
Révélations d'outre-tombe
La Revue des hautes études
La Revue du monde invisible
La Revue de psychologie expérimentale
La Revue scientifique et morale du spiritisme
La Revue spirite
La Revue spiritualiste
La Revue théosophique
La Ruche spirite (Bordeaux)
Le Sauveur des peuples (Bordeaux)
Le Spiritisme
Le Spiritisme kardéciste
La Table parlante
La Tribune psychique
L'Union magnétique
La Vérité (Lyon)
La Vie posthume (Marseille)
Le Voile d'Isis

Other Published Primary Sources

A. T. "Les Tables tournantes, origine et découverte de ce phénomène." Bordeaux: l'auteur, 1853.

Almignana, Abbé A. *Du Somnambulisme, des tables tournantes, et des mediums, considérés dans leurs rapports avec la théologie et la physique.* Paris: Dentu, 1854.

d'Anglemont, Arthur. *Omnithéisme, Dieu dans la science et dans l'amour.* 6 vols. Paris: Comptoir d'Edition, 1891–1896.

Audouard, Olympe. *Les Mondes des esprits ou la vie après la mort.* Paris: Dentu, 1874.

Auguez, Paul. *Les Elus de l'avenir, ou le progrès realisé par le christianisme.* Paris: Dentu, 1857.

——. *Les Manifestations des esprits, réponse à M. Viennet.* Paris: Dentu, 1857.

Babinet, Jacques. "Les Sciences occultes au XIXe siècle, les tables tournantes et les manifestations prétendues surnaturelles considérées au point de vue des principes qui servent de guide dans les sciences d'observation." *La Revue des deux mondes.* 24, no. 6. (1854): 510–532.

Balzac, Honoré de. *Séraphîta.* Paris: L'Harmattan, 1995 [1835].

Bautain, Abbé Louis Eugène Marie [Un Ecclésiastique, pseud.]. "Avis aux chrétiens sur les tables tournantes et parlantes." Paris: Devarenne, 1853.

Bergson, Henri. "Le Rêve." In *Mélanges.* Paris: Presses Universitaires de France, 1972.

Bersot, Ernest. *Mesmer et le magnétisme animal.* Paris: Hachette, 1854.

Bez, Auguste. *Miracles de nos jours.* Bordeaux: l'auteur, 1864.

Blavatsky, Helena Petrovna. *Isis Unveiled: A Master-Key to the Mysteries of Ancient and Modern Science and Theology.* 2 vols. New York: Bouton, 1893.

Blech, Charles. *Contribution à l'histoire de la Société Théosophique en France.* Paris: Editions Adyar, 1933.

Bois, Jules. *Le Miracle moderne.* Paris: Société d'Editions littéraires et artistiques, 1907.

Bourreau, J. B. "Comment et pourquoi je suis devenu spirite." Niort: Favre, 1864.

Breton, André. *Œuvres Complètes.* 2 vols. (Paris: Gallimard 1992).

Burlet, Philibert. *Du Spiritisme considéré comme cause d'aliénation mentale.* Lyon: Richard, 1863.

Cahagnet, Louis-Alphonse. *Magnétisme. Arcanes de la vie future dévoilés, où l'existence, la forme, les occupations de l'âme après sa separation du corps prouvés par plusieurs années d'expériences au moyen de huit somnambules extatiques.* 3 vols. Paris: Germer-Baillière, 1848–1854.

——. *Sanctuaire du spiritualisme, étude de l'âme humaine et de ses rapports avec l'univers, d'après le somnambulisme et l'extase.* Paris: Germer-Baillière, 1850.

Carion, Henri. *Lettres sur l'évocation des esprits.* Paris: Dentu, 1853.

Caston, Alfred de. *Tartuffe spirite, roman de mœurs contemporaine.* Paris: Librairie Centrale, 1865.

Caudemberg, Girard de. *Le Monde spirituel ou science chrétienne de communiquer intimement avec les puissances célestes et les âmes heureuses.* Paris: Dentu, 1857.

Chevillard, Alphonse. *Etudes expérimentales sur certains phénomènes nerveux et solution rationelle du problème spirite, troisième edition revue, corrigée et précédée par un aperçu sur le magnétisme animal.* Paris: Dentu, 1875.

Chevreul, Ernest. *De la baguette divinatoire, du pendule dit explorateur et des tables tournantes, au point de vue de l'histoire, de la critique, et de la méthode expérimentale.* Paris: Mallet-Bachelier, 1854.

Coates, James. *Photographing the Invisible: Practical Studies in Spirit Photography, Spirit Portraiture, and other Rare but Allied Phenomena.* London: L. N. Fowler, 1911.

Collignon, Emilie. "L'Education maternelle, conseils aux mères de famille, le corps et l'esprit." Paris: Ledoyen, 1864.

Committee of the London Dialectical Society. *Report on Spiritualism of the committee of the London Dialectical Society, together with the Evidence, Oral and Written, and a Selection from the Correspondence.* London: Longmans, 1871.

Compte rendu du congrès spirite et spiritualiste international. Paris: Librairie Spirite, 1890.

Compte rendu du congrès spirite et spiritualiste international, tenu à Paris du 16 au 27 septembre 1900. St. Amand: Imprimerie Daniel-Chambon, 1902.

Comte, Auguste. *Discours sur l'esprit positif.* Paris: J. Vrin, 1995 [1844].

Condat, J. J. [Chapelot, pseud.]. *Réflexions sur le spiritisme, les spirites et leur contradicteurs.* Paris: Didier, 1863.

Constant, Alphonse Louis. See Lévi, Éliphas.

Crookes, William. *Researches in the Phenomena of Spiritualism, together with a Portion of his Presidential Address Given before the British Association, 1898, and an Appendix by Sir A. Conan Doyle.* London: Psychic Bookshop, 1926.

——. "Actualité, William Crookes, ses notes sur des recherches faites dans le domaine des phénomènes appelés spirites, pendant les années 1870–1873, publiées par le Quarterly." Translated by Samuel Chinnery and Jane Jaick. Afterword by Pierre-Gaëtan Leymarie. Paris: Librairie Spirite, 1874.

Davies, Charles Maurice. *Mystic London, or Phases of Occult Life in the British Metropolis.* New York: Lovell, Adam, Wesson and Co., 1875.

Delaage, Henri. *L'Eternité dévoilée, ou vie future des âmes après la mort.* Paris: Dentu, 1854.

——. *Le Monde prophétique, ou moyen de connaître l'avenir, suivi de la biographie du somnambule Alexis.* Paris: Dentu, 1853.

Delanne, Gabriel. *L'Ame est immortelle, demonstration expérimentale de l'immortalité.* Paris: Editions de la B.P.S., 1923.

——. *Les Apparitions matérialisées des vivants et des morts.* 2 vols. Paris: Leymarie, 1909.

——. *Le Phénomène spirite, témoignages des savants.* Paris: Leymarie, 1909.

——. *Recherches sur la médiumnité.* Paris: Editions de la B.P.S., 1923.

Denis, Léon. *Après la mort, exposé de la doctrine des esprits, solution scientifique et rationnelle des problèmes de la vie et de la mort, nature et destinée de l'être humain, les vies successives.* Paris: Jean Meyer, n.d. [1890].

——. *Christianisme et spiritisme.* Paris: Jean Meyer, 1946.

——. *Dans l'invisible, spiritisme et médiumnité.* Paris: Librairie des sciences psychiques, 1911.

——. *Le Génie celtique et le monde invisible.* Paris: Jean Meyer, 1927.

——. *La Grande énigme, dieu et l'univers.* Paris: Jean Meyer, 1960.

——. *Jeanne d'Arc medium.* Paris: Leymarie, n.d.

——. *Le Monde invisible et la guerre.* Paris: Librairie des sciences psychiques, 1919.

——. *Le Problème de l'être et la destinée.* Paris: Leymarie, n.d.

Desprez, Julien Florien Félix, archevêque de Toulouse, and Valentin Tournier. "Instruction pastorale sur le spiritisme par Mgr. l'Archevêque de Toulouse, suivie d'une refutation par M. V. Tournier." Paris: Librairie Spirite, 1875.

"Discours prononcés pour l'anniversaire de la mort d'Allan Kardec, inauguration du monument." Paris: Librairie spirite, 1870.

Doyle, Sir Arthur Conan. *The Case for Spirit Photography.* London: Hutchinson, n.d.

——. *The History of Spiritualism.* 2 vols. New York: G.H. Doran Co., 1926.

Dozon, Henri. "Révélations d'outre-tombe. Espoir et résignation. Semez. Règle de société." Paris: Ledoyen, 1862.

Dufaux, Ermance. *La Vie de Jeanne d'Arc, dictée par elle-même.* Paris: Ledoyen, 1858.

Dumas, Alexandre. *Mémoires d'un médecin, Joseph Balsamo.* 2 vols. Paris: Legrand et Crouzet, n.d.

E. Dentu. Paris: Dentu, 1884.

Edoux, Evariste. *Spiritisme Pratique, appel des vivants aux esprits des morts.* Lyon: Librairie moderne, 1863.

Encausse, Gérard. See Papus.

"Examen raisonné des prodiges récents d'Europe et d'Amérique, notamment des tables tournantes et répondantes, par un philosophe." Paris: Vermot, 1853.

Flammarion, Camille [Hermès, pseud.]. *Des Forces naturelles inconnues, à propos des phénomènes produits par les frères Davenport.* Paris: Didier, 1866.

——. *Les Forces naturelles inconnues.* Paris: Flammarion, 1907.

——. *Les Habitants de l'autre monde, révélations d'outre-tombe.* 2 vols. Paris: Ledoyen, 1862.

——. *Mémoires biographiques et philosophiques d'un astronome.* Paris: Flammarion, 1911.

Flournoy, Théodore. *From India to the Planet Mars: A Case of Multiple Personality with Imaginary Languages.* Edited by Mireille Cifali. Translated by Daniel Vermilye. Preface by C. G. Jung. Intro. by Sonu Shamdasani. Princeton: Princeton University Press, 1994.

Fourier, Charles. *Le Nouveau monde amoureux.* Edited by Simone Debout-Oleskiewicz. Paris: Anthropos, 1967.

Fropo, Berthe. "Beaucoup de lumière." Paris: Imprimerie polyglotte, 1884.

Garnier, Adolphe. "Sauvons le genre humain, par Victor Hennequin." Paris: Paul Dupont, 1854.

Gasparin, Agénor de. *Des Tables tournantes, du surnaturel en général et des esprits.* 2 vols. Paris: Dentu, 1854.

Gaudon, Jean, ed. *Ce que disent les tables parlantes: Victor Hugo à Jersey.* Paris: Pauvert, 1963.

Gautier, Judith. *Le Collier des jours, le second rang du collier.* Paris: Renaissance du livre, 1909.

Gautier, Théophile. *Spirite, nouvelle fantastique.* In *L'Œuvre fantastique II—Romans.* Paris: Bordas, 1992 [1866].

Geley, Gustave. *L'Etre subconscient.* Paris: Alcan, 1898.

Gentil, J. A. "L'Ame de la terre et les tables parlantes, ou Sauvons le genre humain, ouvrage examiné au point de vue magnétique de l'influence des besoins sur le moral." Paris: l'auteur, 1854.

Gibier, Paul. *Analyse des choses, essai sur la science future.* Paris: Dentu, 1890.

——. *Spiritisme, Fakirisme occidental.* Paris: Durville, n.d. [1886].

Goupy, Louis. *L'Ether, l'électricité et la matière, deuxième édition de Quœare et invenies.* Paris: Ledoyen, 1854.

——. "Phénomènes de spiritualisme à expliquer." Argenteuil: Worms et cie., 1857.

Grange, Lucie. *La Mission du nouveau-spiritualisme, Lettres de l'esprit de Salem-Hermès, communications prophétiques.* Paris: chez l'auteur, 1896.

Grasset, Joseph. *Le Psychisme inférieur.* Paris: Marcel Rivière, 1913.
——. *Le Spiritisme devant la science.* Paris: Masson, 1904.
Guaïta, Stanislas de. *Au Seuil du mystère.* Paris: Durville, 1915 [1886].
——. *Le Temple de Satan.* Paris: Trédaniel, 1994 [1895].
Guénon, René. *L'Erreur spirite.* Paris: M. Rivère, 1923.
——. *Le Théosophisme, histoire d'une pseudo-religion.* Paris: Nouvelle Librairie Nationale, 1921.
Guibert, Joseph Hippolyte, évêque de Viviers. "Lettre pastorale de monseigneur l'évêque de Viviers, au clergé de son diocèse, sur le danger des expériences des tables parlantes." Privas: Guiremand, 1853.
Guldenstubbé, Baron L. de. *Pneumatologie positive et expérimentale, la réalité des esprits et le phénomène merveilleux de leur écriture directe.* Paris: A. Franck, 1857.
Guyomar, Dr. *Etude de la vie intérieure ou spirituel chez l'homme.* Paris: Delahaye, 1865.
Hennequin, Victor. *Sauvons le genre humain.* Paris: Dentu, 1853.
——. *La Religion.* Paris: Dentu, 1854.
Henry, Victor. *Le Langage Martien.* Paris: Maisonneuve, 1901.
Home, Daniel Dunglas. *Incidents in My Life.* 2 vols. New York: Carleton, 1863.
Houat, L. T. *Etudes et séances spirites, morale, philosophie, medicine, psychologie.* Paris: Ledoyen, 1863.
Hugo, Adèle. *Le Journal d'Adèle Hugo, deuxième volume, 1853,* edited by Frances Vernor Guille. Paris: Minard, 1971.
Hugo, Victor. *La Légende des siècles.* Paris: Gallimard, 1955.
Huguet, Hilarion. *Spiritomanes et spiritophobes, etude sur le spiritisme.* Paris: Dentu, 1875.
Jacob, Alexandre-André [Alexandre Erdan, pseud.] *La France mistique, tablau des excentricités religieuses de ce tems.* 2 vols. Paris: Coulon-Pineau, 1855.
Jacolliot, Louis. *Le Spiritisme dans le monde, l'initiation et les sciences occultes dans l'Inde et chez tous les peuples d'antiquité.* Geneva: Slatkine, 1988 [1875].
Janet, Pierre. *L'Automatisme psychologique, essai de psychologie expérimentale sur les formes inférieures de l'activité humaine.* Paris: CNRS, 1973.
——. "Revue générale, spiritisme contemporain." *Revue philosophique* 33 (1892): 413–442.
——, Théodule Ribot, et al. *IVe Congrès international de psychologie, tenu à Paris, du 20 au 26 août 1900, compte rendu des séances et texte des mémoires.* Paris: Alcan, 1901.
Kardec, Allan [pseud. of Hippolyte Léon Dénizard Rivail]. *La Genèse selon le spiritisme.* Montreal: Editions Select, 1980 [1868].
——. *Le Livre des Esprits, contenant les principes de la doctrine spirite.* Paris: Dervy, 1996 [1860].
——. *Le Livre des Médiums, ou guide des mediums et des évocateurs.* Paris: Dervy, 1978 [1861].
——. *Voyage spirite en 1862.* Paris: Vermet, n.d. [1862].
——. *Œuvres posthumes.* Edited by Pierre-Gaëtan Leymarie. Introduction by André Dumas. Paris: Dervy, 1978.

Larousse, Pierre. *Grand dictionnaire universel du XIXe siècle.* Geneva: Slatkine, 1982.
Lefebvre, H. "La Danse des tables, pochade en un acte." Lyon: Vingtrinier, 1853.
Legas, L. "La Photographie spirite et l'analyse spectrale comparées." Paris: l'auteur, 1875.
Lévi, Eliphas [pseud. of Alphonse-Louis Constant]. *Dogme et rituel de la haute magie.* 2 vols. Paris: Editions Niclaus, 1948.
——. *Histoire de la magie, avec une exposition Claire et précise de ses procédés, de ses rites et de ses mystères.* Paris: Trédaniel, n.d. [1860].
——. *La Science des esprits, révélation du dogme secret des kabalistes, esprit occulte des évangiles, appreciation des doctrines et des phénomènes spirites.* Paris: Trédaniel, n.d. [1865].
Leymarie, Marina, ed. *Le Procès des spirites.* Paris: Librairie Spirite, 1875.
Littré, Emile. "Des Tables parlantes et des esprits frappeurs." *La Revue des deux mondes* 26, no. 3 (1856): 847–872.
Louisy, Paul. *Lumière! Esprits et tables tournantes, révélations médianimiques.* Paris: Garnier Frères, 1854.
Lussan, B.J.B. *Quelques pages sur le spiritisme.* Toulouse: Chauvin, 1865.
M. J. B. "Lettres sur le spiritisme écrites à des ecclésiastiques." Paris: Ledoyen, 1864.
Malgras, J. *Les Pionniers du spiritisme en France.* Paris: Librairie des sciences psychiques, 1906.
Marouseau, Jean-Baptiste. *Réfutation de la doctrine spirite au point de vue religieux.* Paris: Raveau d'Artois, 1865.
Matignon, Ambroise, S. J. *Les Morts et les vivants, entretiens sur les communications d'outre tombe.* Paris: Adrien le Clère, 1862.
——. *La Question du surnaturel, ou la grâce, le merveilleux, le spiritisme au XIXe siècle.* Paris: Adrien Le Clère, 1863.
Matter, Jacques. *Saint-Martin, le philosophe inconnu.* Le Tremblay: Diffusion Rosicrucienne, 1992 [1845].
Michelet, Victor-Emile. *Les Compagnons de la hiérophanie, souvenirs du mouvement herméttiste à la fin du XIXe siècle.* Paris: Dorbon, 1937.
Mirville, Jules Eudes, marquis de. *Pneumatologie, des esprits et de leurs manifestations diverses.* Paris: H. Vrayet de Surcy, 1863 [1853].
Monckhoven, D. v. *Traité générale de photographie.* Paris: Masson, 1873.
Monnier, Cécile. *Je Suis vivant.* Paris: Leymarie, 1920.
——. *Lettres de Pierre.* Intr. by Jean Prieur. 7 vols. Paris: Fernand Lenore, 1980.
Montplaisir, Camille de. *Qu'est-ce que le spiritisme?* Paris: Girard et Josserand, 1863.
Mousseaux, Henri Roger Gougenot des, Chevalier. *La Magie au dix-neuvième siècle, ses agents, ses vérités, ses mensonges.* Paris: Henri Plon, 1863 [1853].
Myers, Frederic W. H. *Human Personality and Its Survival of Bodily Death.* 2 vols. London: Longman, 1903.
——. *La Personnalité humaine, sa survivance, ses manifestations supranormales.* Translated by S. Jankelevitch. Paris: Alcan, 1906.

Nampon. *Du Spiritisme.* Paris: Girard et Josserand, 1863.

Nichols, Thomas Low. *Phénomènes des frères Davenport et leurs voyages en amérique et en angleterre, accompagné de notes et d'opuscules sur la doctrine spirite.* Translated by Mme Bernard Derosne. Paris: Didier, 1865.

Noeggerath, Rufina. *La Survie, sa réalité, sa manifestation, sa philosophie, échos de l'au-delà.* Paris : Flammarion, 1897.

Nordmann, M. *Le Livre des Esprits spiritualistes.* Paris: Patissier, 1863.

Nus, Eugène. *Choses de l'autre monde.* Paris: Dentu, 1880.

Olcott, Henry Steele. *Old Diary Leaves: The Only Authentic History of the Theosophical Society, Fourth Series, 1887–1892.* Adyar: Theosophical Publishing House, 1931.

Papus [pseud. of Gérard Encausse]. *Martinésisme, Willermosisme, Martinisme et Franc- Maçonnerie.* Paris: Chamuel, 1899.

——. *Traité élémentaire de magie pratique.* Paris: Chamuel, 1893.

——. *Traité Elémentaire de science occulte.* Paris: Albin Michel, n.d. [1887].

——. *Traité méthodique de magie pratique.* Paris: Dangles, n.d.

Parsevel-Deschênes, Georges de. *Gardeneur, histoire d'un spirite.* Paris: Librairie du Petit Journal, 1866.

Pauwels, Louis, and Jacques Bergier. *The Morning of the Magicians.* Translated by Rollo Myers. New York: Stein and Day, 1963.

Piérart, Zéphyre-Joseph. "La Vérité sur les Davenport." Paris: Dentu, 1865.

Reynaud, Jean. *Philosophie religieuse, terre et ciel.* Paris: Furne, 1864.

Ribot, Théodule. *Les Maladies de la personnalité.* Paris: Alcan, 1921.

Richet, Charles. *Traité de métapsychique.* Paris: Alcan, 1922.

Rivail, Hippolyte Léon Dénizard. See Kardec, Allan.

Rosenbach, P. "Etude critique sur le mysticisme moderne." *Revue philosophique* 34 (1892): 113–158.

Roubaud, Félix. "La Danse des tables dévoilée, expériences de magnétisme animal pour s'amuser en société, manière de faire tourner une bague, un chapeau, une montre, une table, et même jusqu'aux têtes des expérimentateurs et celles des spectateurs." Paris: l'auteur, 1853.

——. *La Danse des tables, phénomènes physiologiques démontrés.* 2e éd. Paris: Librairie nouvelle, 1853.

Roustaing, Jean-Baptiste. *Spiritisme Chrétien, ou révélation de la révélation, les quatre Evangiles suivis des commandements, expliqués en esprit et en vérité par les évangélistes assistés des apôtres, Moïse.* 3 vols. Paris: Librairie centrale, 1866.

——. "Les Quatre Evangiles de J.B. Roustaing, réponse à ses critiques et à ses adversaires." Bordeaux: J. Durand, 1882.

Saint-Yves d'Alveydre, Alexandre. *Mission des Juifs.* Paris: Calmann Lévy, 1884.

"Sermons sur le spiritisme prêchés à la cathédrale de Metz les 27, 28, et 29 mai 1863 par le R.P. Letierce de la Compagnie de Jésus, refutes par un spirite de Metz." Paris: Didier, 1863.

Sidgwick, Mrs. Henry. "On Spirit Photographs, a Reply to Mr. A. R. Wallace." *Proceedings of the Society for Psychical Research* 7 (1891–1892).

Silas, Ferdinand. "Instruction explicative des tables tournantes, d'après les publications allemandes, américaines, et les extraits des journaux allemands, français et américains." Intro. by Henri Delaage. Paris: Dentu, 1853.

Simon, Gustave, ed. *Chez Victor Hugo: les tables tournantes de Jersey.* Paris: Stock, 1980.

Sinnett, A. P. *Le Monde occulte, hypnotisme transcendant en orient.* Intro. and translated by Félix-Krishna Gaboriau. Paris: Carré, 1887.

Société scientifique du spiritisme. "Fictions et insinuations, réponse à la brochure 'Beaucoup de Lumière.' " Paris: Librairie des études psychologiques, 1884.

Taylor, J. Traill. *The Veil Lifted: Modern Developments in Spirit Photography.* London: Whittaker, 1894.

Un Badaud [pseud.]. *Coup d'oeil sur la magie au XIXe siècle.* Paris: Dentu, 1891.

Un Capitaine [pseud]. *Spiritisme élémentaire, théorique et pratique. Faits spirites et entretiens familiers d'outre-tombe contenant la théorie de l'évocation des esprits ou des âmes des morts (d'après les écrits d'Allan Kardec).* Paris: Ledoyen, 1862.

Vacquerie, Auguste. *Les Miettes de l'histoire.* Paris: Pagnerre, 1863.

Vitoux, Georges. *Les Coulisses de l'au-delà.* Paris: Chamuel, 1901.

Secondary Sources

Abend, Lisa. "Specters of the Secular: Spiritism in Nineteenth-Century Spain." *European History Quarterly* 34, no. 4 (2004): 507–534.

Acquaviva, Sabino Samele. *The Decline of the Sacred in Industrial Society.* Translated by Patricia Lipscomb. Oxford: Blackwell, 1979.

Agulhon, Maurice. *The French Republic, 1879–1992.* Translated by Antonia Nevill. Oxford: Blackwell, 1993.

Allen, James Smith. *Popular French Romanticism: Authors, Readers and Books in the 19th Century.* Syracuse, NY: Syracuse University Press, 1981.

André, Marie-Sophie, and Christophe Beaufils. *Papus, biographie, la belle époque de l'occultisme.* Paris: Berg, 1995.

Ariès, Philippe. *The Hour of Our Death.* Translated by Helen Weaver. New York: Knopf, 1981.

Auspitz, Katherine. *The Radical Bourgeoisie: The* Ligue de l'Enseignement *and the Origins of the Third Republic.* Cambridge: Cambridge University Press, 1982.

Baquiast, Paul. *Une Dynastie de la bourgeoisie republicaine, les Pelletan.* Paris: L'Harmattan, 1996.

Beafuils, Christophe. *Joséphin Péladan, essai sur une maladie du lyrisme.* Grenoble: Millon, 1993.

Beecher, Jonathan. *Charles Fourier: The Visionary and His World.* Berkeley: University of California Press, 1986.

——. *Victor Considerant and the Rise and Fall of French Romantic Socialism.* Berkeley: University of California Press, 2001.

Bellanger, Claude, et al. *Histoire générale de la presse française.* 4 vols. Paris: Presses Universitaires de France, 1972.

Bénichou, Paul. *Le Temps des prophètes.* Paris: Gallimard, 1977.

Berenson, Edward. *The Trial of Madame Caillaux.* Berkeley: University of California Press, 1992.

Bergé, Christine. *L'Au-delà et les Lyonnais, mages, médiums et francs-maçons du XVIIIe au XXe siècle.* Lyon: LUGD, 1995.

——. *La Voix des esprits, ethnologie du spiritisme.* Paris: Métaillié, 1990.

Blackbourn, David. *Marpingen: Apparitions of the Virgin Mary in a Nineteenth-Century German Village.* New York: Vintage, 1993.

Boisset, Yves-Fred. *A la Rencontre de Saint-Yves d'Alveydre et de son œuvre.* Intro. by Robert Amadou. 2 vols. Paris: Sepp, 1996.

Boring, Edwin G. *A History of Experimental Psychology.* 2nd ed. New York: Appleton-Century-Crofts, 1950.

Boy, Daniel. "Les Français et les para-sciences, vingt ans de mesures." *Revue Française de sociologie* 43, no. 1 (January-March 2002): 35–45.

Brandon, Ruth. *The Spiritualists: The Passion for the Occult in the Nineteenth and Twentieth Centuries.* New York: Knopf, 1983.

Braude, Anne. *Radical Spirits: Spiritualism and Women's Rights in Nineteenth-Century America.* Boston: Beacon Press, 1989.

Brower, Matthew Brady. "The Fantasms of Science: Psychical Research in the French Third Republic, 1880–1935." Ph.D. diss., Rutgers University, 2005.

Brown, Diana DeGroat. *Umbanda: Religion and Politics in Urban Brazil.* New York: Columbia University Press, 1986.

Brown, Michael F. *The Channeling Zone: American Spirituality in an Anxious Age.* Cambridge, MA: Harvard University Press, 1997.

Brunet, Georges. *Le Mysticisme social de Saint-Simon.* Paris: Les Presses Françaises, 1922.

Campbell, Colin. "The Cult, the Cultic Milieu, and Secularization." In *The Cultic Milieu: Oppositional Subcultures in an Age of Globalization,* edited by Jeffrey Kaplan and Heléne Lööw. Walnut Creek, CA: AltaMira Press, 2002.

Capron, Eliab Wilkinson. *Modern Spiritualism: Its Facts and Fanaticisms, Its Consistencies and Contradictions.* Boston: Bela Marsh, 1855.

Carlisle, Robert B. *The Proffered Crown: Saint-Simonianism and the Doctrine of Hope.* Baltimore: Johns Hopkins University Press, 1987.

Carroy, Jacqueline, and Régine Plas. "The Origins of French Experimental Psychology: Experiment and Experimentalism." *History of the Human Sciences* 9, no. 1 (1996): 73–84.

Chadwick, Owen. *The Secularization of the European Mind in the Nineteenth Century.* Cambridge: Cambridge University Press, 1975.

Champion, Françoise. "La 'Nébuleuse mystique-ésoterique': une décomposition du religieux entre humanisme révisité, magique, psychologique." In *Le Défi magique*, ed. François Laplantine and Jean-Baptiste Martin. Lyon: Presses Universitaires de Lyon, 1994.

Champion, Françoise, and Martine Cohen. "Recompositions, decompositions. Le renouveau charismatique et la nébuleuse mystique-ésotérique depuis les années soixante-dix." In *Le Débat, histoire, politique, société* 75 (May–August 1993): 81–89.

Charlton, D. G. *Positivist Thought in France during the Second Empire, 1852–1870*. Westport, CT: Greenwood Press, 1959.

——. *Secular Religions in France, 1815–1870*. London: Oxford University Press, 1963.

Chertok, Léon, and Isabelle Stengers. *Le Cœur et la raison, l'hypnose en question de Lavoisier à Lacan*. Paris: Payot, 1989.

Cholvy, Gérard, and Yves-Marie Hilaire. *Histoire religieuse de la France, 1800–1880*. Paris: Privat, 2000.

——. *Histoire religieuse de la France contemporaine*. 3 vols. Paris: Privat, 1988.

Christian, William A., Jr. *Visionaries: The Spanish Republic in the Reign of Christ*. Berkeley: University of California Press, 1999.

Crabtree, Adam. *From Mesmer to Freud: Magnetic Sleep and the Roots of Psychological Healing*. New Haven: Yale University Press, 1993.

Darnton, Robert. *Mesmerism and the End of the Enlightenment in France*. Cambridge, MA: Harvard University Press, 1968.

Daston, Lorraine, and Peter Galison. "The Image of Objectivity." *Representations* 40 (Fall 1992): 81–129.

Dingwall, Eric John. *Some Human Oddities: Studies in the Queer, the Uncanny, and the Fanatical*. London: Home and Van Thal, 1947.

Edelman, Nicole. *Histoire de la voyance et du paranormal du XVIIIe siècle à nos jours*. Paris: Seuil, 2006.

————. *Voyantes, guérisseuses et visionnaires en France, 1785–1914*. Paris: Albin Michel, 1995.

Eliade, Mircea. "Cultural Fashions and History of Religions." In *Occultism, Witchcraft, and Cultural Fashions: Essays in Comparative Religions*. Chicago: University of Chicago Press, 1976.

Ellenberger, Henri. *The Discovery of the Unconscious: The History and Evolution of Dynamic Psychiatry*. New York: Basic Books, 1970.

Encausse, Philippe. *Sciences occultes, ou 25 années d'occultisme occidental, Papus, sa vie et son œuvre*. Paris: Editions OCIA, 1949.

Faivre, Antoine. *Accès de l'ésotérisme occidental*. 2 vols. Paris: Gallimard, 1986.

Fenech, Georges. *Face aux sectes: politique, justice, Etat*. Intro. by Alain Vivien. Paris: Presses Universitaires de France, 1999.

Fuentès, Patrick, and Philippe de la Cortadière. *Camille Flammarion*. Paris: Flammarion, 1994.

Garçon, Maurice. *La Justice contemporaine, 1870–1932*. Paris: Grasset, 1933.

Gauchet, Marcel. *The Disenchantment of the World.* Translated by Oscar Burge. Princeton: Princeton University Press, 1997.

Gauld, Alan. *The Founders of Psychical Research.* London: Routledge and Kegan Paul, 1968.

——. *A History of Hypnotism.* Cambridge: Cambridge University Press, 1992.

Geertz, Clifford. *The Interpretation of Cultures.* New York: Basic Books, 1973.

Godwin, Joscelyn. "The Beginnings of Theosophy in France." London: Theosophical History Center, 1989.

——. *The Theosophical Enlightenment.* Albany: State University of New York Press, 1994.

Goldsmith, Barbara. *Other Powers: The Age of Suffrage, Spiritualism, and the Scandalous Victoria Woodhull.* New York: Knopf, 1998.

Guérard, Albert Leon. *French Prophets of Yesterday: A Study of Religious Thought under the Second Empire.* London: T. Fisher Unwin, 1913.

Gutek, Gerald Lee. *Pestalozzi and Education.* New York: Random House, 1968.

Gutierez, Grégrory. "Le Discours du réalisme fantastique: la revue Planète." Mémoire de maîtrise de Lettres modernes spécialisées, Université Sorbonne—Paris IV, 1998.

Hacking, Ian. *Rewriting the Soul: Multiple Personality and the Sciences of Memory.* Princeton: Princeton University Press, 1995.

Hammer, Olav. *Claiming Knowledge: Strategies of Epistemology from Theosophy to the New Age.* Leiden: Brill, 2001.

Hanegraaff, Wouter J. *New Age Religion and Western Culture: Esotericism in the Mirror of Secular Thought.* Albany: State University of New York Press, 1998.

Harris, Ruth. *Lourdes: Body and Spirit in the Secular Age.* London: Penguin, 1999.

Harvey, David Allen. *Beyond Enlightenment: Occultism and Politics in Modern France.* DeKalb: Northern Illinois University Press, 2005.

Hess, David. *Samba in the Night: Spiritism in Brazil.* New York: Columbia University Press, 1994.

——. *Spirits and Scientists: Ideology, Spiritism and Brazilian Culture.* University Park: Pennsylvania State University Press, 1991.

Hughes, H. Stuart. *Consciousness and Society: The Reorientation of European Social Thought, 1890–1930.* New York: Knopf, 1958.

Introvigne, Massimo. "Ordeal by Fire: The Tragedy of the Solar Temple." *Religion* 25 (1995): 269–273.

Johnson, Christopher H. *Utopian Communism in France: Cabet and the Icarians, 1839–1851.* Ithaca: Cornell University Press, 1974.

Jonas, Raymond. *France and the Cult of the Sacred Heart, an Epic Tale for Modern Times.* Berkeley: University of California Press, 2000.

Jones, Caroline A., and Peter Galison. *Picturing Science, Producing Art.* London: Routledge, 1998.

Kaufman, Suzanne K. *Consuming Visions: Mass Culture and the Lourdes Shrine.* Ithaca: Cornell University Press, 2004.

Kselman, Thomas. *Death and the Afterlife in Modern France.* Princeton: Princeton University Press, 1993.

——. *Miracles and Prophecies in Nineteenth-Century France.* New Brunswick: Rutgers University Press, 1983.

Lachapelle, Sofie. "Attempting Science: The Creation and Early Development of the Institut Métapsychique International in Paris, 1919–1931." *Journal of the History of Behavioral Sciences* 41, no. 1 (2005): 1–24.

——. "A World Outside Science: French Attitudes Toward Mediumistic Phenomena, 1853–1931." Ph.D. diss., University of Notre Dame, 2002.

L'Alouette, Jacqueline. *La Libre pensée en France, 1848–1940.* Paris: Albin Michel, 1997.

Langlois, Claude. *Le Catholicisme au féminin, les congrégations françaises à supérieure générale au XIXe siècle.* Paris: Cerf, 1984.

Laplantine, François, and Marion Aubrée. *La Table, le livre et les esprits, naissance, évolution et actualité du mouvement social spirite entre France et Brésil.* Paris: J.C. Lattès, 1990.

Laplantine, François, and Jean-Baptiste Martin, eds. *Le Défi magique.* 2 vols. Lyon: Presses Universitaires de Lyon, 1994.

Laqueur, Thomas. "Why the Margins Matter: Occultism and the Making of Modernity." *Modern Intellectual History* 3, no. 1 (2006): 111–135.

Laurant, Jean-Pierre. *L'Esotérisme chrétien en France au XIXe siècle.* Lausanne: Editions de l'âge d'homme, 1992.

Le Bras-Chopard, Armelle. *De L'Egalité dans la différence, le socialisme de Pierre Leroux.* Paris: Presses de la Fondation Nationale des Sciences Politiques, 1986.

Le Maléfan, Pascal. *Folie et spiritisme, histoire du discours psychopathologique sur la pratique du spiritisme, ses abords et ses avatars (1850–1950).* Paris: L'Harmattan, 1999.

Lefebvre, Anne-Marie. "Spirite de Théophile Gautier, étude historique et littéraire." Thèse de doctorat. Sorbonne troisième cycle, 1978.

Luckmann, Thomas. *The Invisible Religion: The Problem of Religion in Modern Society.* New York: Macmillan, 1967.

Luhrmann, T. H. *Persuasions of the Witch's Craft: Ritual Magic and Witchcraft in Present Day England.* Oxford: Blackwell, 1989.

Manuel, Frank. *The New World of Henri Saint-Simon.* Cambridge, MA: Harvard University Press, 1956.

Marois, André. *Olympio: The Life of Victor Hugo.* New York: Harper and Bros., 1956.

Matlock, Jann. "Ghostly Politics." *Diacritics* 30, no. 3 (Fall 2000): 53–71.

McCann, Brigitte. *Raël, journal d'une infiltrée.* Outremont, QC: Stanké, 2004.

McCauley, Elizabeth. *A. A. E. Disderi and the Carte de Visite Portrait Photograph.* New Haven: Yale University Press, 1985.

McManners, John. *Death and the Enlightenment: Changing Attitudes to Death in Eighteenth-Century France.* Oxford: Oxford University Press, 1981.

McPhee, Peter. *A Social History of France, 1780–1880.* New York: Routledge, 1992.

Méheust, Bertrand. *Un Voyant prodigieux: Alexis Didier, 1826–1886.* Le Plessis-Robinson: Institut Synthélabo, 2003.

——. *Somnambulisme et médiumnité.* 2 vols. Le Plessis-Robinson: Institut Synthélabo, 1999.

Micale, Mark S. "On the 'Disappearance' of Hysteria: A Study in the Clinical Destruction of a Diagnosis." In *Isis* 84, no. 3 (September, 1993): 496–526.

Michelat, Guy. "Parasciences, sciences et religion." In *Le Débat, histoire, politique, société* 75 (May-August 1993): 90-100.

Moore, R. Laurence. *In Search of White Crows: Spiritualism, Parapsychology and American Culture.* New York: Oxford University Press, 1977.

Mucchielli, Laurent. "Aux Origines de la psychologie universitaire en France (1870–1900): enjeux intellectuels, contexte politique, réseaux et stratégies d'alliance autour de la *Revue philosophique* de Théodule Ribot." *Annals of Science* 55 (1998): 263–289.

Nathan, Michel. *Le Ciel des Fouriéristes, habitants des étoiles et réincarnation de l'âme.* Lyon: Presses Universitaires de Lyon, 1981.

Nicolas, Serge. *Histoire de la psychologie française, naissance d'une nouvelle science.* Paris: In Press Editions, 2002.

Nicolas, Serge, and Agnes Charvillat. "Théodore Flournoy (1854–1920) and Experimental Psychology: Historical Note." *American Journal of Psychology* 111, no. 2 (Summer 1998): 279–294.

Nord, Philip. *The Republican Moment: Struggles for Democracy in Nineteenth-Century France.* Cambridge, MA: Harvard University Press, 1995.

Oppenheim, Janet. *The Other World: Spiritualism and Psychical Research in England, 1850–1914.* Cambridge: Cambridge University Press, 1985.

Owen, Alex. *The Darkened Room: Women, Power, and Spiritualism in Late Nineteenth-Century England.* London: Virago, 1989.

——. *The Place of Enchantment: British Occultism and the Culture of the Modern.* Chicago: University of Chicago Press, 2004.

Pauwels, Louis, and Jacques Bergier. *The Morning of the Magicians.* Trans. Rollo Myers. New York: Stein and Day, 1963.

Parot, Françoise. "Le Banissement des esprits, naissance d'une frontière institutionnelle entre spiritisme et psychologie." *Revue de synthèse* 115, no. 3–4 (July–Dec. 1996): 417–443.

Pike, Sarah. *New Age and Neopagan Religions in America.* New York: Columbia University Press, 2004.

Pincus-Witten, Robert. *Occult Symbolism in France: Joséphin Péladan and the Salons de la Rose-Croix.* New York: Garland, 1976.

Poovey, Mary. *A History of the Modern Fact: Problems of Knowledge in the Sciences of Wealth and Society.* Chicago: University of Chicago Press, 1998.

Prothero, Stephen. *The White Buddhist: The Asian Odyssey of Henry Steel Olcott.* Bloomington: Indiana University Press, 1996.

Rémond, René. *L'Anticléricalisme en France de 1815 à nos jours.* Paris: Fayard, 1999.

Richardson, James T., and Massimo Introvigne. " 'Brainwashing' Theories in European Parliamentary and Administrative Reports on 'Cults' and 'Sects.' " In *Journal for the Scientific Study of Religion* 40, no. 2 (June 2001): 143–168.

Roudinesco, Elisabeth. *La Bataille de cent ans.* 2 vols. Paris: Seuil, 1986.

Sausse, Henri. *Biographie d'Allan Kardec.* Intro. by Léon Denis. Paris: Jean Meyer, 1927.

Sedgwick, Mark. *Against the Modern World: Traditionalism and the Secret Intellectual History of the Twentieth Century.* Oxford: Oxford University Press, 2004.

Sentes, Bryan, and Susan Palmer. "Presumed Immanent: The Raëlians, UFO Religions, and the Postmodern Condition." *Nova Religio* 4, no. 1 (Oct. 2000): 86–105.

Shapin, Steven. *A Social History of Truth: Civility and Science in Seventeenth-Century England.* Chicago: University of Chicago Press, 1994.

Sharp, Lynn L. "Echoes from the Beyond: Purgatory and Catholic Communication with the Dead." Paper delivered at the thirty-first annual conference of the Western Society for French History, Newport Beach, CA, Oct. 31, 2003.

——. *Secular Spirituality: Reincarnation and Spiritism in Nineteenth Century France.* Lanham, MD: Lexington Books, 2006.

Skultans, Vieda. *Intimacy and Ritual: A Study of Spiritualism, Mediums and Groups.* London: Routledge, 1974.

Stock-Morton, Phyllis. *Moral Education for a Secular Society: The Development of* Morale Laïque *in Nineteenth-Century France.* Albany: State University of New York Press, 1988.

Stoczkowski, Wiktor. *Des hommes, des dieux et des extraterrestres, ethnologie d'une croyance moderne.* Paris: Flammarion, 1999.

Sword, Helen. *Ghostwriting Modernism.* Ithaca: Cornell University Press, 2002.

Tocquet, Robert. *Les Pouvoirs secrets de l'homme, le bilan du paranormal.* Intro. by Louis Pauwels. Paris: Editions les Productions de Paris, 1963.

Treitel, Corinna. *A Science for the Soul: Occultism and the Genesis of the German Modern.* Baltimore: Johns Hopkins University Press, 2004.

Turner, Frank M. *Between Science and Religion: The Reaction to Scientific Naturalism in Late Victorian England.* New Haven: Yale University Press, 1974.

Vernette, Jean. *Jésus dans la nouvelle religiosité.* Paris: Desclée, 1987.

Viatte, Auguste. *Les Sources occultes du romantisme.* 2 vols. Paris: Champion, 1928.

——. *Victor Hugo et les illuminés de son temps.* Montreal: Editions de l'arbre, 1922.

Vovelle, Michel. *La Mort et l'occident de 1300 à nos jours.* Paris: Gallimard, 1983.

Walton, Whitney. *Eve's Proud Descendants: Four Women Writers and Republican Politics in Nineteenth-Century France.* Stanford: Stanford University Press, 2000.

Washington, Peter. *Madame Blavatsky's Baboon: A History of the Mystics, Mediums and Misfits Who Brought Spiritualism to America.* New York: Schocken, 1996.

Weber, Eugen. *France Fin-de-siècle.* Cambridge, MA: Harvard University Press, 1986.

Weber, Max. "The Social Psychology of the World Religions." In *From Max Weber: Essays in Sociology.* Trans. and ed. Hans Heinrich Gerth and Charles Wright Mills. Oxford: Oxford University Press, 1958.

Weisberg, Barbara. *Talking to the Dead: Kate and Maggie Fox and the Rise of Spiritualism.* New York: HarperCollins, 2004.

Wernick, Andrew. *Auguste Comte and the Religion of Humanity: The Post-Theistic Program of French Social Theory.* Cambridge: Cambridge University Press, 2001.

White, Hayden. *Metahistory: The Historical Imagination in Nineteenth-Century Europe.* Baltimore: Johns Hopkins University Press, 1973.

Wilson, David B. "On the Importance of Eliminating *Science* and *Religion* from the History of Science and Religion: The Cases of Oliver Lodge, J. H. Jeans and A. S. Eddington." In *Facets of Faith and Science,* ed. Jitse M. van der Meer, 1:27–48. Lanham, MD: University Press of America, 1996.

Winter, Alison. *Mesmerized: Powers of Mind in Victorian Britain.* Chicago: University of Chicago Press, 1998.

Yates, Frances A. *Giordano Bruno and the Hermetic Tradition.* Chicago: University of Chicago Press, 1998.

Index

Italic page numbers refer to illustrations.